MATHEMATICAL EXPLORATIONS

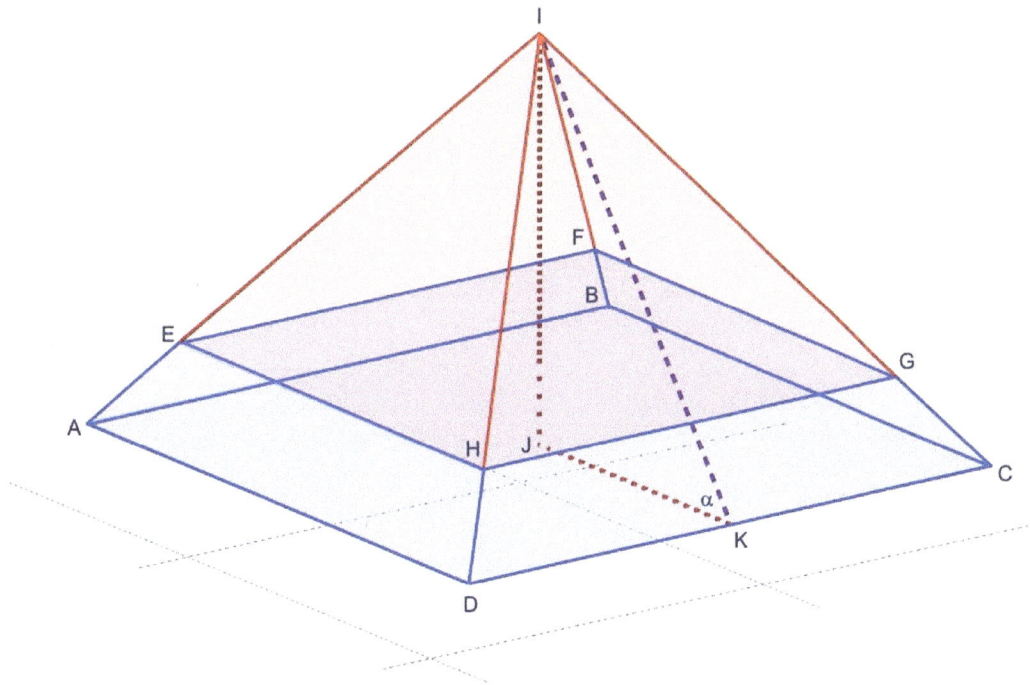

JAMES R WARREN

BLOXWICH
2022

First Published in the United Kingdom in 2022 by Midland Tutorial Productions

First Edition: 1 August 2022

ISBN 978 1 7396296 6 3

Printed and Bound by IngramSpark

Midland Tutorial Productions Publishers
31 Victoria Avenue
Bloxwich
Walsall
WS3 3HS
United Kingdom

MATHEMATICAL EXPLORATIONS

An Album of Research Reporrts

First Edition

James R Warren

MIDLAND TUTORIAL PRODUCTIONS
BLOXWICH

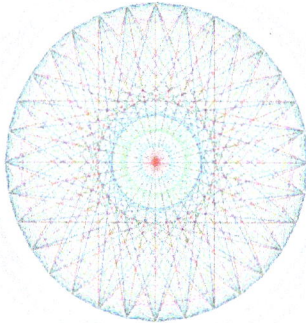
MATHEX WARREN

To The Glory of
The Loving God

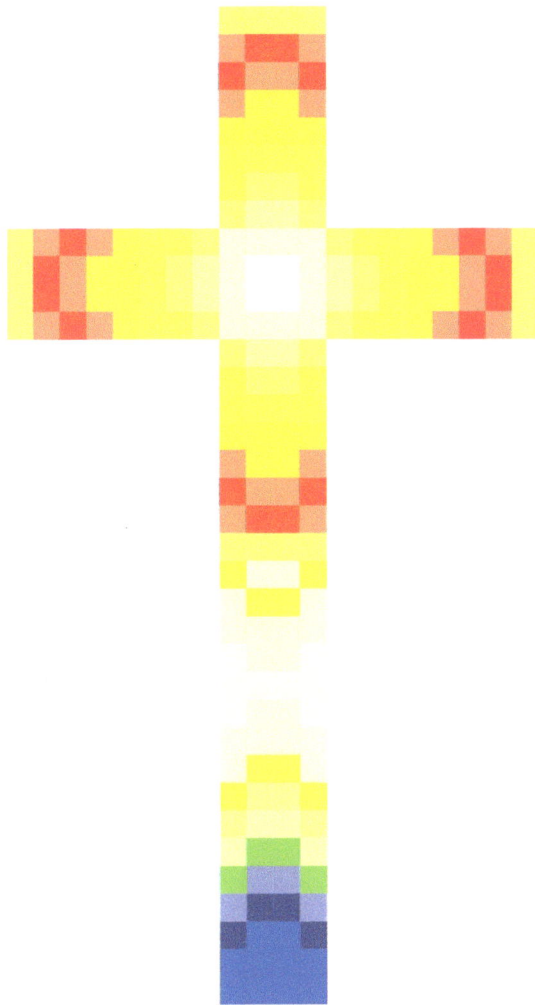

Who Made Our Minds Free

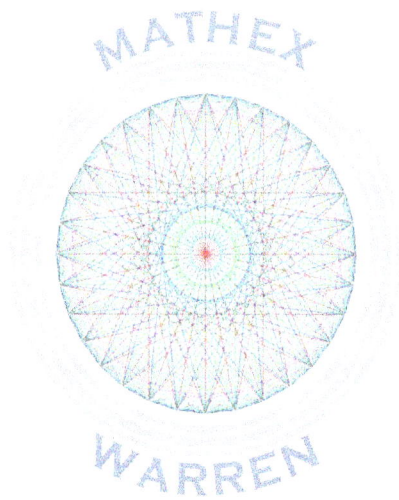

TABLE OF CONTENTS

Page

CHAPTER ONE

Histogram Interpolation

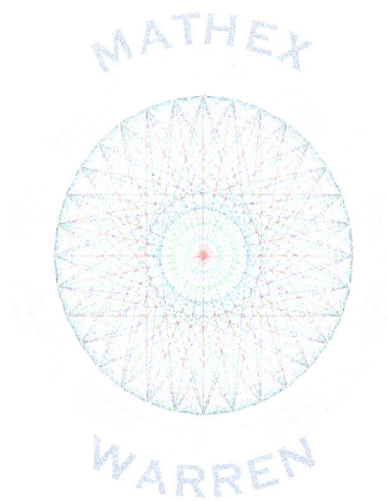

The Reconfiguration of Grouped Frequency Data Binning
As Mediated by Certain Fitting Functions

by
James R Warren BSc MSc PhD PGCE

PART ONE
THE GROUPED MEAN AND GROUPED STANDARD DEVIATION
OF GROUPED DATA FREQUENCIES

By way of example consider the 2005 and 2015 Age Structure data for the UK as provided by Cassie Hayter at the United Kingdom Office for National Statistics[1].

The relevant data specified by Sex and Age is tabulated in Part Two.

There is a very large number of ways in which such data may be summarised as Grouped Frequency Distributions.

One such possibility is to consolidate on Sex and construct a fourteen-interval Grouped Frequency Distribution scheme as presented by Nomisweb[2].

NOMISWEB UK AGE STRUCTURE

LB	UB	Mid Point	Caption	Count	Cumulate
0	4	2	Age 0 to 4	4027092	4,027,092
5	7	6	Age 5 to 7	2413398	6,440,490
8	9	8.5	Age 8 to 9	1541048	7,981,538
10	14	12	Age 10 to 14	3549251	11,530,789
15	15	15	Age 15 to 15	727076	12,257,865
16	17	16.5	Age 16 to 17	1513008	13,770,873
18	19	18.5	Age 18 to 19	1583789	15,354,662
20	24	22	Age 20 to 24	4294683	19,649,345
25	29	27	Age 25 to 29	4441234	24,090,579
30	44	37	Age 30 to 44	12759825	36,850,404
45	59	52	Age 45 to 59	13146744	49,997,148
60	64	62	Age 60 to 64	3501719	53,498,867
65	74	69.5	Age 65 to 74	6339752	59,838,619
75	84	79.5	Age 75 to 84	3745558	63,584,177
85	89	87	Age 85 to 89	969591	64,553,768

$$\Sigma = 64553768$$

Table 1.1
UK Age Structure GFD
According to the Irregular Nomis Presentation

Note that this Grouped Frequency Distribution (GFD) *is constructed from the Cassie Hayter ONS year-wise data* and differs in detail from the GFD on the Nomis website.

Our requirement is to express this same sex-consolidated ONS source data in nine GFD intervals (i.e. nine bins or buckets). That is, we wish to address the GFD shown in Table 1.2:-

DECADES AGE STRUCTURE

LB	UB	Mid Point	Caption	Actual Count	Actual Cumulate
0	9	4.5	Age 0 to 9	7981538	7981538
10	19	14.5	Age 10 to 19	7373124	15354662
20	29	24.5	Age 20 to 29	8735917	24090579
30	39	34.5	Age 30 to 39	8460461	32551040
40	49	44.5	Age 40 to 49	8930109	41481149
50	59	54.5	Age 50 to 59	8515999	49997148
60	69	64.5	Age 60 to 69	7116308	57113456
70	79	74.5	Age 70 to 79	4887212	62000668
80	89	84.5	Age 80 to 89	2553100	64553768
			Totals	64553768	355124008
			Means	7172641	39458223
			SDs	2007116	19494881

Table 1.2
UK Age Structure GFD
According to a Regular Decades Presentation

Note that, as required by simple data conservation, the Total of the Group Frequencies, Σf, is identically 64553768 in each distribution: The instantaneous Census Population.

For a GFD, the Estimated Grouped Mean, $\bar{x}$, is given by:-

$$\bar{x} = \frac{\sum fx}{\sum f}$$

Equation 1.1

where f is the Frequency (Count) in a particular Group or Bin, and x is the Abscissal Interval Mid-Point of that Group.

Also for a GFD, the Standard Deviation is:-

$$\sigma = \sqrt{\frac{\sum fx^2}{\sum f} - \bar{x}^2}$$

Equation 1.2

Let GFD1 be the First Configuration of GFD bins, and GFD2 be the Second. Then, *provided that GFD1 and GFD2 exhaust the identical data set*, the following relations apply:-

$$\bar{x}_{GFD1} = \bar{x}_{GFD2}$$
Equation 1.3

and:-

$$\sigma_{GFD1} = \sigma_{GFD2}$$
Equation 1.4

In the case of statistical engines of (re)distribution such as regression equations, Equations 1.3 and 1.4 are likely to be approximate rather than exact, but a well-chosen engine will act within the data limits of error.

GFD Moments Tableaus for the Nomis and Decade Groupings

Table 1.3 presents the GFD statistical solution tableau for Equations 1.3 and 1.4:-

GROUPED FREQUENCY MOMENTS (all data <u>ONS</u> sourced)

fx_{nomis}	fx^2_{nomis}	fx_{decade}	fx^2_{decade}	
8054184	16108368	35916921	161626145	
14480388	86882328	106910298	1550199321	
13098908	111340718	214029967	5243734179	
42591012	511092144	291885905	10070063705	
10906140	163592100	397389851	17683848347	
24964632	411916428	464121946	25294646030	
29300097	542051785	459001866	29605620357	
94483026	2078626572	364097294	27125248403	
119913318	3237659586	215736950	18229772275	
472113525	17468200425			
683630688	35548795776			
217106578	13460607836			
440612764	30622587098			
297771861	23672862950			
84354417	7338834279			
2553381538	135271158393	2549090996	134964758762	**Totals**
39.55433767		39.48787305		**Grouped Means**
	23.04202280		23.05302366	**Grouped SDs**
		0.16803370		**%SpDef($g\mu_{nomis}, g\mu_{decade}$)**

Table 1.3
UK Age Structure GFD
Grouped Frequency Tableau and Moments
For the Nomis and Decade GFD Patterns

It is clear that, within tolerable error, Equations 1.3 and 1.4 are satisfied and that data values are conserved upon re-binning.

PART TWO
THE RE-DISTRIBUTION OF GROUPED DATA

In our functional re-distribution (re-binning) example we will treat of the 2015 Population data for the United Kingdom as assessed by the UK Office of National Statistics (ONS). This year-of-age data is listed in Table 2.1.

In our context of the re-assignment of ONS 2015 year-wise Population Data we have to consider three arrangements:-

(1) The "original" ONS n = 90 Sub-populations by current personal Age.

We will call this the Data Pair Series $\{a_i,\pi_i\}$ where a_i (treated as independent variable x) is the Age of Person; and π_i is the Number of Persons at that Age.

(2) The NomisWeb pattern n = 15 Sub-populations from the ONS data (<u>NOT</u> the published Nomis data).

We will call this the Data Pair Series $\{UBnomis_i, p_i\}$ where $UBnomis_i$ is the Upper Bound Value (treated as independent variable x) is the Grouped Age Data; and p_i is the Number of Persons in that Age Range.

Grouped Data Statistics employ the Range Mid-Points, $MPnomis_i$, defined as:-

$$MPnomis_i = LBnomis_i + \frac{UBnomis_{i+1} - LBnomis_i}{2}$$

Equation 2.1

(3) The Decades pattern n = 9 Sub-populations from the ONS data (NOT the published Nomis data).

We will call this the Data Pair Series $\{UBdec_i, P_i\}$ where $UBdec_i$ is the Upper Bound Value (treated as independent variable x) is the Grouped Age Data; and P_i is the Number of Persons at that Age.

Grouped Data Statistics employ the Range Mid-Points, $MPdec_i$, defined as:-

$$MPdec_i = LBdec_i + \frac{UBdec_{i+1} - LBdec_i}{2}$$

Equation 2.2

Decades-pattern Data may be:-
(a) Actual
  Gathered by partitioned summations of the ONS data $\{a_i,\pi_i\}$.
(b) Sextic
  Estimated by Sextic Polynomial Regression of Cumulates of the Actual Decade-wise Binned Data Cumulates $\{UBdec_i,C_i\}$.
(c) 14-Degree
  Estimated by Fourteen-Degree Polynomial Regression of the Actual Decade-wise Binned Data Cumulates $\{UBdec_i,C_i\}$.

Table 2.1
UK Populations

UK POPULATIONS

Age	Male 2015	Female 2015	2015
0	398539	378230	776769
1	403020	383105	786125
2	413548	393156	806704
3	427117	406959	834076
4	421505	401913	823418
5	413235	394805	808040
6	408987	390552	799539
7	412479	393340	805819
8	399245	379516	778761
9	389988	372299	762287
10	375072	357559	732631
11	368574	351456	720030
12	358439	341255	699694
13	352759	336249	689008
14	361698	346190	707888
15	373264	353812	727076
16	385048	365519	750567
17	391202	371239	762441
18	403985	383393	787378
19	410770	385641	796411
20	414063	393000	807063
21	431038	409522	840560
22	433886	418170	852056
23	446514	439399	885913
24	462222	446869	909091
25	455341	439199	894540
26	446764	440630	887394
27	448965	448203	897168
28	434633	442025	876658
29	443500	441974	885474
30	443494	443595	887089
Totals	12728894	12248774	24977668
Means	410609.48	395121.7419	805731.226

UK POPULATIONS

Age	Male 2015	Female 2015	2015
31	431902	434663	866565
32	434309	439796	874105
33	432286	441050	873336
34	437099	443329	880428
35	437332	443039	880371
36	420414	424046	844460
37	392107	395566	787673
38	387285	388680	775965
39	392441	398028	790469
40	399874	406065	805939
41	405771	413492	819263
42	425003	431616	856619
43	443067	450901	893968
44	454825	468750	923575
45	444995	457361	902356
46	455570	469734	925304
47	456082	469923	926005
48	464148	474510	938658
49	461581	476841	938422
50	465548	480584	946132
51	461048	475891	936939
52	452399	464911	917310
53	443862	453646	897508
54	429088	438360	867448
55	411759	420874	832633
56	402345	412272	814617
57	393423	403022	796445
58	378415	388841	767256
59	364674	375037	739711
60	350167	362126	712293
61	349645	363721	713366
Totals	13078464	13366675	26445139
Means	421885.94	431183.0645	853069

UK POPULATIONS

Age	Male 2015	Female 2015	2015
62	345,319	358,762	704,081
63	333,812	349,821	683,633
64	335,678	352,668	688,346
65	340,422	360,410	700,832
66	348,975	367,768	716,743
67	366,002	387,904	753,906
68	395,680	418,856	814,536
69	303,869	324,703	628,572
70	292,273	315,276	607,549
71	290,461	313,740	604,201
72	267,272	294,594	561,866
73	236,273	265,034	501,307
74	210,093	240,147	450,240
75	215,612	247,443	463,055
76	210,995	244,995	455,990
77	202,447	237,101	439,548
78	188,015	225,678	413,693
79	174,775	214,988	389,763
80	162,277	203,186	365,463
81	145,722	188,243	333,965
82	133,787	178,566	312,353
83	124,667	172,237	296,904
84	112,442	162,382	274,824
85	98,571	150,471	249,042
86	84,424	133,863	218,287
87	71,190	118,573	189,763
88	61,372	105,902	167,274
89	51,143	94,082	145,225
Totals	6103568	7027393	13130961
Means	217984.57	250978.3214	468962.893

64553768 UK 2015 Population Total

Formation of Grouped Data Bins

NomisWeb Pattern Data Group Formation
Bin Content
The NomisWeb Groups are of irregular age-range widths as shown in Table 2.2:-

NOMISWEB UK AGE STRUCTURE

LB	UB	Mid Point	Caption	Count	Cumulate
0	4	2	Age 0 to 4	4027092	4,027,092
5	7	6	Age 5 to 7	2413398	6,440,490
8	9	8.5	Age 8 to 9	1541048	7,981,538
10	14	12	Age 10 to 14	3549251	11,530,789
15	15	15	Age 15 to 15	727076	12,257,865
16	17	16.5	Age 16 to 17	1513008	13,770,873
18	19	18.5	Age 18 to 19	1583789	15,354,662
20	24	22	Age 20 to 24	4294683	19,649,345
25	29	27	Age 25 to 29	4441234	24,090,579
30	44	37	Age 30 to 44	12759825	36,850,404
45	59	52	Age 45 to 59	13146744	49,997,148
60	64	62	Age 60 to 64	3501719	53,498,867
65	74	69.5	Age 65 to 74	6339752	59,838,619
75	84	79.5	Age 75 to 84	3745558	63,584,177
85	89	87	Age 85 to 89	969591	64,553,768
			Totals	64553768	443426216

Table 2.2
NomisWeb UK Age Structure

Bin Content, $Bnomis_i$, is computed using the partial series summations:-

$$Bnomis_i = \sum_{i=LB}^{UB} p_i$$

Equation 2.3

where LB and UB are respectively values of ONS a_i that define the primitive series year-wise Population Estimates.
Cumulates
The relevant Cumulate, $Cnomis_i$, Series is defined by:-

$$Cnomis_j = \sum_{l=0}^{j} \sum_{i=LB_j}^{UB_j} p_i = \sum_{k=0}^{UB_j} \pi_k$$

Equation 2.4

For the re-assignment of counts to new configurations of grouped distributions a very large number of approaches suggest themselves.

I have essayed or considered a small subset of these, and in particular some methods that are sophistications of the Algebraic Polynomial Equation (APE).

Among the advantages of the APE is that it is by definition continuous; it is integrable and multiply-differentiable in all parts; it is simple and versatile; and it is amenable to many types of development.

It has disadvantages: The APE is not necessarily realistic from a scientific point of view. Far more seriously it is *vulnerable to dramatic interpunctual excursion* which may yield ludicrously-extreme dependent outcomes between data points, or even the entire de-routing of the functional trajectory at extrema. This problem is called The Runge Phenomenon and a bugbear, to a greater or a lesser extent, of all APE-based functional fitments. The Runge Phenomenon may be tolerable for simpler polynomial equation fitments to few data pairs, but becomes a serious limitation to higher-degree approximations of larger data swarms. So unwise applications of distributive functions, especially in demographic or political theory, may lead us to conclusions that are the very opposite of fact.

In general, the Algebraic Polynomial Equation may be specified by:-

$$P_n(x) = c_0 + c_1 x + c_2 x^2 + \cdots + c_n x^n = \sum_{i=0}^{n} c_i x^i$$

Equation 2.5

where $P_n(x)$ is a Polynomial Equation (a function of Independent Variable x) or Order n; and c_0 ... c_n are Coefficients to be determined.

An APE of the First Degree has $c_1 x$ as its highest term and describes a straight line curve: It is a Linear Equation of $n = 1$. The Quadratic Equation is $n = 2$; the Cubic $n = 3$; the Quartic $n = 4$; and so on. Only APEs of $n<6$ have *analytic* solutions.

The Degree of an APE can only be as high as one less than the number of data pairs to be fitted, i.e. $n-1$. Accordingly, I had reason to study the case of $n = 14$, because that is one less than the number of Bin Upper Bounds in the NomisWeb interpretation of the UK Population Distribution.

In this research I tried two basic methods of fitting an Algebraic Polynomial Equation to the pseudo-continuous Cumulative Data Series $\{UB_i, C_i\}$: The Lagrange Interpolation Polynomial (LIP), and the Polynomial Regression (PolyReg). A PolyReg is a mathematical statistical process that is essentially a sophistication of Multiple Regression. The aim of both methods is to define the set of Coefficients c_0 ... c_n, in a scientifically-tenable way, or in other words so that the resulting functional equation accurately mirrors the data relationship.

The Lagrange Interpolation Polynomial is defined as:-

$$p_n(x) = \sum_{j=0}^{n} y_j \cdot \mathcal{L}_{n,j}(x) = \sum_{j=0}^{n} y_i \prod_{k=0,k\neq j}^{n} \frac{x - x_k}{x_j - x_k}$$

Equation 2.6

where $p_n(x)$ is the Lagrange Interpolation Polynomial; y_i is the ith. Independent Variable (ex data); and $\mathcal{L}_{n,j}(x)$ is the Lagrange Polynomial. Lambers[3] shows how a LIP can be reduced to an ordinary but unique APE.

Clearly, the LIP process in its pure Equation 2.6 form is an order n^2 process, but fast methods are feasible, notwithstanding their manifest dangers.

The LIP provides an "exact" rendezvous of the fitted function with every data point, but as we have remarked it can really go a wandering and hike a real holiday elsewhere.

Whilst the LIP function is beguilingly accurate for the ages of minority Table 2.3 shows how LIP fails utterly to represent the Cumulate Function (CF) beyond the bin whose UB age is 19: The curve goes into a dramatic nose-dive.

I used the dCode[4] LIP tool, but I imply no mis-programming by the French authors: The failure of LIP is intrinsic to its mathematics, not the fault of any man.

Accordingly, I abandoned Lagrange Interpolation as a potential engine of grouped frequency distribution.

Actual Cumulate	Synthetic Cumulate	SpDef%
4027092	4.027E+06	-0.006536
6440490	6.441E+06	-0.003250
7981538	7.980E+06	0.014483
11530789	1.151E+07	0.215084
12257865	1.222E+07	0.303761
13770873	1.369E+07	0.559618
15354662	1.521E+07	0.954704
19649345	1.908E+07	2.912757
24090579	2.232E+07	7.349777
36850404	1.310E+07	64.461006
49997148	-1.104E+08	320.858919
53498867	-2.216E+08	514.215288
59838619	-6.666E+08	1213.977954
63584177	-1.645E+09	2687.039407
64553768	-2.464E+09	3917.366171
-4.982E+09	8.730E+03	**Total**
	582.014610	**Mean**

Table 2.3
The Utter Failure of 15-point Lagrange Fitment to
NomisWeb-Style Binning Upper Bounds

I am told it is no use trying Neville's Method or any other fashionable technique of "exact" interpolation, because these are affected by The Runge Phenomenon to a greater or lesser degree.

Now Polynomial Regression is guaranteed never to visit a given data-point but it may approach very near; and if used sensibly it tends to smooth the fitted curve (i.e. the fitted APE); and to suppress the effects of The Runge Phenomenon.

Bearing these facts in mind, I also tried higher-order Polynomial Regressions on the ONS data as distributed in the NomisWeb pattern: The graphical Sextic Regression (n = 6) furnished by EXCEL®, and also an n-1 = 14 Polynomial Regression mediated by Arachnoid[5]. As with my comparison of LIP to manually-distributed reality, I assessed the viably of the several fits using both point-wise and summative calculations of Percentage Specific Defect defined in this way:-

$$\%Sp.Def. = 100 \left(\frac{y_{theo} - y_{meth}}{y_{theo}} \right)$$
Equation 2.7

where %Sp.Def. is the local Percentage Specific Defect; y_{theo} is the actual (Fiducial) Dependent Variable outcome and y_{meth} is the Dependent outcome according to the method or function under test.

Before we progress to further discussion of the distribution function fitments we may briefly touch upon the Verhulstian Logistic Regression paradigm traditional in demographic work if only to eliminate it in our context.

An APE-dependant Logistic Regression is designed to reflect the allegedly sigmoidal character of population cumulates and may be defined as:-

$$L_n(x) = \frac{K}{1 + e^{-P_n(x)}}$$
Equation 2.8

where $L_n(x)$ is the Logistic Function value at x; K is a Scaling Constant; and $P_n(x)$ is a fitted APE. From a pedantically technical point of view, K is redundant and could be replaced with unity, but K is often of course convenient, especially in contexts like ours where the magnitudes of the independent and dependent variables differ by several orders of magnitude.

As illustrated in Figure 2.1, study of the plotted cumulates attaching to any of the ONS data binnings shows that our UK relationship of Cumulate Population to Age is quasi-linear except in the asymptotic reach of the highest ages. Therefore Logistic Regression is a potentially-troublesome complication of what is a thoroughly non-classical population structure. Therefore, I decided to stick with simple APE models.

Regression Coefficients

The aim of any APE-related procedure is to identify reliable term coefficients. Table 2.4 presents those for both the Sextic (EXCEL®) and the 14-Degree (Arachnoid®) polynomial regressions, whose Coefficients of Determination R^2 are respectively 0.99994396 and 0.99999981. Both values suggest analytic determinacy, but of course this is hardly a pragmatic assumption for social or demographic data, and we shall see that these Coefficients of Determination are betrayed by functional behaviours in detail.

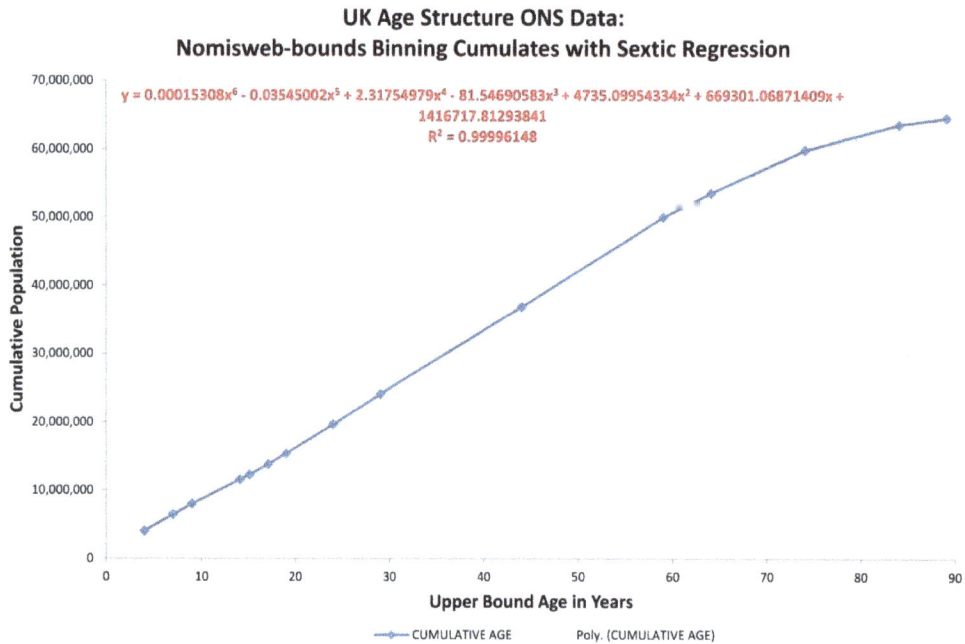

UK Age Structure ONS Data:
Nomisweb-bounds Binning Cumulates with Sextic Regression

$y = 0.00015308x^6 - 0.03545002x^5 + 2.31754979x^4 - 81.54690583x^3 + 4735.09954334x^2 + 669301.06871409x + 1416717.81293841$
$R^2 = 0.99996148$

Figure 2.1
The Linear Habit of the UK Population Cumulate

SEXTIC REGRESSION		14-DEGREE REGRESSION	
R^2	0.99994396	R^2	0.99999981
c_0	892033.53889465	c_0	2.84268698350006E+05
c_1	780526.33928680	c_1	1.01286616389361E+06
c_2	-2779.22236347	c_2	-2.12772728157855E+04
c_3	157.65489954	c_3	7.21896482074150E+02
c_4	-1.54234634	c_4	-1.15503512867146E+02
c_5	-0.00491915	c_5	1.31725410291557E+01
c_6	0.00005887	c_6	-5.92713919669502E-01
		c_7	9.96880278061684E-03
		c_8	8.38961893426772E-05
		c_9	-6.22853879518785E-06
		c_{10}	1.04351634082902E-07
		c_{11}	-7.53542172363245E-10
		c_{12}	1.08085074798085E-12
		c_{13}	1.50255828913638E-14
		c_{14}	-6.18507199159515E-17

Table 2.4
Regression Coefficients

My implementation of EXCEL® is, in practice, restricted to a 14-figure presentational accuracy and accordingly the full precision of the Arachnoid computation has been slightly reduced.

Re-Distribution

The posterior re-distribution of population counts into a re-configured sequence of bins, for example the fifteen NomisWeb bins to our nine decade-spanning buckets, is in essence the reverse of the prior formation of n Prior Cumulates, C_i.

Because we employ the mediation of an APE distributor function we may define the new Posterior Cumulate values D_i using:-

$$D_j(x) = P_n(x)$$
Equation 2.9

for all the UB_j Upper Bound values in the posterior binning.

Having the m D_j to hand for the m new buckets we may now specify the posterior 1 ... m-1 Bin Counts E_j in terms of:-

$$E_{j+1} = D_{j+1} - D_j$$
Equation 2.10

where:-

$$E_0 = D_0$$
Equation 2.11

Ideally, a function and its trajectory that fits the paired data may visit each plotted point; be smoothly continuous of steady first and subsequent derivative; and everywhere be steady, economical and rational.

Fortunately or otherwise, precision is as elusive as exactitude unattainable. For in such fallacy knowledge is shown itself impossible: Without shortcomings and mutabilities progress also would elude us.

PART THREE
THE BEHAVIOR OF
POLYNOMIAL BINNING FUNCTIONS

Having cogent sets of Polynomial Term Coefficients, c_{term}, for both the Sextic and Fourteen-Degree PolyReg fitments we are in a position to use the respective Algebraic Polynomials $P_6(x)$ and $P_{14}(x)$ to establish the posterior Decade Grouped Frequency bucket series, which if desired may be plotted as histograms or other convenient diagrams.

Without immense effort, or the concomitant vagaries of that, we can test the fidelity of the results by using point-wise and summative Population Specific Defect, as well as (summative) Grouped Frequency Means and Standard Deviations.

Table 3.1 presents the results for the Actual Decade Bin Counts from the ONS annual age data for combined males and females in the year 2015, and parallel grouped frequency distributions for the Sextic and 14-Degree binnings.

Inspection discloses a few salient facts. The Census Population $\Sigma\pi = 64553768$ is conserved only in the case of the final Actual Cumulate of the decade wise binning. Both the $P_6(x)$ and $P_{14}(x)$ re-distributions compromise count to some little extent; the former to the loss of 49097 individuals and the latter to the loss of 5.

Bin-wise Percent Specific Defect for $P_6(x)$ varies between around -5.7 to 0.4 and the mean PSD is 0.9737, call it one percent. One the other hand, bin-wise PSD for the higher-degree $P_{14}(x)$ system is remarkably tight at around -0.02 to 1.74 for ages less than 59, but shows catastrophic failure for the older age ranges to 17.4185 in the 80-89 bucket. This is quite counter-intuitive and re-emphasises the caution we must apply in the treatment of higher-order APEs: For even PolyReg is not immune to The Runge Phenomenon.

As in life, it is all swings and roundabouts. You lose fewer men and women, but pay with less knowledge of the losses at the older ages.

The respective differences in Standard Deviation reflect, at 3.3498 for the Sextic and 8.5908 for the 14-Degree, this greater uncertainty of the expensive, higher-order, processes.

Grouped Moments

Table 3.2 shows the Grouped Frequency Momental Statistics for the Actual, Sextic and 14-Degree Grouped Frequency Distributions (GFDs), as well as summative PSDs for the latter.

The near-identities of the Grouped Means on the one hand for all three distributions, and the similar near-identity for the Grouped Standard Deviations indicate the broad Conservation of Data and the overall efficacy of the PolyReg process. Once again, however, disparities are in the detail.

Population Specific Defect (PSD) is always an effective discriminant of discrepancy: For the Sextic $P_6(x)$ fitment it is 0.20075333, and for the 14-Degree $P_{14}(x)$ it is 0.13329815. It is a big difference, but as we have seen we cannot say that the higher-degree fit is one and a half times better. That would be an abuse of the statistics, and an abuse of thought itself.

DECADES AGE STRUCTURE

LB	UB	Mid Point	Caption	Actual Count	Actual Cumulate	Sextic Synthetic Count	Sextic Synthetic Cumulate	Bin-Wise Count SpDef%	14-Degree Synthetic Count	14-Degree Synthetic Cumulate	Bin-Wise Count SpDef%
0	9	4.5	Age 0 to 9	7981538	7981538	7796205	7796205	2.3220	7957101	7957101	0.3062
10	19	14.5	Age 10 to 19	7373124	15354662	7793473	15589679	-5.7011	7402046	15359148	-0.3923
20	29	24.5	Age 20 to 29	8735917	24090579	8288585	23878264	5.1206	8737975	24097122	-0.0236
30	39	34.5	Age 30 to 39	8460461	32551040	8774225	32652489	-3.7086	8487325	32584448	-0.3175
40	49	44.5	Age 40 to 49	8930109	41481149	8894342	41546831	0.4005	8774313	41358761	1.7446
50	59	54.5	Age 50 to 59	8515999	49997148	8377970	49924801	1.6208	8639180	49997941	-1.4465
60	69	64.5	Age 60 to 69	7116308	57113456	7081612	57006413	0.4876	6683721	56681662	6.0788
70	79	74.5	Age 70 to 79	4887212	62000668	5031630	62038043	-2.9550	5763723	62445385	-17.9348
80	89	84.5	Age 80 to 89	2553100	64553768	2466628	64504671	3.3869	2108387	64553773	17.4185
			Totals	64553768	355124008	64504671	354937397	0.9737	64553773	355035341	5.4335
			Means	7172641	39458223	7167186	39437489	0.1082	7172641	39448371	0.6037
			SDs	2007116	19494881	1992772	19494339	3.3498	2039431	19510220	8.5908

Table 3.1
Decade Age Structure

GROUPED MOMENTS (all data ONS sourced)

Actual fx_{nomis}	Actual fx^2_{nomis}	Sextic fx_{ons}	Sextic fx^2_{ons}	14-Degree fx_{ons}	14-Degree fx^2_{ons}	
8054184	16108368	35082925	157873161	35806956	161131303	
14480388	86882328	113005364	1638577773	107329673	1556280253	
13098908	111340718	203070343	4975223413	214080379	5244969280	
42591012	511092144	302710765	10443521396	292812726	101020039039	
10906140	163592100	395798215	17613020585	390456929	17375333341	
24964632	411916428	456599343	24884664219	470835312	25660524489	
29300097	542051785	456763982	29461276840	431100030	27805951955	
94483026	2078626572	374856452	27926805668	429397377	31990104611	
119913318	3237659586	208430051	17612339320	178158736	15054413220	
472113525	17468200425					
683630688	35548795776					
217106578	13460607836					
440612764	30622587098					
297771861	23672862950					
84354417	7338834279					
2553381538	135271158393	2546317441	134713302376	2549978118	134950747491	**Totals**
39.55433767	23.04202280	39.47493102	23.02513646	39.50161247	23.02475680	**Grouped Means** **Grouped SDs**
			0.20075333		0.13329815	$\%\mathbf{SpDef}(g\mu_{actual}, g\mu_{Pn(x)})$

Table 3.2
Grouped Moments for the
Actual and Synthetic Decade Binnings

Re-Binning Departure from Truth

We adumbrated above the way in which re-distributions of data compromise accuracy whilst increasing precision, often in beguiling but misleading ways that vary between different *parts* of the new distribution. Earlier, we discussed at some length the intrinsic weaknesses of some sophisticated mechanisms and their potential for disastrous error, such as completely to confound the naïve black-box user.

In this final part of our study we shall look a little more closely at Departure from Truth.

Table 3.3 isolates those parts of Table 3.1 that treat of bin-wise Population Specific Defect (PSD) and adds summative calculations of PSD which are respectively 0.0760558876 and -0.0000074833 for $P_6(x)$ and $P_{14}(x)$. The four order of magnitude difference in absolute PSD might convince of the utter and decisive superiority of higher-order determinations, but we have already seen that in this case at least the lower-order distribution is generally the more accurate.

RE-BINNING DEPARTURE FROM TRUTH

SEXTIC Bin-Wise Count SpDef%	14-DEGREE Bin-Wise Count SpDef%	
2.3220	0.3062	
-5.7011	-0.3923	
5.1206	-0.0236	
-3.7086	-0.3175	
0.4005	1.7446	
1.6208	-1.4465	
0.4876	6.0788	
-2.9550	-17.9348	
3.3869	17.4185	
0.9737	5.4335	**Totals**
0.1082	0.6037	**Means**
3.3498	8.5908	**SDs**
0.0760558876	-0.0000074833	**%SpDef (Means)**

Table 3.3
The Re-Binning Departures from Truth

PART FOUR
GRAPHICAL COMPARISONS

Comparative diagrams illustrate the disparities and differences we have discussed in a dramatic way.

First of all, let us compare the raw year-by-year histogram of the raw ONS yearly UK Population Data as forwarded by Ms Cassie Hayter (Figure 4.1) with the general conformation of the Actual ONS-based Nomisweb-boundaries (Figure 4.2) histogram and the Actual ONS-based Decade-boundaries histogram (Figure 4.3):-

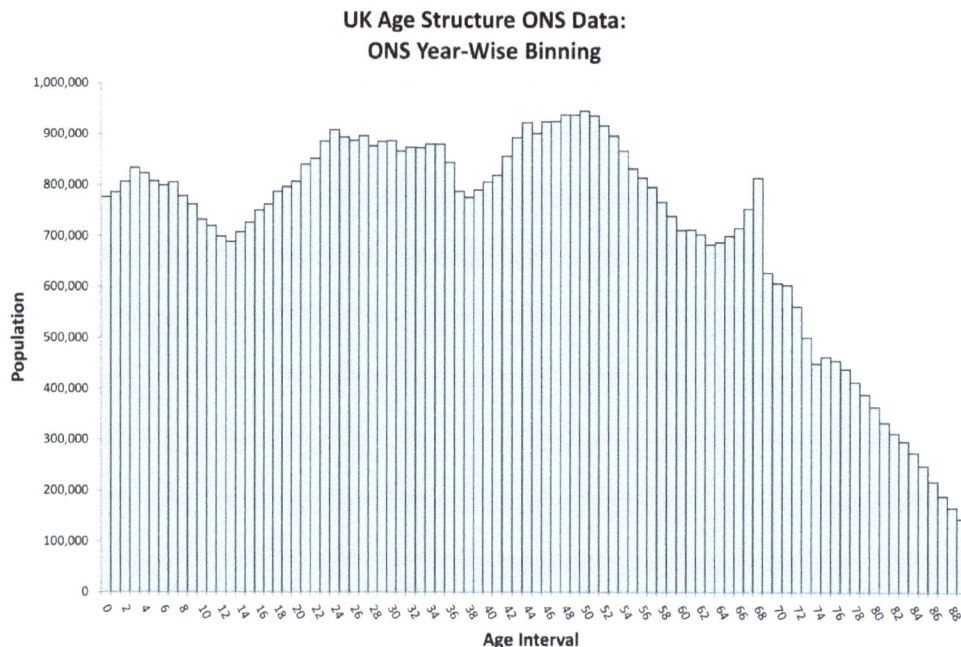

UK Age Structure ONS Data:
ONS Year-Wise Binning

Figure 4.1
The Raw ONS Data Population Histogram

Though the Figure 4.2 Nomisweb-pattern conformation is something similar to the given ONS data appearance, the Figure 4.3 Decades binning looks nothing like either. And yet the descriptive statistics confirm that this is all the same data, yielding the same information.

It shows how deceptive appearances can be.

This point is perhaps reinforced by Figure 4.4 that illustrates a comparison of the Nomisweb-pattern Actual data bin populations with the decade-wise $P_6(x)$ and $P_{14}(x)$ renditions. The drastic smoothing of the polynomial regression decade binnings implies a drastic loss of information, and yet this is belied by the Data Conservation we have already observed.

The regressions vaguely agree except that the $P_{14}(x)$ grouped frequency curve is a little erratic.

Finally, in that latter context of the two polynomial regression fitments we may illustrate Departure from truth with Figure 4.5. Total agreement with the fiducial should follow the zero abscissa. We can see that the Sextic $P_6(x)$ is erratic but consistently-so along its course, whereas the 14-Degree $P_{14}(x)$ fitment is faithful until Age 59 and then fluctuates wildly.

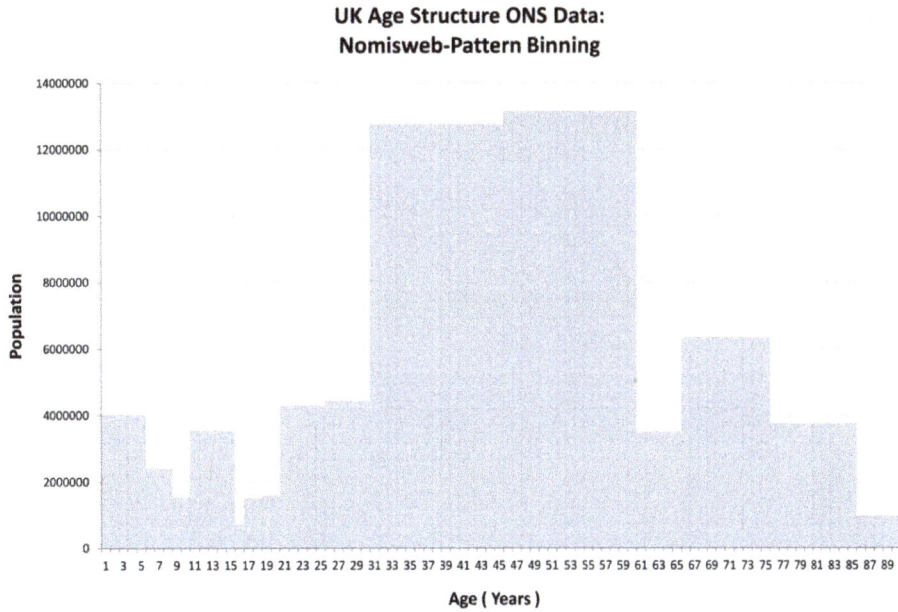

Figure 4.2
The Actual-Data Nomisweb-boundaries UK Population Histogram

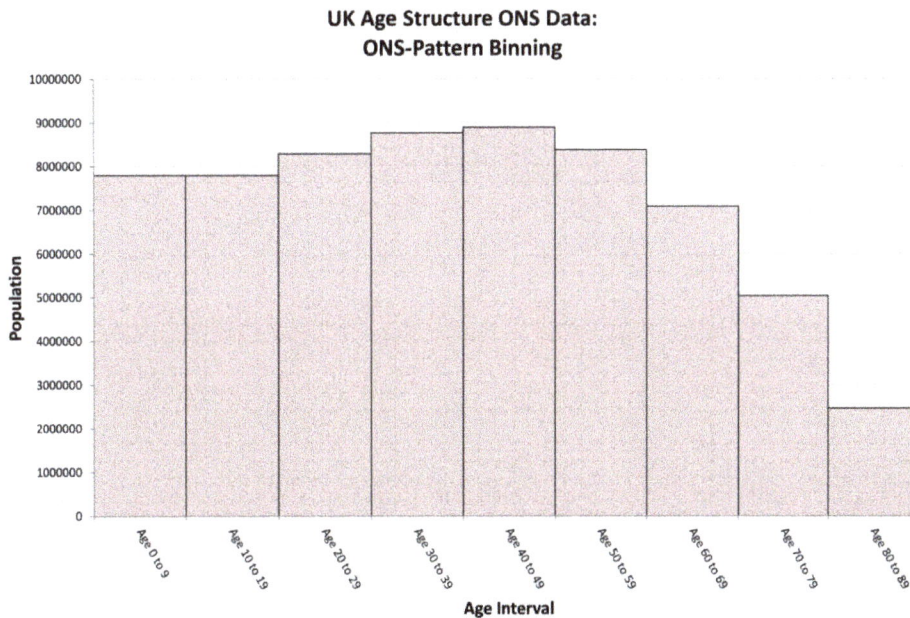

Figure 4.3
The Actual-Data Decades-boundaries UK Population Histogram

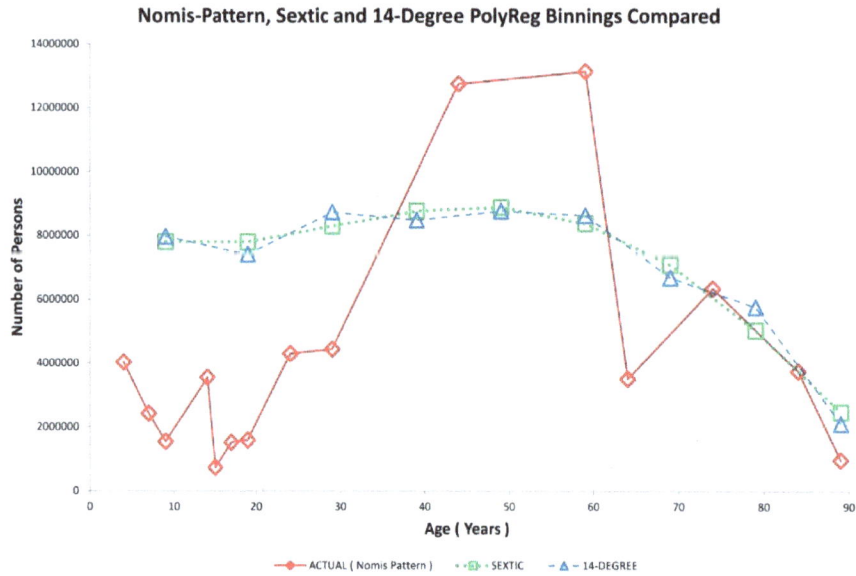

Figure 4.4
The Actual-Data according to Nomisweb Binning
In Comparison to
Sextic and 14-Degree Age Upper Bound versus Cumulate
Decade-wise Binnings

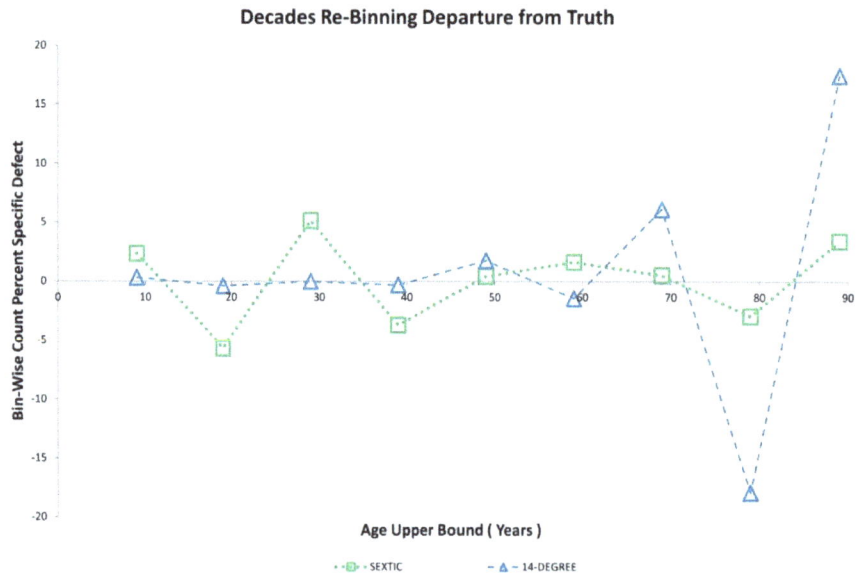

Figure 4.5
The Fidelity of the Fitted Sextic and 14-Degree Regression
Distributor Functions
In Terms of PSD Departure from Truth

References

1   ONS
The United Kingdom Office for National Statistics
1 Drummond Gate
Pimlico
London
SW1V 2QQ

Population Estimates for UK, England and Wales,
Scotland and Northern Ireland: mid-2012 to mid-2016
MYE series worksheets (EXCEL®)
Compiled by Laura Todd and Cassie Hayter
© Crown copyright 2018

https://www.ons.gov.uk/peoplepopulationandcommunity/
populationandmigration/
populationestimates/datasets/
populationestimatesforukenglandandwalesscotlandandnorthernireland

2   NomisWeb
Official Labour Market Statistics

Nomis
Durham University
Department of Geography
Lower Mountjoy
South Road
Durham
DH1 3LE
UK

https://www.nomisweb.co.uk

3   Lambers

James V Lambers
MAT 772: Numerical Analysis for Computational Science
August 23 2016
The University of Southern Mississippi

Jim Lambers
MAT 772
Fall Semester 2010-11
Lecture 5 Notes

http://www.math.usm.edu/lambers/mat772/book772.pdf

4   dCode
Lagrange Interpolating Polynomial Calculator
Source : https://www.dcode.fr/lagrange-interpolating-polynomial
© 2018 dCode

https://www.dcode.fr/lagrange-interpolating-polynomial

5   Arachnoid
Polynomial Regression Data Fit
Copyright © 2013, P. Lutus
This is the JavaScript version of PolySolve (07.20.2013)

https://arachnoid.com/polysolve/

CHAPTER TWO

Pyramid Planning

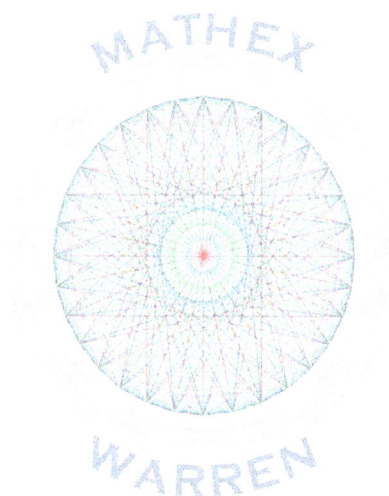

The Drafting and Ground Geometry of the Ancient Egyptian Pyramids

by
James R Warren BSc MSc PhD PGCE

PART ONE
THE MENSURATION OF THE REGULAR RIGHT PYRAMID

<u>Pyramid Anatomy</u>

Mathematically, a regular right pyramid is any Euclidean ideal solid whose base is a regular n-sided polygon and which has an apex perpendicularly above the polygon center.
Figure 1.1 illustrates the conformation for the case of a classical square pyramid:-

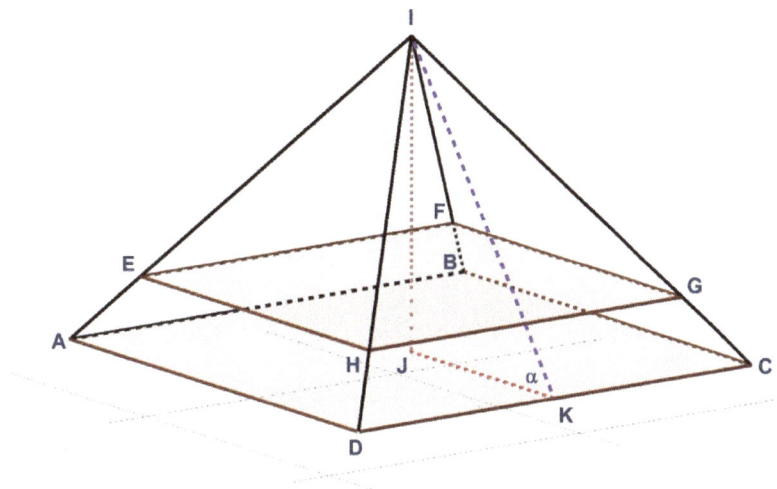

Figure 1.1
Square Pyramid Conformation

A, B, C and D are vertices that define the Base of the pyramid. The centre of the Base, which is a regular tetragon (square), sits at J. The normal, JI, connects the Base to the Apex at I. The Height, h, is constituted by that perpendicular IJ; whilst the Apothemal Radius, b, is exemplified by JK which is perpendicular to the Basel Edge DC. The hypotenuse IK constitutes the Slant Height f.

The Basal Edges AB, BC, CD and DA are, each of length S, so that it follows that DK and KC are each S/2 in length.

Angle AJD is β which is equivalent to 2π/n, where n is the number of Inclined Sides of the Pyramid.

Knowledge of n, S, and h (or alternatively f) wholly defines a regular right pyramid.

The plane EFGH is parallel to the Base ABCD and is the Medial Plane that divides the volume of the whole pyramid in half. Thus the solid figure ABCDEFGH constitutes a Frustum of the Pyramid. It is what Egyptologists might describe as a mastaba, a distinct truncated pyramidal structure that was also constructed by the Ancients.

Figure 1.2 clarifies the division of the equal volumes with a blue tinted frustum surmounted by a red sub-pyramid:-

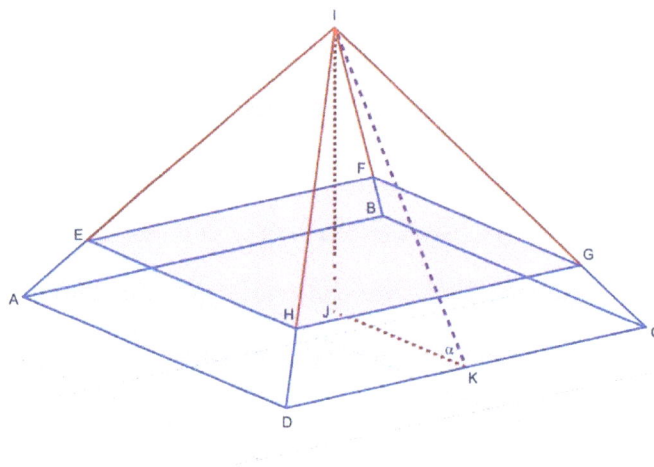

Figure 1.2
Square Pyramid Frustum and Sub-Pyramid

For further clarification, Figure 1.3 illustrates the assignment of notation to the various lineaments of a pyramid, and to the angle β=2π/n:-

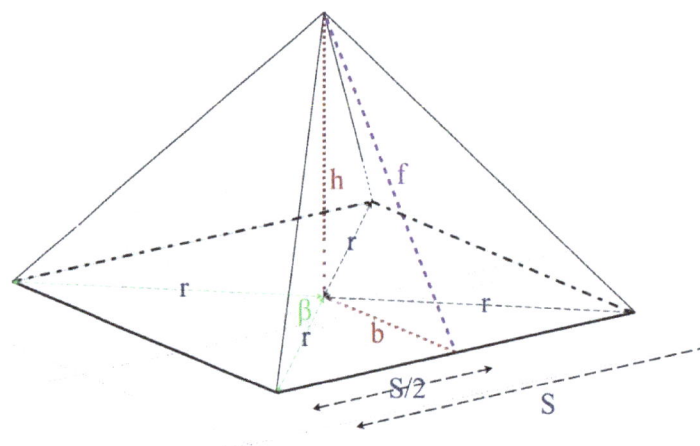

Figure 1.3
Pyramid Notation

Analytic Geometry

We have seen that the fundamental Vertex Interangle β is given by:-

$$\beta = \frac{2\pi}{n}$$
Equation 1.1

where π is the Ludolphine Constant and n is the Number of Sides of the Basal Regular Polygon. It follows immediately that:-

$$\frac{\beta}{2} = \frac{\pi}{n}$$
Equation 1.2

a fact that may assist simplifications.

By Pythagoras it follows that Edge Length s (not the Basal Edge Length, S) is given by:-

$$s = \sqrt{f^2 + \left(\frac{S}{2}\right)^2} = \sqrt{f^2 + \frac{S^2}{4}}$$
Equation 1.3

whilst the Basal Vertex Radius, r, is similarly yielded by:-

$$r = \sqrt{\frac{S}{2(1 - \cos(\beta))}}$$

$$= \sqrt{S^2 \cdot \csc^2\left(\frac{\beta}{2}\right)}$$

$$= \frac{S}{2}\csc\left(\frac{\beta}{2}\right)$$

$$= \frac{S}{2}\csc\left(\frac{\pi}{n}\right)$$
Equation 1.4

Similarly the Basal Apothemal Radius, b, is given by:-

$$b = \sqrt{r^2 - \frac{S^2}{4}}$$

$$= \sqrt{\frac{S^2}{4}\csc^2\left(\frac{\pi}{n}\right) - \left(\frac{S^2}{4}\right)}$$

$$= \sqrt{\frac{S^2}{4}\left(csc^2\left(\frac{\pi}{n}\right) - 1\right)}$$

$$= \frac{S}{2}\sqrt{cot^2\left(\frac{\pi}{n}\right)}$$

$$= \frac{S}{2}cot\left(\frac{\pi}{n}\right)$$

Equation 1.5

If the Slant Height f is known and we wish to compute the Apical Height h we may employ:-

$$h = \sqrt{f^2 - b^2}$$

$$= \sqrt{f^2 - \left[r^2 - \frac{S^2}{4}\right]}$$

$$= \sqrt{f^2 - \left[\frac{S}{2(1-\cos(\beta))} - \frac{S^2}{4}\right]}$$

$$= \sqrt{f^2 - \frac{S^2}{2}\left[\frac{1}{(1-\cos(\beta))} - \frac{1}{2}\right]}$$

Equation 1.6

As we shall see, Equation 1.6 is not necessarily in its most reduced terms.
The Figure Radius, R, is invariably bigger than either r or b and may be expressed as:-

$$R = b + f$$

$$= \sqrt{\frac{S^2}{4}csc^2\left(\frac{\pi}{n}\right) - \left(\frac{S^2}{4}\right)} + f$$

Equation 1.7

Again, R is not necessarily reduced to its simplest terms.

Planar Forms of Pyramid Sides

Sometimes children are encouraged to model regular pyramids and other ideal solids by drafting side outlines on stiff card; cutting the outlines; and then folding along the basal edges to form a three-dimensional example.
Many pedagogues consider that gives youngsters a natural and practical appreciation of solid geometry.

We may speculate that the Ancients may have approached pyramid planning in a similar fashion by perhaps drawing to scale upon papyrus or parchment, and then replicating the design upon prepared ground, perhaps with ropes, pegs and water levels.

Further to explore this concept we may examine three planar expressions of simple pyramids.

Case A

Case A is a regular tetrahedron with of course four equilaterally triangular sides whose six edges are of four units (forty graphical millimeters).

Figure 1.4 illustrates the planar representation:-

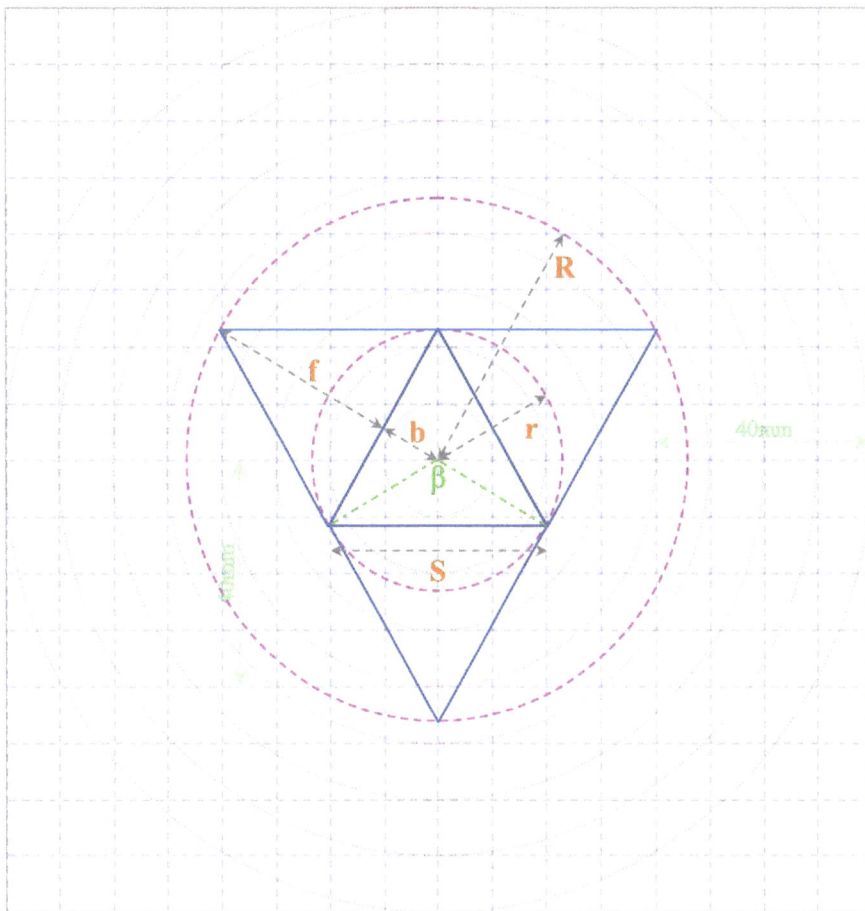

Figure 1.4
Planar Form of a Regular Tetrahedron

The nature of the Figure Radius, R, is now clear and the student is invited mentally to fold the three outer triangles along the outer lines of the central triangle and bring the three outermost R-contingent vertices to meet at an apex orthogonal to the plane of the paper.

Case B

Case B is a square pyramid of an aspect common to several Ancient Egyptian pyramid mausolea. The side edge to height ratio S/h is 1½, β is of course π/2, and accordingly if S = 5 units then h = 3.333'.

The layout is seen in Figure 1.5:-

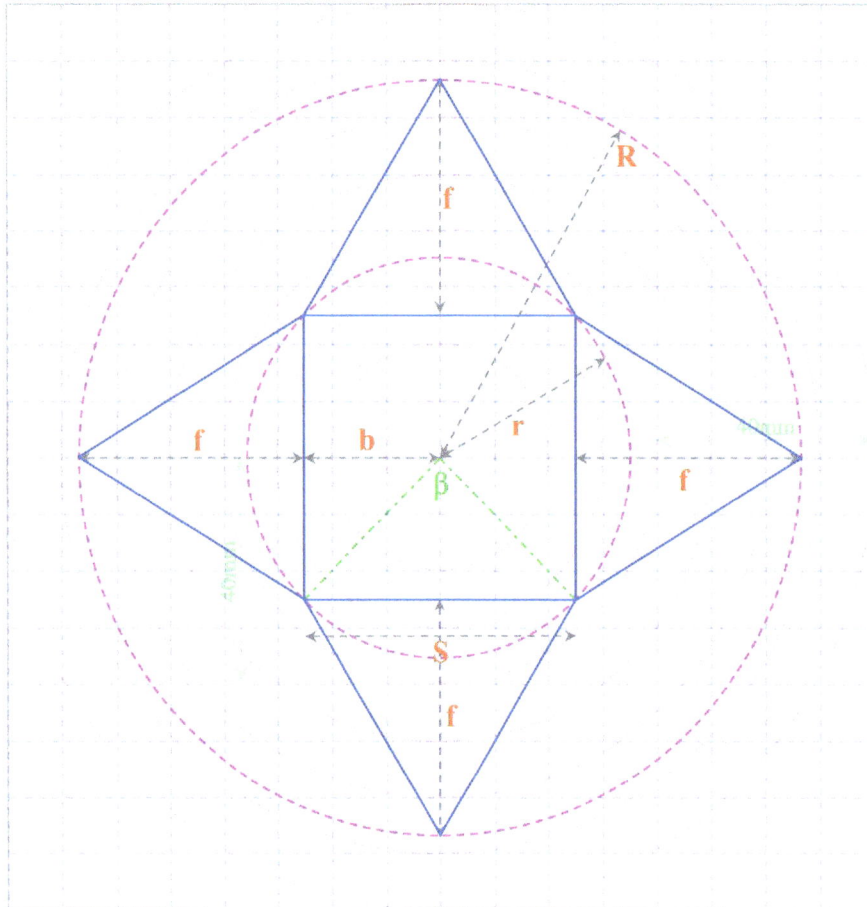

Figure 1.5
Planar Form of a Typical Square Pyramid

Case C

Case C concerns a regular pentagonal right pyramid for which the ratio S/h is specified by:-

$$\frac{S}{h} = \frac{1 + \sqrt{5}}{2}$$

Equation 1.8

In other words S/h is the Ratio of Phidias.
If we again choose S = 5 (for convenience: then h = 3.09017 approximately).
The planar form is shown in Figure 1.6.

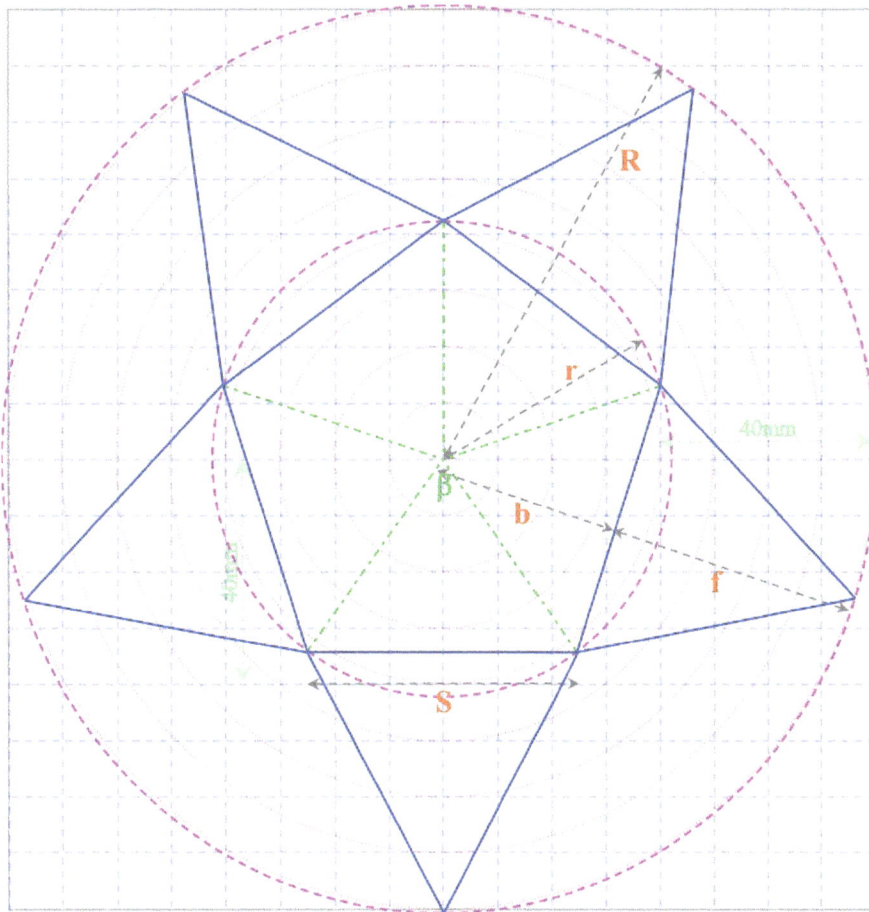

Figure 1.6
Planar Form of a Regular Pentagonal Pyramid
of Phidian S/h

The Surface Area of Pyramids

    The Side Separation Model

    The Side Separation Model envisions a right pyramid as comprising its base, which is treated separately; to which are added n identical triangular sloping sides.
    This conception is of course intuitively obvious, but does not necessarily lead to simple or computationally-efficient formulae.
    We may summarise the Side Separation Model of Pyramid Surface Area in these terms:-

$$A = B + \frac{Pf}{2}$$
Equation 1.9

where A is the Pyramid Surface Area, B is the Area of the Base, and P is the Perimeter of the Base.

Elementary analysis provides the Base Area as:-

$$B = n.\frac{1}{2}r^2.sin(\beta) = \frac{nSb}{2}$$
$$= \frac{n}{2}.\frac{S}{2(1-cos(\beta))}.sin(\beta)$$
Equation 1.10

whilst the Perimeter is:-

$$P = nS$$
Equation 1.11

giving the Area of the n Sloping Sides, A_k, as:-

$$A_k = \frac{nSf}{2}$$
Equation 1.12

Therefore the Surface Area A becomes:-

$$A = \frac{n}{2}.\frac{S}{2(1-cos(\beta))}.sin(\beta) + \frac{nSf}{2}$$
Equation 1.13

A little wrangling yields:-

$$A = \frac{nS^2}{2}\left[\frac{1}{2tan\left(\frac{\beta}{2}\right)} + \frac{\sqrt{h^2 + \frac{S^2}{2}\left(\frac{1}{1-cos(\beta)} - \frac{1}{2}\right)}}{S}\right]$$
Equation 1.14

Equation 1.14 may further be simplified to:-

$$A = \frac{nS^2}{4}\left[cot\left(\frac{\pi}{n}\right) + \frac{\sqrt{4h^2 + S^2.cot^2\left(\frac{\pi}{n}\right)}}{S}\right]$$

Equation 1.15

I essayed further simplifications using a symbolic processor (Wolfram Alpha®) but without success.

The Deltoid ("Kite") Model

An alternative conception of pyramid surface area is suggested by treating the planar form as a radial array of n deltoids ("kites"), each deltoid being composed of a base sectoral triangle attached along a base side to a slant-side triangle of altitude f. Both triangles are isosceles. The geometry is illustrated in Figure 1.7:-

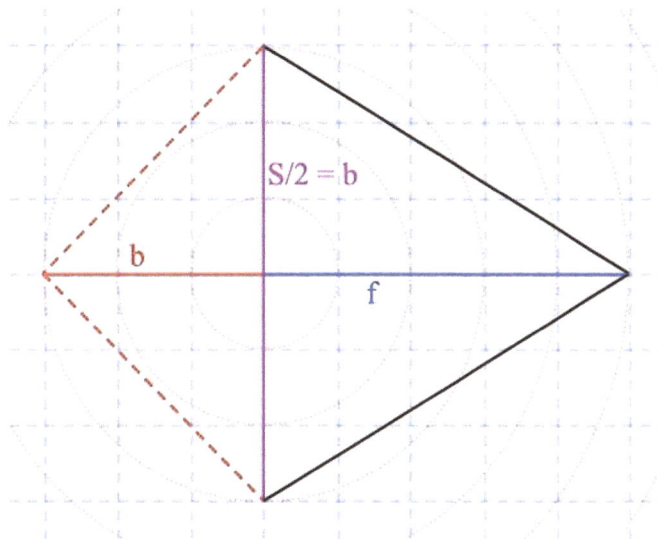

Figure 1.7
A Component Deltoid

The deltoid of Figure 1.7 represents the easternmost of four identical planar couples that fold to form a square pyramid of n=4, S=6, h=4 and f=5.
Inspection readily reveals that the Area of a Deltoid, A_{delt}, is:-

$$A_{delt} = \frac{S}{2}(b + f)$$

Equation 1.16

from which it is clear that the Surface Area of the whole pyramid is:-

$$A = nA_{delt} = \frac{nS}{2}(b + f)$$
Equation 1.17

Substitution for b and minor re-arrangement gives:-

$$A = \frac{n}{2} \cdot \left[\frac{S^2}{2} \cdot \cot\left(\frac{\pi}{n}\right) + Sf \right]$$
Equation 1.18

If Pyramid Height h is available Equation 1.18 may be re-expressed as:-

$$A = \frac{n}{2} \cdot \left[\frac{S^2}{2} \cdot \cot\left(\frac{\pi}{n}\right) + S\sqrt{h^2 + b^2} \right]$$
Equation 1.19

which by substitution for b² gives:-

$$A = \frac{n}{2} \cdot \left[\frac{S^2}{2} \cdot \cot\left(\frac{\pi}{n}\right) + S\sqrt{h^2 + \frac{S^2}{4} \cdot \cot^2\left(\frac{\pi}{n}\right)} \right]$$
Equation 1.20

By combining the outcomes of the Side Separation and the Deltoid Models we may disclose this identity:-

$$\frac{n}{2} \cdot \left[\frac{S^2}{2} \cdot \cot\left(\frac{\pi}{n}\right) + S\sqrt{h^2 + \frac{S^2}{4} \cdot \cot^2\left(\frac{\pi}{n}\right)} \right] = \frac{nS^2}{4}\left[\cot\left(\frac{\pi}{n}\right) + \frac{\sqrt{4h^2 + S^2 \cdot \cot^2\left(\frac{\pi}{n}\right)}}{S} \right]$$
Equation 1.21

Which reduces to:-

$$\cot^2\left(\frac{\pi}{n}\right) = \frac{2}{1 - \cos\left(\frac{2\pi}{n}\right)} - 1$$
Equation 1.22

The Volume of Pyramids

The Volume, V, of a regular right pyramid is given by:-

$$V = \frac{hn}{6} \cdot \frac{S^2}{2 \tan\left(\frac{4\pi}{n}\right)}$$

Equation 1.23

which may be re-expressed as:-

$$V = \frac{1}{12} \cdot nhS^2 \cdot \cot\left(\frac{4\pi}{n}\right)$$

Equation 1.24

Even using my symbolic processor I was unable further to simplify Equation 1.24.

PART TWO
SOME STATISTICAL FEATURES OF THE
SURVIVING TWENTY-THREE ANCIENT EGYPTIAN PYRAMIDS

By "surviving" we mean those pyramids that have come down to us in such a degree of preservation that their gross dimensions may reliably be assessed.

Table 2.1 presents key Age, Quality and Dimensions data for these structures ranked in descending order of my Second Quality Factor, Q_2.

Age

Most of our twenty-three pyramids are of Old Kingdom date between 4600 and 3800 years BP or approximately 2686–2181 BC.

This is essentially a Chalcolithic or Copper Age era of intermediate technology between the Neolithic Age of stone tools and early agriculture, and a later Bronze Age of established non-ferrous alloy tool technologies. Whilst it is likely that much of Old Kingdom quarrying and manufacture may have been executed with legacy lithic tools, much fine work may have been accomplished using hardened copper, probably natural or artificial arsenical copper alloy.

I know little about Old Kingdom drawing or computational techniques, except that they must have been very impressive.

Simple regression studies disclosed no significant relations between Pyramid Ages and any of: Quality Factor Q_2; Apothemal Slope, α; Height, h; or Edge Ratio, ρ.

Quality Factors

Statistical Quality Factors were computed in order to assess the precision with which the individual pyramids had been constructed, which was of course a function of design, materials and workmanship, but also by implication the degree to which the architects' "intention" was realised.

Accordingly, Quality Factors depend upon the degree of agreement between Computed (or Theoretical) metrics and the corresponding Measured data, as expressed in terms of Specific Defects.

A (Percentage) Specific Defect, %Sp.Def., is defined by:-

$$\%Sp.Def(X,Y) = 100\left(\frac{X-Y}{X}\right)$$
Equation 2.1

where X is a Theoretical Value and Y is the corresponding Measured or Empirical Value.

Value.

Specifically, we are interested in:-
(a) $CSA\alpha_{comp}$ vs. $SA\alpha_{meas}$
        Computed Apothemal Slope versus Measured Apothemal Slope
(b) CV_{comp} vs. WV_{meas}

Computed Pyramid Volume versus Measured Pyramid Volume

These paired data give rise respectively to the Specific Defects %Sp.Def.α and %Sp.Def.V

The First Quality Factor

The First Quality Factor, Q_1, is defined as:-

$$Q_1 = -ln|\%Sp.Def.\alpha \times \%Sp.Def.V|$$
Equation 2.2

A major draw-back of Q_1 is that it is undefined if either specific defect is zero.

The Second Quality Factor

The Second Quality Factor, Q_2, is defined as:-

$$Q_2 = ln[(\%Sp.Def.V + \%Sp.Def.\alpha \times \%Sp.Def.V)^2]$$
Equation 2.3

Q_2 is robust under perfection and reasonably stable at values between about fifteen and thirty.

Dimensions

In the context of Egyptian pyramids the Radial Interangle β is not interesting because it is invariably π/2, since all such pyramids are (or were meant to be) square.

The key dimensions of interest (measured in meters) are:-
  (a)   Side Length, S
  (b)   Height, h
  (c)   Arris Length, s

The Arris Length is that of the Sloping Edge between the faces of the pyramid.

Additionally, these dimensionless ratios were computed:-
  (1)   Edge Ratio, S/s=ρ
  (2)   Side Height Ratio, S/h

Side Length and Arris Length

The edges S and s show a very strong linear correlation for the 23 pyramids. The Coefficient of Determination was +0.98572957. The fitted regression equation is:-

$$s = 4.48702020 + 0.89871354S$$
Equation 2.4

or in rough terms:-

$$s = 4.5 + 0.9S$$
Equation 2.5

The fitted line is shown in Plot 2.1

Sorted Serial	Pyramid Modern Name	Age, Ω, at 2019AD	Pyramid Quality Factor (2)	Computed Slope Angle, α, (radians)	Height, h, (meters)	Edge Ratio S/s, ρ	Side Length, S, (meters)	Arris Length, s, (meters)	Side Height Ratio, S/h
1	The Great Pyramid of Giza	4572.5	29.52914	0.9056108	146.6	1.050441	230	218.9556	1.568895
2	Pyramid of Khafre	4572.5	29.218	0.9272952	143.5	1.028992	215.25	209.1854	1.5
3	Red Pyramid	4572.5	28.68521	0.7621465	105	1.172188	220	187.6832	2.095238
4	Bent Pyramid	4572.5	28.05646	0.8406183	105	1.109787	188	169.4019	1.790476
5	Pyramid of Senusret III	3898	25.14482	0.978369	78	0.975048	105	107.687	1.346154
6	Pyramid of Amenemhat III	3898	25.04632	0.9600704	75	0.994937	105	105.5344	1.4
7	Pyramid of Neferirkare	4438.5	24.91561	0.7994817	54	1.143705	105	91.80686	1.944444
8	Pyramid of Menkaure	4572.5	24.73624	0.9026059	65.5	1.053347	103.4	98.16328	1.578626
9	Pyramid of Senusret I	3898	24.64854	0.8621701	61.25	1.090909	105	96.25	1.714286
10	Pyramid of Hawara	3898	24.41372	0.8351309	58	1.11447	105	94.21518	1.810345
11	Pyramid of Senusret II	3898	24.2634	0.7421181	48.6	1.186603	106	89.33062	2.18107
12	Pyramid of Amenemhet I	3916	23.54071	0.9186249	55	1.03767	84	80.9506	1.527273
13	Pyramid of Nyuserre	4438.5	23.26376	0.9127205	51.68	1.043502	79.9	76.5691	1.546053
14	Pyramid of Djedkare-Isesi	4438.5	23.17672	0.9288208	52.5	1.02745	78.75	76.53124	1.495238
15	Pyramid of Teti	4282	23.17672	0.9272952	52.5	1.028992	78.5	76.40272	1.495238
16	Pyramid of Pepi I	4282	23.17672	0.9272952	52.5	1.028992	78.75	76.53124	1.5
17	Pyramid of Pepi II	4282	23.17672	0.9272952	52.5	1.028992	78.75	76.53124	1.5
18	Pyramid of Merenre	4282	23.17672	0.9272952	52.5	1.028992	78.75	76.53124	1.5
19	Pyramid of Sahure	4438.5	22.95547	0.8734478	47	1.080718	78.75	72.86825	1.675532
20	Pyramid of Userkaf	4438.5	22.76805	0.9286026	49	1.027671	73.3	71.32633	1.495918
21	Pyramid of Unas	4438.5	21.53233	0.979447	43	0.973856	57.75	59.30035	1.343023
22	Pyramid of Khendjer	3809	20.8858	0.9581861	37.35	0.996949	52.5	52.66068	1.405622
23	Pyramid of Ibi	4189.5	17.70562	0.9272952	21	1.028992	31.5	30.6125	1.5
		4262	24.2258	0.897911	65.5209	1.054487	106.0370	99.7839	1.6051 Mean
		268.9129	2.6510	0.062526	30.6169	0.056484	52.9509	47.9309	0.2187 PSD
		6.3095	10.9428	6.963463	46.7284	5.356494	49.9363	48.0348	13.6240 PCV(%)

Table 2.1
Age, Quality and Dimensions Data for the Twenty-Three Egyptian Pyramids

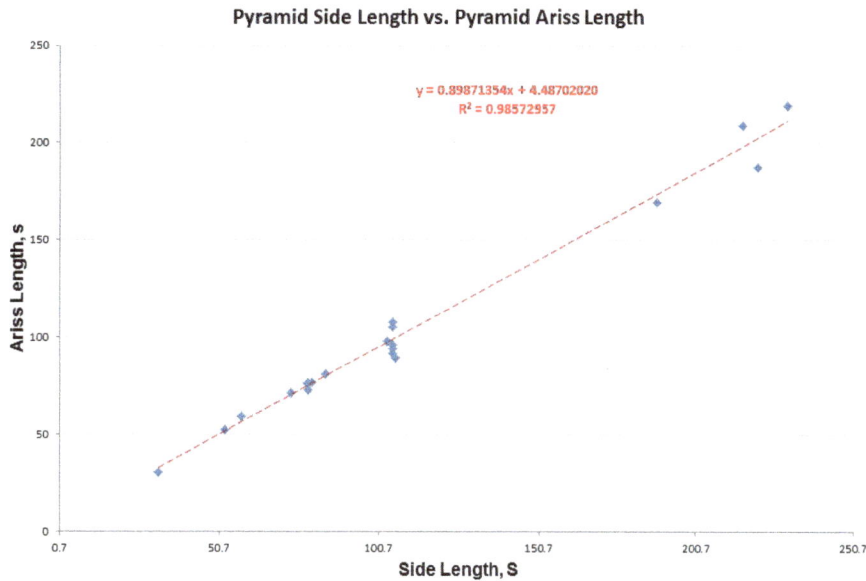

Plot 2.1
The Tight Correlation of Egyptian Pyramid Side Length with Arris Length

Histograms of Dimensionless Ratios

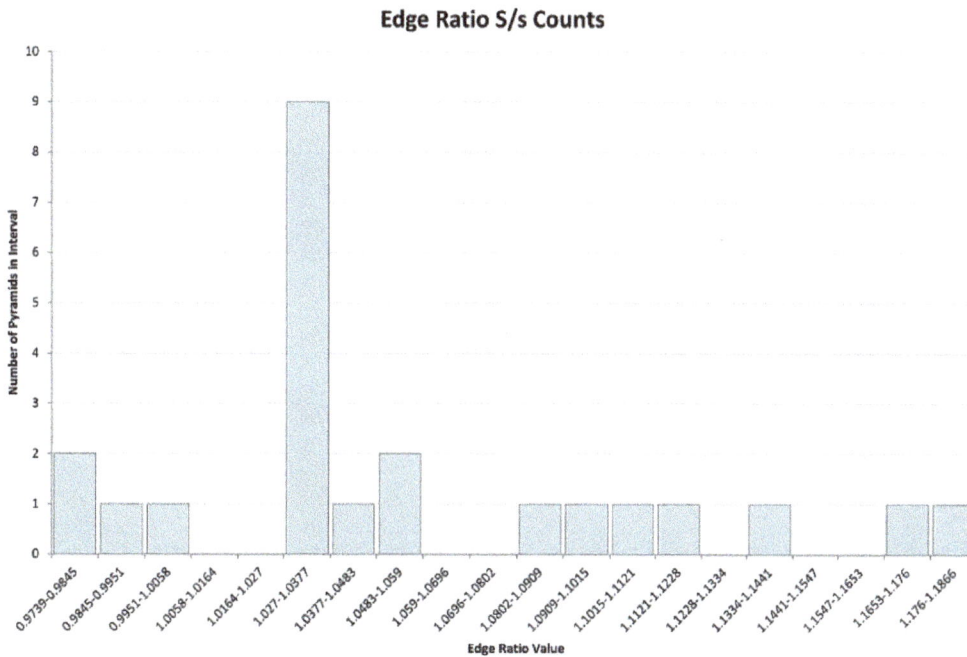

Plot 2.2
A Histogram of Pyramid Basal Edge Length/Arris Edge Length Ratios

Side Ratio S/h Counts

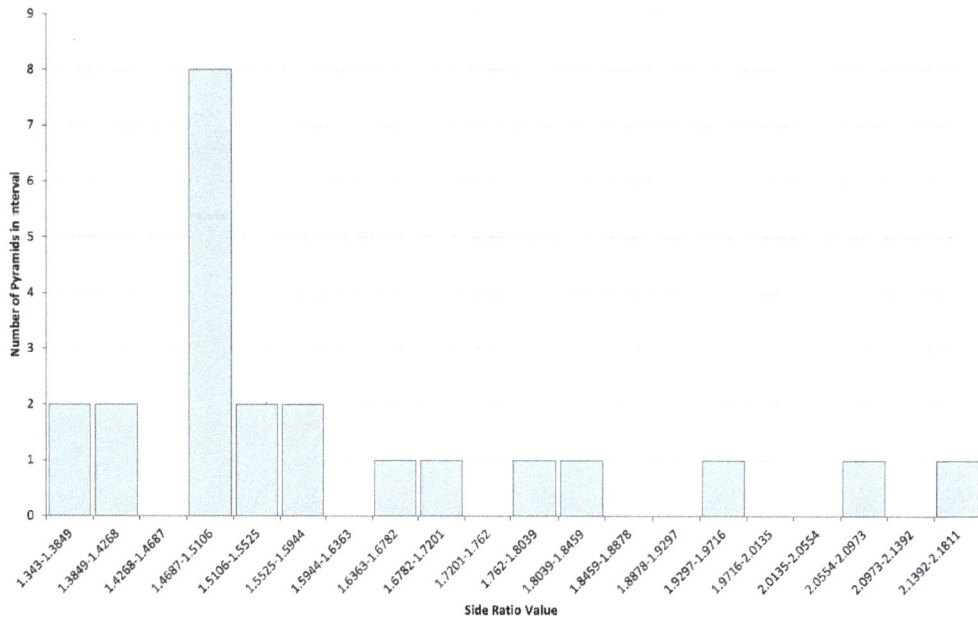

Plot 2.3
A Histogram of Pyramid Basal Edge Length/Height Ratios

Sine Ratio h/s Counts

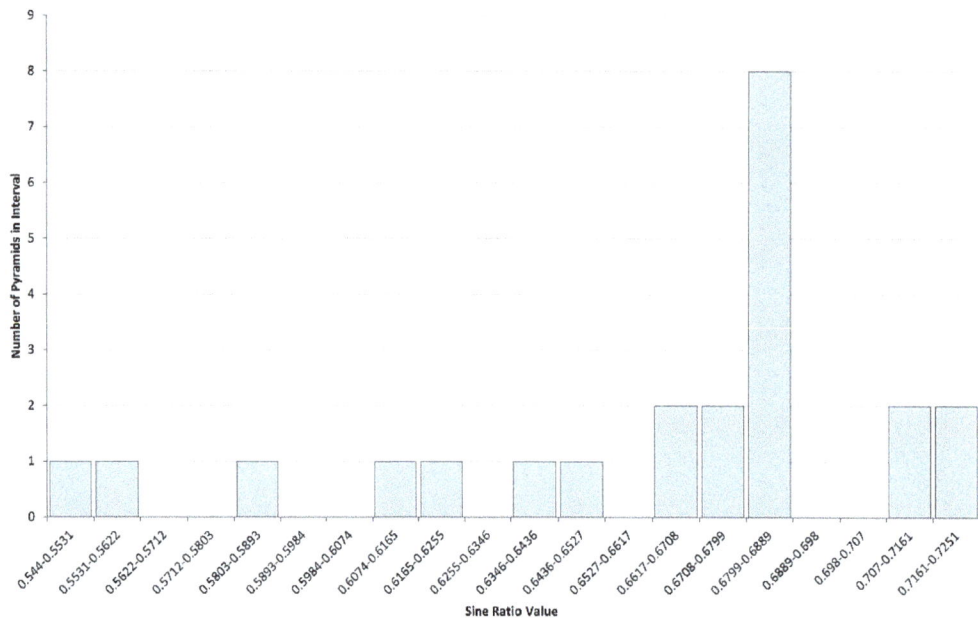

Plot 2.4
A Histogram of Pyramid Height/Arris Edge Length Ratios

<u>Distributions Analysis</u>

There are clear modal values of pyramid proportions that betray the intentions of their designers. In particular, there are modes for the dimensionless ratios S/s, S/h and h/s, amongst which are obvious mathematical linkages.

The precise values of these three modes are clarified in Table 2.2.

SELECTED LOCAL GROUPED FREQUENCY STATISTICS

	Edge Ratio, S/s	Side Ratio, S/h	Major Sine, h/s
Triad Lower Bound	1.016405	1.42682713	0.6708
Triad Upper Bound	1.048317	1.55253443	0.697962
Group Width	0.010637	0.04190244	0.009054
Group Count	20	20	20
Submodal Interval Midpoint, x_0	1.021724	1.44777834	0.675327
Modal Interval Midpoint, x_1	1.032361	1.48968078	0.684381
Supermodal Interval Midpoint, x_2	1.042998	1.53158321	0.693435
Submodal Group Frequency, f_0	0	0	2
Modal Group Frequency, f_1	9	8	8
Supermodal Group Frequency, f_2	1	2	0
Σf	10	10	10
Σfx	10.33425	14.9806127	6.825699
Σfx^2	10.67977	22.4446849	4.659148
Triad Grouped Mean	1.033425	1.49806127	0.68257
Triad Grouped SD	0.003191	0.01676097	0.003622
Modal Group Lower Bound	1.027042	1.46872956	0.679854
Modal Group Width	0.010637	0.04190244	0.009054
Mode	1.032674	1.49267381	0.683734

Table 2.2
Resolution of Dimensionless Ratio Modes for Egyptian Pyramids

It can be seen that modal S/h is very near to 1.5 and the majority of the pyramids seemed to have "aimed" at this value as a design basis.

Notwithstanding that, the biggest and perhaps the oldest of the pyramids, The Great Pyramid of Giza, seems to be based upon S/h of $\pi/2$ (1.570796) because its actual S/h is 1.568895.

The percentage specific defect involved is +0.121042% so the Giza proportion is unlikely to be fortuitous.

There is, however, a suspicion that the pyramid designers of succeeding centuries either made genuine mistakes about the "required" S/h standard, or else found $\pi/2$ inconvenient, because S/h=1.5 became the norm. The percentage specific defect of this value relative to $\pi/2$ is +4.507%.

The motives for selecting either value for S/h, if indeed they are chosen as opposed to being the automatic outcome of methods, are obscure. We will investigate this problem in Part Three.

The Greater Ratio of Phidias, (sometimes called the Golden Mean, the Golden Section or something), Φ, is defined by:-

$$\Phi = \frac{1 + \sqrt{5}}{2}$$

Equation 3.1

which has a value of approximately 1.61803398875
The Lesser Ratio of Phidias, ϕ, is defined by:-

$$\phi = \Phi - 1$$
Equation 3.2

whose value is about 0.61803398875

ϕ and Φ are transcendental numbers whose mantissas are comprised of an infinite number of apparently random digits and hence never terminate. Like other transcendentals they are defined algorithmically. The Ratios of Phidias exhibit a number of remarkable recurrence relations and have long been employed in architecture and other branches of aesthetics.

Also the Ratio of a Circle Circumference to its Diameter, also known as The Ludolphine Constant, π, is a *different* transcendental number related to elementary geometry and has a value adjacent to 3.14159265359.

Both π and ϕ manifest in natural creatures: Who are the architecture of God.

We may note that:-

$$\phi\Phi = 1$$
Equation 3.3

and:-

$$\sqrt{1 + \frac{\pi^2}{16}} * \sqrt{1 - \frac{\frac{\pi^2}{16}}{1 + \frac{\pi^2}{16}}} = 1$$

Equation 3.4

which of course implies that:-

$$\sqrt{1+\frac{\pi^2}{16}} * \sqrt{1-\frac{\frac{\pi^2}{16}}{1+\frac{\pi^2}{16}}} = \phi\Phi$$

Equation 3.5

Therefore, though they are transcendental π and Φ are related and may presumably be approximated in terms of each other.

It is such approximations that we wish to examine because the architecture of the Ancient Egyptian pyramids, and in particular the Great Pyramid of Cheops is said to relate π and φ.

Such approximations may be said to be paraphidian properties in that they are numerically very near to the Lesser Phidian Ratio or its complementary Greater Ratio: But not theoretically equivalent. It is my contention that the Ancients could simulate Phidian Proportion with rope-and-peg technologies applied in literal exercises of geometry, as well as in paper-based genotypical planning.

In terms of Figure 1.2 in Part One of this disquisition consider the pyramid half-section illustrated in Figure 3.1:-

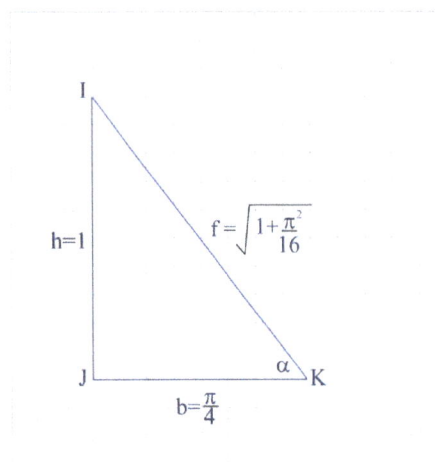

Figure 3.1
The Pyramid Half-Section as a
Right-Angled Triangle

It is sometimes claimed in literature that Angle α is equivalent to φ or that φ = f/h or that the figure IJK is in some other sense Phidian. In particular, this is asserted for the Great Pyramid of Cheops. You can see that this cannot possibly be true since every side is a function of π and integers only, and that therefore Φ and φ are precluded.

But that is not to say that the geometry cannot *approximate* Phidian constants to some degree.

Further to investigate these matters of constructive approximation it may be helpful to unfold the pyramid and represent its edges and faces in two-dimensional planar terms (forgive the pleonasm) as depicted in Figure 3.2 for this case of unit height and Side Length S of $\pi/2$:-

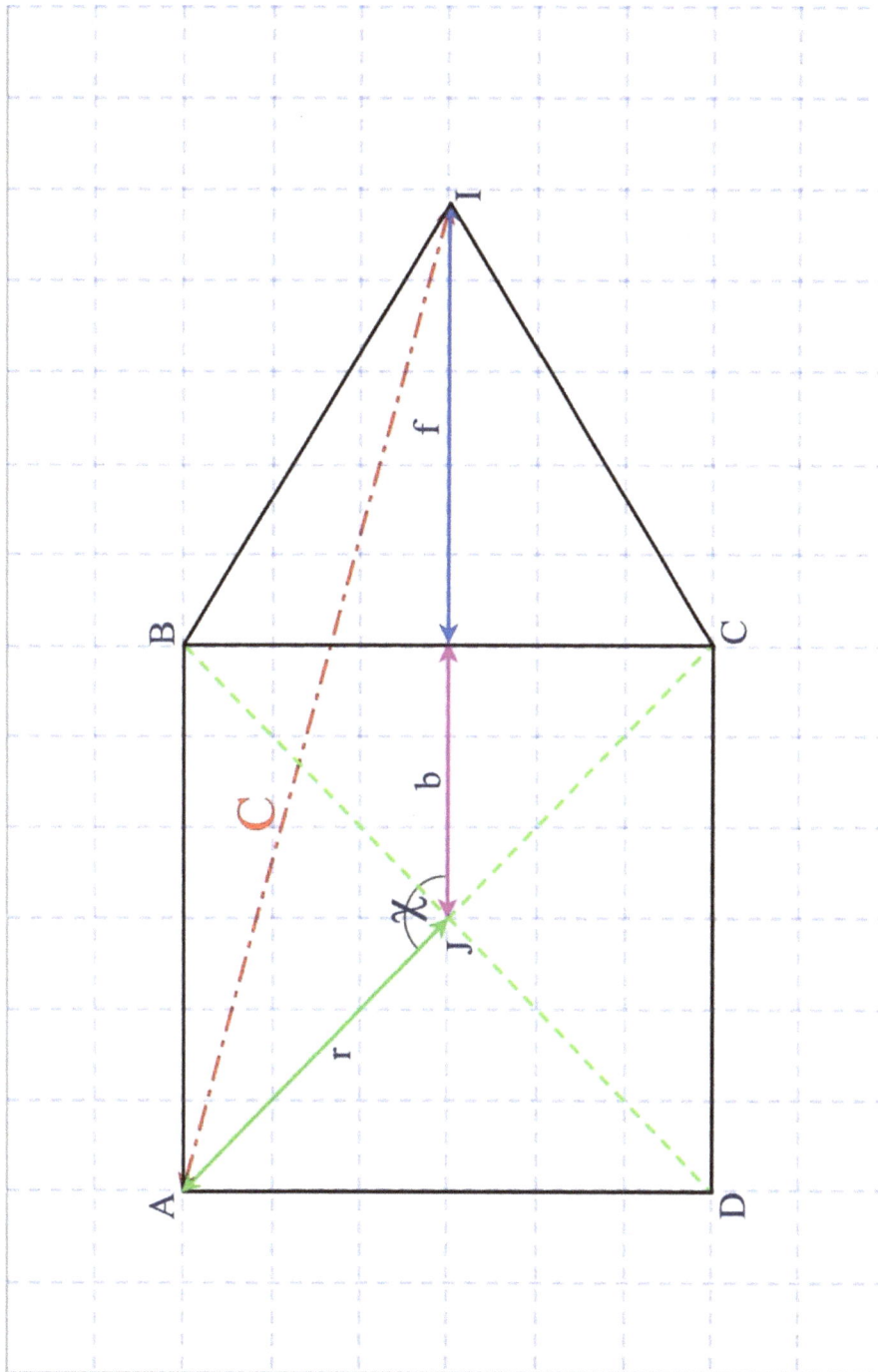

Figure 3.2
The Crossing Line C in the Collapsed Plan of a Paraphidian Pyramid

Our priority is to study the Clipping or Crossing Line between the A and I vertices: The red line of alternating dash lengths marked with the big red "C" (not the black vertex C). We can see that the Vertex Radii, r, are given by:-

$$r = \sqrt{b^2 + b^2} = \sqrt{2b^2} = \sqrt{2\left(\frac{\pi}{4}\right)^2} = \sqrt{\frac{\pi^2}{8}}$$

Equation 3.6

and that the line JI is of length:-

$$b + f = \frac{\pi}{4} + \sqrt{1 + \frac{\pi^2}{16}}$$

Equation 3.7

Also, the included angle χ is visibly $3\pi/4$ radians, which in surd form is:-

$$\cos\frac{3\pi}{4} = -\frac{1}{\sqrt{2}}$$

Equation 3.8

Now by the Cosine Rule the Clipping Line C is determinable as:-

$$C = \sqrt{r^2 + (b+f)^2 - 2r(b+f)\cos\chi}$$

Equation 3.9

which by substitution becomes:-

$$C = \sqrt{\frac{\pi^2}{8} + \left(\frac{\pi}{4} + \sqrt{1 + \frac{\pi^2}{16}}\right)^2 - 2\sqrt{\frac{\pi^2}{8}}\left(\frac{\pi}{4} + \sqrt{1 + \frac{\pi^2}{16}}\right).-\frac{1}{\sqrt{2}}}$$

Equation 3.10

Noting that:-

$$(b+f)^2 = 1 + \frac{\pi^2}{8} + \frac{\pi}{2}\cdot\sqrt{1 + \frac{\pi^2}{16}}$$

Equation 3.11

Equation 3.10 simplifies by:-

$$C = \sqrt{\frac{\pi^2}{8} + 1 + \frac{\pi^2}{8} + \frac{\pi}{2} \cdot \sqrt{1 + \frac{\pi^2}{16}} + \sqrt{2}\sqrt{\frac{\pi^2}{8}\left(\frac{\pi}{4} + \sqrt{1 + \frac{\pi^2}{16}}\right)}}$$

Equation 3.12

which may be re-written:-

$$C = \sqrt{1 + \frac{\pi^2}{4} + \sqrt{\frac{\pi^2}{4}\cdot\left(1 + \frac{\pi^2}{16}\right)} + \sqrt{2}\sqrt{\frac{\pi^2}{8}\left(\frac{\pi}{4} + \sqrt{1 + \frac{\pi^2}{16}}\right)}}$$

$$= \sqrt{1 + \frac{\pi^2}{4} + \sqrt{\frac{\pi^2}{4}\cdot\left(1 + \frac{\pi^2}{16}\right)} + \sqrt{\frac{2\pi^2}{8}\left(\frac{\pi}{4} + \sqrt{1 + \frac{\pi^2}{16}}\right)}}$$

$$= \sqrt{1 + \frac{\pi^2}{4} + \sqrt{\frac{\pi^2}{4}\cdot\left(1 + \frac{\pi^2}{16}\right)} + \sqrt{\frac{\pi^2}{4}\left(\frac{\pi}{4} + \sqrt{1 + \frac{\pi^2}{16}}\right)}}$$

$$= \sqrt{1 + \frac{\pi^2}{4} + \sqrt{\frac{\pi^2}{4}\cdot\left(1 + \frac{\pi^2}{16}\right)} + \frac{\pi}{2}\left(\frac{\pi}{4} + \sqrt{1 + \frac{\pi^2}{16}}\right)}$$

$$= \sqrt{1 + \frac{\pi^2}{4} + \frac{\pi}{2}\cdot\sqrt{1 + \frac{\pi^2}{16}} + \frac{\pi}{2}\left(\frac{\pi}{4} + \sqrt{1 + \frac{\pi^2}{16}}\right)}$$

$$= \sqrt{1 + \frac{\pi^2}{4} + \frac{\pi^2}{8} + \frac{\pi}{2}\cdot\sqrt{1 + \frac{\pi^2}{16}} + \frac{\pi}{2}\sqrt{1 + \frac{\pi^2}{16}}}$$

$$= \sqrt{1 + \frac{\pi^2}{4} + \frac{\pi^2}{8} + \pi\cdot\sqrt{1 + \frac{\pi^2}{16}}}$$

$$= \sqrt{1 + \frac{3\pi^2}{8} + \pi\cdot\sqrt{1 + \frac{\pi^2}{16}}}$$

Equation 3.13

Thence from Equation 3.13 by Wolfram Alpha®:-

$$C = \sqrt{1 + \frac{3\pi^2}{8} + \frac{\pi}{4} \cdot \sqrt{16 + \pi^2}}$$

$$= \frac{1}{2}\sqrt{\frac{1}{2} \cdot \left(8 + 3\pi^2 + 2\pi\sqrt{16 + \pi^2}\right)}$$

Equation 3.14

Then continuing manually:-

$$C = \frac{1}{2} \cdot \sqrt{\frac{1}{2}\sqrt{8 + 3\pi^2 + 2\pi\sqrt{16 + \pi^2}}}$$

$$= \frac{1}{2} \cdot \frac{1}{\sqrt{2}}\sqrt{8 + 3\pi^2 + 2\pi\sqrt{16 + \pi^2}}$$

$$= \frac{1}{\sqrt{8}} \cdot \sqrt{8 + 3\pi^2 + 2\pi\sqrt{16 + \pi^2}}$$

$$= \sqrt{1 + \frac{3\pi^2}{8} + \frac{\pi}{4}\sqrt{16 + \pi^2}}$$

Equation 3.15

To assist further simplifications we shall work with the square of the Clipping Line length for a while:-

$$C^2 = 1 + \frac{3\pi^2}{8} + \frac{\pi}{4}\sqrt{16 + \pi^2}$$

$$= 1 + \frac{3\pi^2}{8} + \sqrt{\frac{\pi^2}{16}}\sqrt{16 + \pi^2}$$

$$= 1 + \frac{3\pi^2}{8} + \sqrt{\pi^2 + \frac{\pi^4}{16}}$$

Equation 3.16

Noting that:-

$$\sqrt{\pi^2 + \frac{\pi^4}{16}} \approx 4$$

Equation 3.17

and that the Percentage Specific Defect function PerSpDef(x,y) is defined by:-

$$PerSpDef(x, y) = 100 \left(\frac{x - y}{x} \right)$$

Equation 3.18

then:-

$$PerSpDef \left(\sqrt{\pi^2 + \frac{\pi^4}{16}}, 4 \right) = -0.13253618$$

Equation 3.19

We may propose a first approximant of C as:-

$$C_A = \sqrt{5 + \frac{3\pi^2}{8}}$$

Equation 3.20

C_A is one of the simplest and best of the approximants of C for the Case E of the Paraphidian Regular right pyramid where r is driven by h = 1 and b = π/4.
For that system PerSpDef(C,C_A) is -0.0304378.
It is also the case that:-

$$C_B = \sqrt{1 + 6\phi + \pi \sqrt{1 + \phi}}$$
$$= \sqrt{1 + 6\phi + \pi f}$$
$$= \sqrt{1 + 6\phi + \pi \sqrt{1 + \frac{\pi^2}{16}}}$$

Equation 3.21

Which, for a PerSpDef(C,C_B) of -0.04082906 provides a practical linkage of the Phidian and the Ludolphine within the Clipping Line.
Also:-

$$C_C = \sqrt{5 + 6\phi}$$

Equation 3.22

and:-

$$1 + \pi \sqrt{1 + \frac{\pi^2}{16}} \approx 5$$

Equation 3.23

$$\phi \approx \sqrt{\left(\frac{4}{\pi}\right)^2 + 1}$$

Equation 3.24

A selection of these techtonically potentially useful approximations that relate the Phidian and Ludolphine constants, and other relevant parameters, are listed with their Defects in Appendix One.

PART FOUR
THE ANALYTIC GEOMETRY OF CLASSICAL PARAPHIDIAN PYRAMIDS
(CASE E)

In this Part we wish to examine the archetype of Ancient Egyptian pyramids exemplified by the Great Pyramid of Giza (sometimes called the Great Pyramid of Khufu or Cheops).

The Great Pyramid of Giza is the largest and one of the oldest of the monuments that have come down to us and is the only survivor of The Seven Wonders of the Ancient World discussed by Diodorus Siculus.

To be particular, this species of right square pyramid is driven by a unitary height (i.e. h = 1) and a Side Length S of $\pi/2$. It immediately follows that b = $\pi/4$ and that Apothemal Slant Height L $\equiv$ f is given by:-

$$f = \sqrt{1^2 + b^2} = \sqrt{1 + \left(\frac{\pi}{4}\right)^2} = \sqrt{1 + \frac{\pi^2}{16}} \approx 1.27155428$$

Equation 4.1

The Basal Side Interangle, β, is of course:-

$$\beta = \frac{2\pi}{n} = \frac{2\pi}{4} = \frac{\pi}{2}$$

Equation 4.2

The Vertex Radius of the Base (i.e. the radius of the circumscribed circle), r, is given by:-

$$r = \sqrt{\frac{S^2}{2(1 - \cos(\beta))}} = \frac{1}{2}\sqrt{S^2 . \csc\left(\frac{\beta}{2}\right)^2} = \frac{S}{2} \cdot \csc\left(\frac{\beta}{2}\right) = \frac{S}{2} \cdot \csc\left(\frac{\pi}{n}\right)$$

Equation 4.3

For the case of a square pyramid n = 4 and accordingly:-

$$r = \sqrt{\frac{\pi^2}{8}}$$

Equation 4.4

By the Pythagorean Theorem:-

$$b = \sqrt{r^2 - \frac{S^2}{4}} = \sqrt{\frac{S^2}{4} \cdot \csc^2\frac{\pi}{n} - \frac{S^2}{4}} = \sqrt{\frac{S^2}{4} \cdot \left(\csc^2\frac{\pi}{n} - 1\right)} = \frac{S}{2} \cdot \sqrt{\cot^2\frac{\pi}{n}} = \frac{S}{2} \cdot \cot\left(\frac{\pi}{n}\right)$$

Equation 4.5

For Case E b does of course compute as $\pi/4$.

Equation 4.1 for Apothemal Slope Height f may formally be developed as:-

$$f = \sqrt{h^2 + b^2} = \sqrt{h^2 + \frac{S^2}{2(1 - cos\beta)} - \frac{S^2}{4}} = \sqrt{h^2 + \frac{S^2}{2} \cdot \left(\frac{1}{1 - cos\beta} - \frac{1}{2}\right)}$$

Equation 4.6

which for this Case E computes as the expression in Equation 4.1.

The Arris Length, s, and the Pyramid Height, h, are also yielded by arrangements of the Pythagorean Theorem in terms of the other figure dimensions.

Arris Length s is given by:-

$$s = \sqrt{f^2 + \frac{S^2}{4}}$$

Equation 4.7

which for Case E is:-

$$s = \sqrt{1 + \frac{\pi^2}{8}} \approx 1.49455697$$

Equation 4.8

Pyramid Height, h, known to be unity by definition, may be elaborated as:-

$$h = \sqrt{f^2 - b^2} = \sqrt{f^2 - \left(r^2 - \frac{S^2}{4}\right)} = \sqrt{f^2 - \left(\frac{S^2}{2(1 - cos\beta)} - \frac{S^2}{4}\right)}$$

$$= \sqrt{f^2 - \frac{S^2}{2}\left(\frac{1}{1 - cos\beta} - \frac{1}{2}\right)}$$

Equation 4.9

A Putative Modus of Disposition

It is known that the Ancient Egyptians developed an elaborate rope-based technology that was mastered and applied by a class of harpedonaptai or rope-stretchers. If we are of Grecian inclination we may call their science tonotechnics (τόνος, rope: τέχνη craft).

There are an infinite array of procedures for using ropes-and-post (or equivalently compass and straight-edge) for drawing the plans of a pyramid upon a flat surface. As always in this Fallen life our choice depends upon our prejudice, convenience and know-how, but especially the exigencies of economy.

Figure 4.1 shows a likely possible plan:-

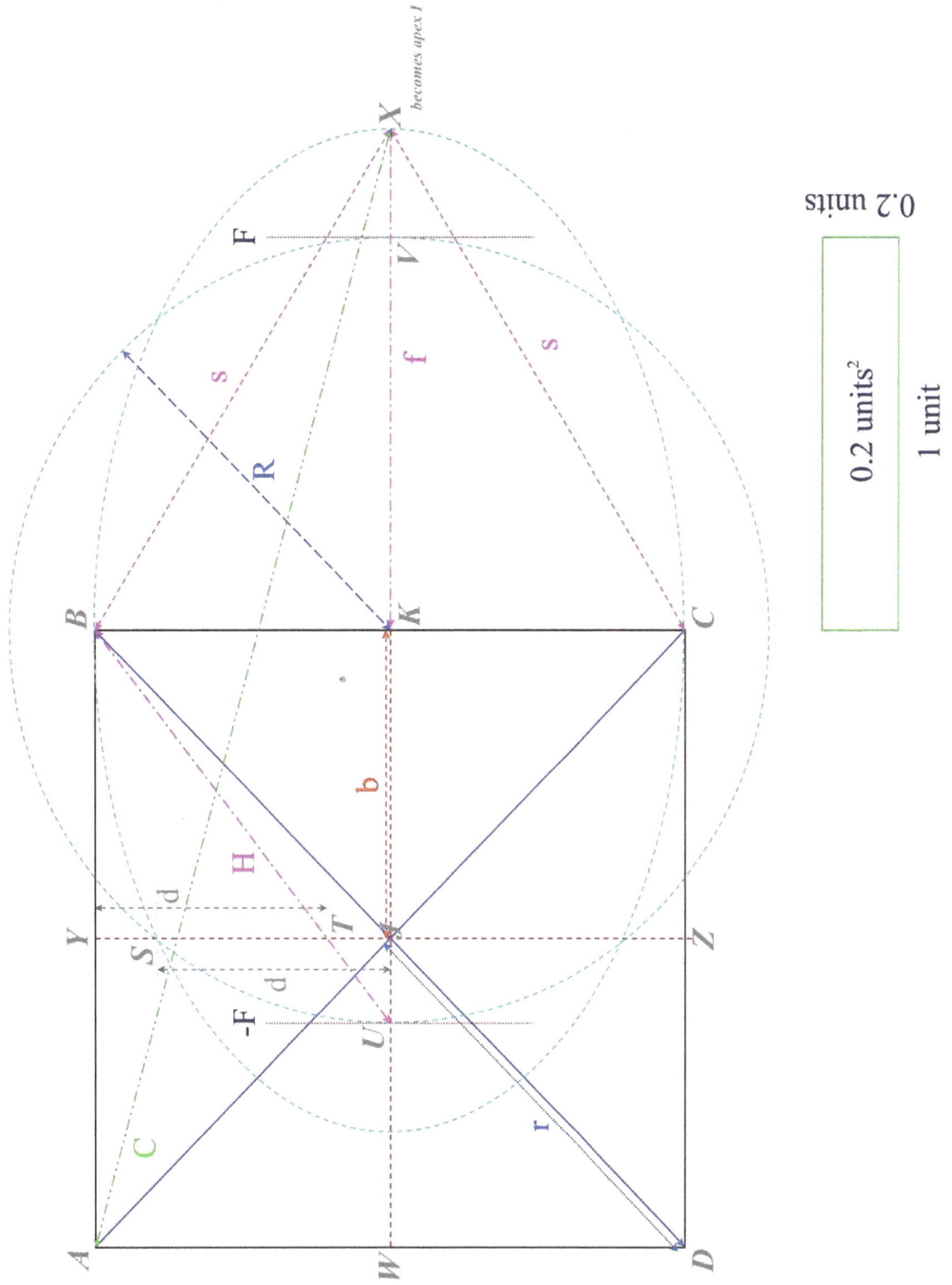

Figure 4.1
A Likely Planar Representation of a Paraphidian Pyramid

Convenient Functions

At this juncture we may define a few useful functions.

Firstly, our usual Percentage Specific Defect metric for the relative error of theoretical and empirical quantities nominally equivalent:-

$$Sp.\,Def\left(x_{theo},y_{emp}\right) = 100\left(\frac{x_{theo} - y_{emp}}{x_{theo}}\right)$$

Equation 4.10

Secondly, the Gauss-Kummer Mean, GKh, as:-

$$GKh(x,y) = \frac{(x-y)^2}{(x+y)^2}$$

Equation 4.11

and thirdly the Ramanujan Second Approximator of Elliptical Perimeter, RamaII(x,y) as:-

$$RamaII(x,y) = \pi(x+y)\cdot\left(1 + \frac{3\cdot GKh(x,y)}{10 + \sqrt{4 - 3\cdot GKh(x,y)}}\right)$$

Equation 4.12

Ramanujan II is a very accurate (but not of course exact) estimator of the length of an ellipse circumference. In both the Gauss-Kummer Mean and RamaII the arguments x and y identify with a and b, respectively the Major and Minor Semi-Diameters (semi-axes) of the ellipse.

Auxiliary Elements

A few additional variables and lineaments may conveniently be defined here. Ellipse Eccentricity, e_e, is:-

$$e_e = \sqrt{1 - \frac{b^2}{a^2}}$$

Equation 4.13

The (positive and negative) Ellipse Foci F and –F which with reference to Figure 4.1 constitute the respective line segments **KV** and **UK** are computed by:-

$$F = a.e_e = a\sqrt{1 - \frac{b^2}{a^2}} = \sqrt{1 + \frac{\pi^2}{16}}\sqrt{1 - \frac{b^2}{a^2}} = 1$$

Equation 4.14

from which:-

$$F = a.e_e = a\sqrt{1 - \frac{b^2}{a^2}} = \sqrt{1 + \frac{\pi^2}{16}}\sqrt{1 - \frac{b^2}{a^2}} = \sqrt{1 + \frac{\pi^2}{16}}\sqrt{1 - \frac{\frac{\pi^2}{16}}{1 + \frac{\pi^2}{16}}} = \phi\Phi = 1$$

Equation 4.15

The geometry dictates, of course, that –F is minus unity.
The Focal Hypotenuse, H, is equivalent to the Apothemal Slant Height, f, that is to
say:-

$$H \equiv f = a = \sqrt{1 + b^2} = \sqrt{1 + \frac{\pi^2}{16}}$$

Equation 4.16

The Vertex Radius of the Square Base, r, is:-

$$r = \sqrt{2}b = \sqrt{2} \cdot \frac{\pi}{4} = \frac{\pi}{2\sqrt{2}}$$

Equation 4.17

The Radius of the Circle of Equivalent Area to the Ellipse, R, is:-

$$R = \sqrt{ab} = \sqrt{\frac{\pi}{4}\sqrt{1 + \frac{\pi^2}{16}}} = \frac{1}{2}\sqrt{\pi}\sqrt[4]{1 + \frac{\pi^2}{16}} = \frac{\sqrt{\pi}\sqrt[4]{16 + \pi^2}}{4} = 0.999337977109 \approx 1$$

Equation 4.18

Sp.Def(1,R) is 0.066202289145
The Ellipse Coincidence Height, d, (c.f. segment *YT* on Figure 4.1):-

$$d = \sqrt{R^2 - b^2} = \sqrt{1 - \frac{\pi^2}{16}} = 0.618990892447 \approx \phi$$

Equation 4.19

Another near approximation is:-

$$f \approx \frac{\Phi}{\phi} \cdot \frac{4}{\pi} \cdot \left(1 - \frac{\pi^2}{16}\right) = \left(\frac{\Phi}{\phi}\right)\left(\frac{4}{\pi} - \frac{\pi}{4}\right) = \left(\frac{\Phi}{\phi}\right)\left(\frac{1}{b} - b\right) = (1 + \Phi)\left(\frac{1}{b} - b\right)$$
Equation 4.20

for which:-

$$Sp.\,Def\left(f, (1+\Phi)\left(\frac{1}{b} - b\right)\right) = -0.442847171894$$
Equation 4.21

In terms of areal and circumferential relativities for the ellipse and its equivalent circle given that R is 0.999337977109 as opposed to unity the following defects were computed:-

$$Sp.\,Def\,(2\pi, P_{circle}) = 0.066202289145$$
Equation 4.22

$$Sp.\,Def\,(\pi, A_{circle}) = 0.132360750859$$
Equation 4.23

$$Sp.\,Def\,(P, P_{circle}) = 4.176073101781$$
Equation 4.24

$$Sp.\,Def\,(A, A_{circle}) = 0$$
Equation 4.25

The Loop Length, λ, for drawing the ellipse about the turnstakes at U and V is given by:-

$$\lambda = 2(F + H) = 2 \cdot \left[\sqrt{1 + \frac{\pi^2}{16}} + 1\right] = 4.54310855063$$
Equation 4.26

The Proposed Modus

(1)   A Horizon $WJKX$ would be defined in an exact West-East line probably by astronomical means and a turnstake driven at J, the intended Base Center.

(2)   The $-b$ and $+b$ segments would be defined along $WJKX$, possibly with a $\pi/4$ calibrated cord and turnstakes driven at W and K.

(3)   The perpendicular (i.e. North-South) crossline $YTJZ$ would be defined by bisection: Either by scribing the *vesica piscis* from W and K; or perhaps by trisecting the hemicircle KYW and then bisecting the central arc.

(4)   The Base Corner Vertices $ABCD$ are defined, probably by scribing from each of Y, K, Z and W in order to maximise good-conditioning.

(5)   A second calibrated cord or chain of length f equivalent to H is rotated in tension about a stake at **B** until it touches the horizon line at points **U** and **V** that respectively mark the foci –F and F. Stakes are driven at **U** and **V**.

(6)   A loop rope of length λ is held in tension around posts U and V and pulled about to describe the ellipse with a=**KX** and b=**KB**.

(7)   Behold, **KX**, a segment of the horizon, defines the Apothemal Slope Height, L (or f), which when (mentally) rotated about **BC,** through an angle of approximately π-acos(ϕ) radians, generates the Apex **I** from Point **X**.

PART FIVE
THE ANALYTIC GEOMETRY OF CERTAIN SESQUIMETRIC PYRAMIDS
(CASE B)

In Part Four we discussed the geometry and the possible design principle of Paraphidian pyramids whose ratio of Side Length to Apex Height was $\pi/2:1$, or algebraically-expressed $S/h=\pi/2$. Part Two showed that only the Great Pyramid of Giza (4573BP) and The Pyramid of Menkaure (4573BP) fell indubitably into that class.

There is, however, a much larger group of Sesquimetric Pyramids whose S/h is approximately one and a half. That is, $S/h \approx 1.5$. Among the oldest of these may be the Pyramid of Khafre (4573BP), but the later and mutually-contemporary family of the Pyramids of Teti I, Pepi I, Pepi II and Merenre (4282BP) are surely definite examples.

Neferirkare (4439BP) is a sesquimetric pyramid of a distinct but not unique type. It immediately follows that:-

$$b = \frac{3}{4}$$
Equation 5.1

Amongst the infinite possibilities at this ratio we shall study the sub-cases of:-

(i) Integral Form or the 2-by-5 Paradigm
In this case whilst $b = 0.75$, $f = 1.25$ and so the ratio $f/b \equiv 5/3$.
Accordingly the dividend of Side Length by Arris Length S/s is around 1.02899151.

(ii) The Neferirkare Paradigm
In this form the Apothemal Slant Height f_{Nef} is defined by:-

$$f_{Nef} = \frac{b}{tan\left(\frac{\pi}{6}\right)} = b\sqrt{3}$$
Equation 5.2

whilst the Angle opposite the planar apex X, γ_{Nef}, is given by:-

$$\gamma_{Nef} = atan\left(\frac{b}{f_{Nef}}\right) = 0.5404195$$
Equation 5.3

This fact leads to some interesting ambiguities and confusions, not least because 0.5404195 radians computes to 30.96375653 degrees and not 30 degrees.

<u>General Features</u>

The planimetric geometrical lineaments of sesquimetric pyramids are computable as:-

$$r = \frac{S}{2} csc\left(\frac{\pi}{n}\right) = \sqrt{2b^2} = 1.06066017$$
Equation 5.4

$$b = \frac{S}{2} cot\left(\frac{\pi}{n}\right) = 0.75$$
Equation 5.5

$$f = \sqrt{h^2 + b^2} = 1.25$$
Equation 5.6

$$s = \sqrt{f^2 + \frac{S^2}{4}} = 1.45773797$$
Equation 5.7

$$h = \sqrt{f^2 - b^2} = 1$$
Equation 5.8

$$R = b + f = 2$$
Equation 5.9

$$\frac{R}{r} = \frac{2}{3}\sqrt{8}$$
Equation 5.10

$$\frac{R^2}{r^2} = \frac{4 \times 8}{3} = \frac{32}{9}$$
Equation 5.11

Though of beguiling simplicity, the numerical outcomes of the supposedly-general Equations 5.4 through 5.11 must be treated with extreme caution because as aforenoted there are subtly distinct types of sesquimetric pyramids, to the bafflement of Ancients and Moderns alike.

<u>The Two-by-Three Paradigm</u>

This is a "simple" ellipse-scribed pyramid whose Side Length S = 1.5, h = 1, b = 0.75, and whose a ≡ f $_{35}$ = 1.25. Hence:-

$$\frac{b}{f_{35}} = \frac{3}{5}$$
Equation 5.12

By the Cosine Rule, The Clipping Line, C_{35}, is therefore:-

$$C_{35} = \sqrt{r_{35}^2 + (b + f_{35})^2 - 2r_{35}(b + f_{35}) - \cos\left(\frac{3}{4}\pi\right)}$$

$$= \sqrt{r_{35} + (b + f_{35})^2 + \sqrt{2}r_{35}(b + f_{35})}$$

$$= \sqrt{\frac{9}{8} + \left(\frac{3}{4} + \frac{5}{4}\right)^2 + \sqrt{2}\sqrt{\frac{9}{8}\left(\frac{3}{4} + \frac{5}{4}\right)}}$$

$$= \sqrt{\frac{9}{8} + 2^2 + \sqrt{2}\sqrt{\frac{9}{8}}2}$$

$$= \frac{\sqrt{130}}{4}$$

$$= 2.85043856$$
Equation 5.13

Let the Ellipse Eccentricity. e_e, be defined by:-

$$e_e = \sqrt{1 - \frac{b^2}{a^2}}$$
Equation 5.14

Then the ($\pm$) Ellipse Foci, F_{35}, are given by:-

$$F_{35} = ae_e = a\sqrt{1 - \frac{b^2}{a^2}} = 1$$
Equation 5.15

From which the Focal Radius, H_{35}, an identity of Apothemal Slope Line, f_{35}, is yielded as:-

$$H_{35} = \sqrt{F_{35} + b^2} = 1.25$$
Equation 5.16

For this paradigm the resulting Pyramid Volume, V_{35}, is yielded as:-

$$V_{35} = \frac{n}{6} \cdot \frac{S^2}{2(1 - \cos(\beta))} \cdot \sin\left(\frac{2\pi}{n}\right) \cdot h = 0.75$$

Equation 5.17

whilst the Pyramid Surface Area, A_{35}, is given by:-

$$A_{35} = \frac{n}{2} \cdot \left[\frac{S^2}{2} \cdot \cot\left(\frac{\pi}{n}\right) + S \cdot f_{35}\right] = 1.25$$

Equation 5.18

Geoscribing Loop Length λ is defined as:-

$$\lambda = 2(F + H) = 2(1 + 1.25) = 4.5$$

Equation 5.19

For this variant of the sesquimetric pyramid the Loop Length is very conveniently three times the Pyramid Side Length S, and could just as helpfully be laid-off along three sides of the structure's square base.

The general conformation is shown in Figure 5.1.

One of the candidate modi of disposition that we may propose is:-

The Proposed Modus

(1)   A Horizon *WJKX* would be defined in an exact West-East line probably by astronomical means and a turnstake driven at *J*, the intended Base Center.

(2)   The –b and +b segments would be defined along *WJKX*, possibly with a 3/4 calibrated cord and turnstakes driven at *W* and *K*.

(3)   The perpendicular (i.e. North-South) crossline *YTJZ* would be defined by bisection: Either by scribing the *vesica piscis* from *W* and *K*; or perhaps by trisecting the hemicircle *KYW* and then bisecting the central arc.

(4)   The Base Corner Vertices *ABCD* are defined, probably by scribing from each of *Y, K, Z* and *W* in order to maximise good-conditioning.

(5)   A second calibrated cord or chain of length f equivalent to H is rotated in tension about a stake at *B* until it touches the horizon line at points *U* and *V* that respectively mark the foci –F and F. Stakes are driven at *U* and *V*.

(6)   A loop rope of length λ is held in tension around posts *U* and *V* and pulled about to describe the ellipse with a=*KX* and b=*KB*.

(7)   Behold, *KX*, a segment of the horizon, defines the Apothemal Slope Height, L (or f), which when (mentally) rotated about *BC* generates the Apex *I* from Point *X*.

Such a methodology is not materially different from that of the Paraphidian Pyramid exemplified by the Great Pyramid of Cheops at Giza, except for Phase (5) where the second cord length λ is re-defined for Focal Radius H.

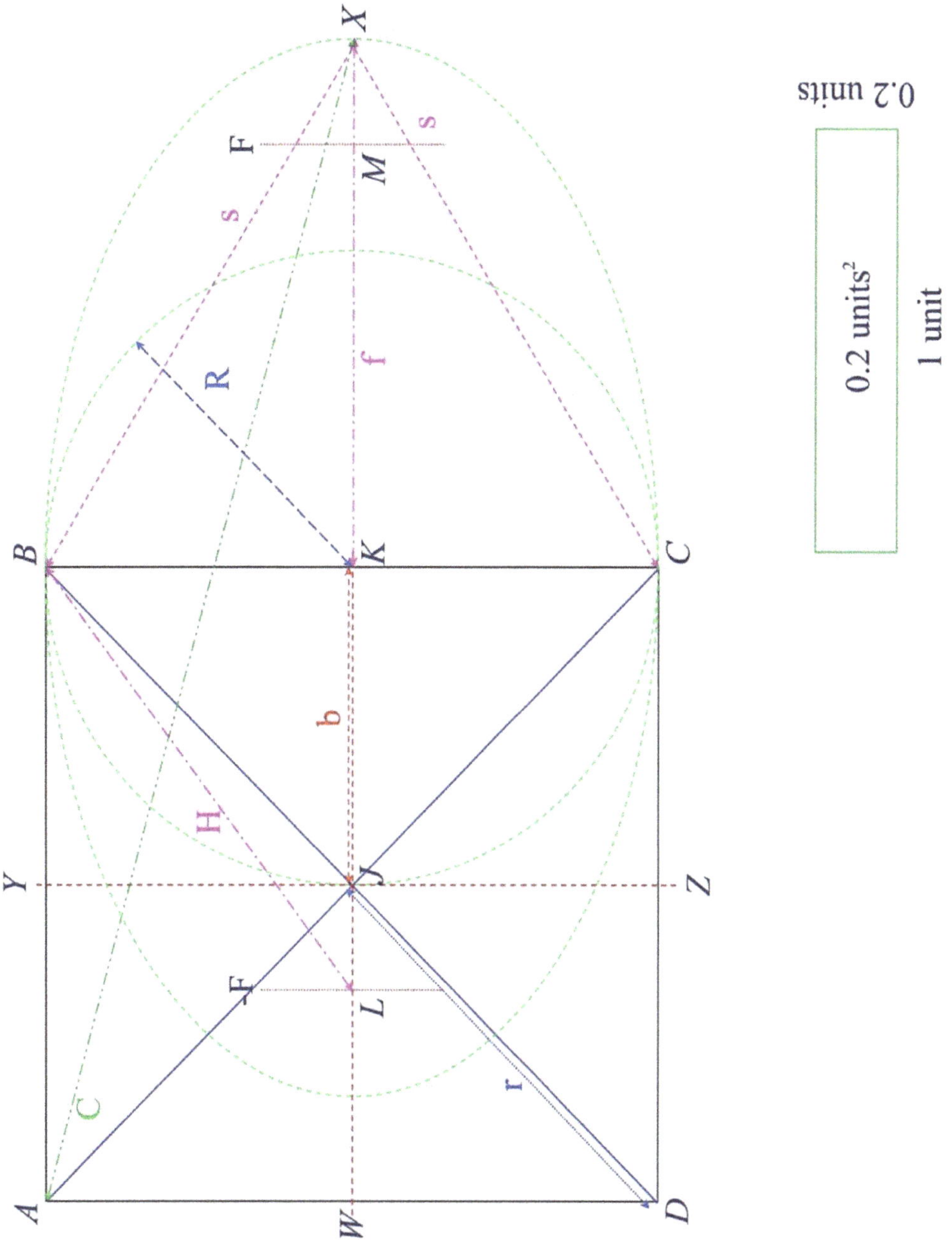

Figure 5.1
A Likely Planar Representation of a Sesquimetric Pyramid
Of the Three-by-Five Type

The Neferirkare Paradigm

This variant is not in any convenient sense an ellipse-scribed pyramid, but nevertheless has Side Length S = 1.5, h = 1, and b = 0.75.

The governing fact of the Neferirkare type is:-

$$f_{Nef} - \frac{b}{tan\left(\frac{\pi}{6}\right)} - b\sqrt{3} - 1.29903810568$$

Equation 5.20

Hence:-

$$\frac{b}{f_{Nef}} = \frac{3}{b\sqrt{3}} = \frac{\sqrt{3}}{b} = 2.30940107676$$

Equation 5.21

By the Cosine Rule, The Clipping Line, C_{Nef}, is therefore:-

$$C_{Nef} = \sqrt{r_{Nef}{}^2 + \left(b + f_{Nef}\right)^2 - 2r_{Nef}\left(b + f_{Nef}\right) - cos\left(\frac{3}{4}\pi\right)}$$

$$= \sqrt{2b^2 + (b + b\sqrt{3})^2 + \sqrt{2}\sqrt{2b^2(b + b\sqrt{3})}}$$

$$= \sqrt{b^2\left(2 + 1 + 2\sqrt{3} + 3 + 2 + 2\sqrt{3}\right)}$$

$$= 2b\sqrt{2 + \sqrt{3}}$$

$$= \frac{6}{4}\sqrt{2 + \sqrt{3}}$$

$$= 2.89777748$$

Equation 5.22

Let the Ellipse Eccentricity. e_e, be defined by:-

$$e_e = \sqrt{1 - \frac{b^2}{a^2}}$$

Equation 5.23

Then the (±) Ellipse Foci, F_{Nef}, are given by:-

$$F_{Nef} = ae_e = a\sqrt{1 - \frac{b^2}{a^2}} = 1.06066017$$

Equation 5.24

From which the Focal Radius, H_{Nef}, an identity of Apothemal Slope Line, f_{Nef}, is yielded as:-

$$H_{Nef} = \sqrt{F_{Nef} + b^2} = 1.29903811$$

Equation 5.25

For this paradigm the resulting Pyramid Volume, V_{Nef}, is yielded as:-

$$V_{Nef} = \frac{n}{6} \cdot \frac{S^2}{2(1 - cos(\beta))} \cdot sin\left(\frac{2\pi}{n}\right) \cdot h = 0.79549513$$

Equation 5.26

whilst the Pyramid Surface Area, A_{Nef}, is given by:-

$$A_{Nef} = \frac{n}{2} \cdot \left[\frac{S^2}{2} \cdot cot\left(\frac{\pi}{n}\right) + S \cdot f_{Nef}\right] = 6.14711432$$

Equation 5.27

Geoscribing Loop Length λ_{Nef} is defined as:-

$$\lambda_{Nef} = 2(F + H) = 2(1.06066017 + 1.29903811) = 4.71939655135$$

Equation 5.28

It is clear enough that for the Neferirkare type the metrical lengths $f \equiv H$ and λ are nowhere near as convenient for ellipse generation as in the three-by-five subcase and that accordingly a more wieldy modus of disposition might be preferred.

Figure 5.2 offers the layout of just such a procedure involving the scribing of two circles rather than one ellipse.

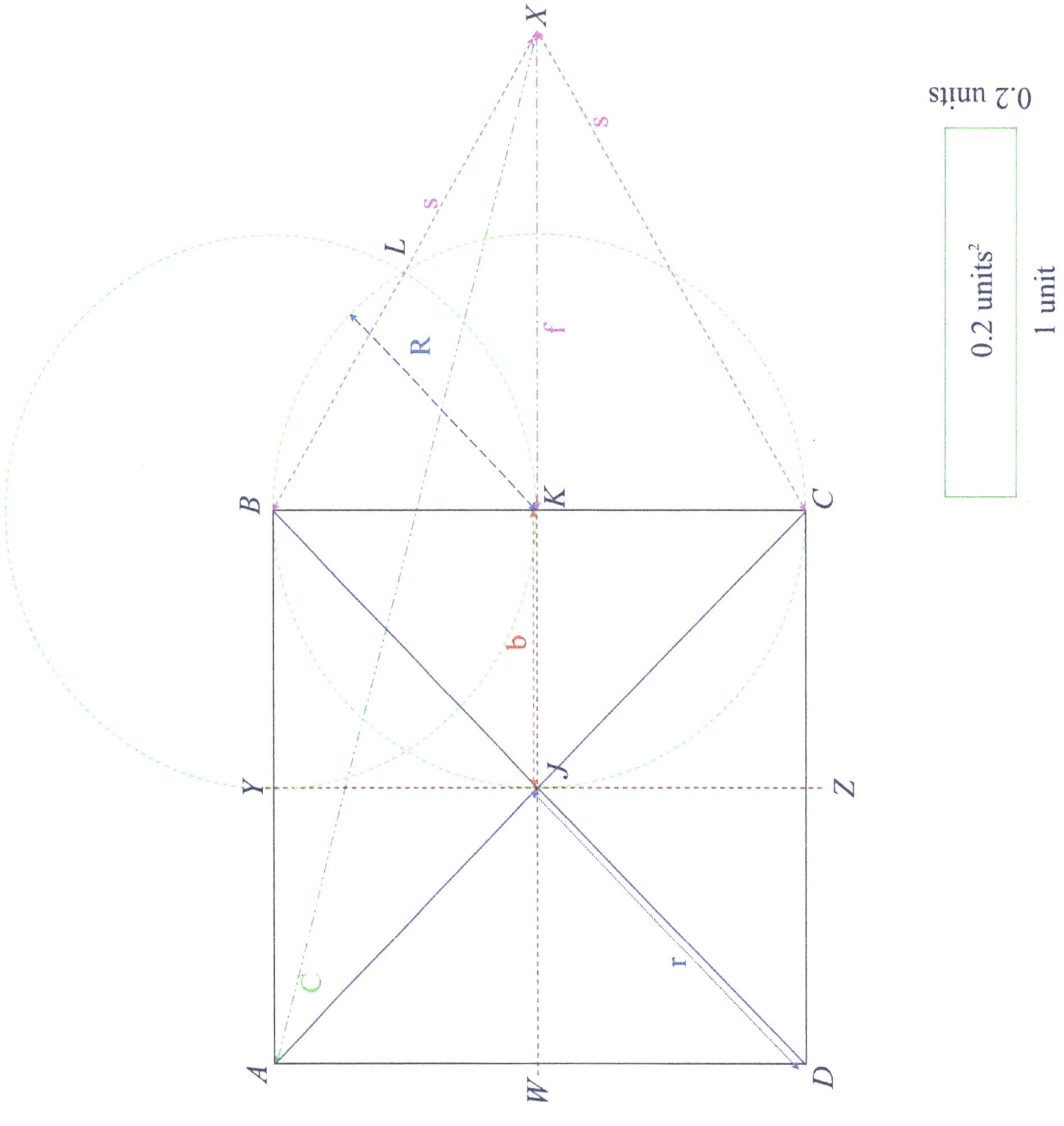

Figure 5.2
A Likely Planar Representation of a Sesquimetric Pyramid
Of the Neferirkare Type

The Proposed Modus

(1)   A Horizon **WJKX** would be defined in an exact West-East line probably by astronomical means and a turnstake driven at **J**, the intended Base Center.

(2)   The –b and +b segments would be defined along **WJKX**, possibly with a 3/4 calibrated cord and turnstake driven at **W** and **K**.

(3)   The perpendicular (i.e. North-South) crossline **YTJZ** would be defined by bisection: Either by scribing the *vesica piscis* from **W** and **K**; or perhaps by trisecting the hemicircle **KYW** and then bisecting the central arc.

(4)   The Base Corner Vertices **ABCD** are defined, probably by scribing from each of **Y**, **K**, **Z** and **W** in order to maximise good-conditioning.

(5)   A tensioned cord of length b is rotated about a turnstake at Vertex **B**.

(6)   That tensioned cord of length b is then rotated about a turnstake at Midpoint **K**. The point where the arcs intersect is **L**.

(7)   A tensioned rope of at least length s is laid from B through L to intersect the horizon at **X**.

(8)   Behold: Distance **KF** is Apothemal Slant Height f. Rotation of **KF** about **BC** brings **K** to identity with the pyramid apex **I**.

PART SIX
CONCLUSIONS

In some senses, there can be no conclusions concerning the Pyramids of the Ancients, for until we meet the latter Elsewhere their secrets shall remain recondite.

Man fears Time, but Time fears the Pyramids

as some forgotten genius perceived, possibly in the twelfth Christian century.

We have seen that the Ancient Egyptians attempted *system* in their planning of the monuments but there are very subtle differences in the geometry of the pyramids that may reflect misunderstandings by Ancient architects who *intended* to emulate their ancestors. After all, though some paraphidian and sesquimetric pyramids are contemporary, and clearly of complex though variant *design*, there are classes of sesquimetric pyramids that are one or two centuries younger and may on purpose have been planned on debased principles, or to ideals of a hermetic or numeromantic character lost to history. Or fire or flood may have destroyed orthodox drawings, and left bereft sons and apprentices who had to find their own ways, whether by the fiat of self-will, or through the forests of consensus.

Square pyramids are rich in right triangles, there being three principals, each replicated four-fold. Of course, plane rotations about the apical axis generate infinite examples. We do not know if the Old Kingdom builders understood the Pythagorean Theorem. Babylonian records of that postdate the age of Khufu by some seven hundred years.

It is not necessarily realistic to think that the Ancients would have entertained Cartesian conceptions of three dimensionality: Hence planar planning of height is credible.

Whatever the Ancient motives, surely beyond the merely valedictory, the true value of these structures was, and is, to fascinate all posterity and inspire the joy and science of their remotest descendants, not merely genetic but more significantly spiritual.

I am afraid I am not one of those who thinks that the Ancient builders were tutored by extra-terrestrial animals or even by God in Person, because for sure none of such agents would have made the elementary errors and forgettings we see in the elaborations of the Ancients or indeed in our own labors, however pressing their commitments elsewhere, or urgent their desires to abscond forever.

When we abstract such parameters as The Ratio of Phidias or The Ludolphine Constant, or indeed any number of incidental transcendentals from our analyses of dumb stones we have to ask ourselves to which extent if any the Ancients *intentionally* incorporated such measures into their work. Or whether these things were natural fallout from their line-and-circle scrivenings, and that the draftsmen would have greeted our questions with incredulity and bemusement if we had asked them regarding the importance of these metrics.

To the modern mind the building of the pyramids was dangerous and futile. The Ancients would have countered that all things are dangerous and the end of our progress is death. The Ancient Egyptians seemed to think (or maybe we libel them) that they could physically instantiate a perpetual elite by embalming corpses and sealing them in impregnable mausolea. Perhaps the Roman Catholic belief in Redemption Through Works is a remote and demotic echo of such ideas. Certainly we Protestants see such as naivety at best; blasphemy at worst. Eternal Life is God's Option, the operation of Divine Grace.

The Pyramids are grand but imperfect, like men and all our works, but as tokens of aspiration they are salutary.

<u>Notation</u>

Numerical Values refer to the putative ideal case of the
Great Pyramid of Giza (Cheops or Khufu)

A	Ellipse Area	
a	Ellipse Major Semi-Diameter	
b	Side Half-Length	($b = \pi/4$)
b	Ellipse Minor Semi-Diameter	
C	The Length of the Plan Clipping (or Crossing) Line	
e_e	Ellipse Eccentricity	
Φ	The Greater Ratio of Phidias	(about 1.618034)
ϕ	The Lesser Ratio of Phidias	(about 0.618034)
F	Focal Length	
F	Distance from Ellipse Center to Focus	(i.e. $\pm\varepsilon$)
H	Focal Hypotenuse	
h	Pyramid Height	($h = 1$)
λ	Ellipse Descriptive Loop Length	
L	Slant Height (normal to Edge S)	
P	Ellipse Perimeter	
Π_A	Pyramid Surface Area	
Π_V	Pyramid Volume	
R	Radius of the Circle that is Equivalent in Area to the Ellipse	
S	Side Length	($S = \pi/2$)
s	Arris Length (non-basal edge)	
Sp.Def(x,y)	The (Percentage) Specific Defect for Exact x against Approximate y	

General References and Bibliography

1   "The Great Pyramid of Khufu (Cheops)"
Franz Löhner and Teresa (Zubi) Zuberbühler
http://www.cheops-pyramide.ch
Copyright: 2006
9 pp

2   "Building the Great Pyramid"
Franz Löhner and Teresa (Zubi) Zuberbühler
https://www.cheops-pyramide.ch/khufu-pyramid/khufu-numbers.html
Copyright: 2006

3   "Geometrical Substantiation of Phi, the Golden Ratio and the Baroque of
       Nature, Architecture, Design and Engineering"
Md Akhtaruzzaman and Amir A Shafie
Department of Mechatronics Engineering
International Islamic University
Kuala Lumpur 53100
Malaysia
International Journal of Arts 2011; 1(1): 1-22
Corresponding Author: akhter900@yahoo.com
http://journal.sapub.org/arts
Copyright: 2011 Scientific and Academic Publishing

4   "Pi and the Great Pyramid"
Ralph Greenburg
Copyright: 2000
https://www.sites.math.washington.edu/~greenber/PyPyr.html
8 pp

5   "The Golden section ratio: Phi"
Dr Ron Knott
http://www.maths.surrey.ac.uk/hosted-sites/R.Knott/Fibonacci/phi.html
Copyright: 1996-2016
19 pp

6   "Engineering the Pyramids"
Dick Parry   21 April 2005
Sutton Publishing   (2004)
ISBN 0-7509-3414-X
192 pp

7   "Sacred Geometry"
           Miranda H Lundy   (author and illustrator)
           Wooden Books Ltd of Glastonbury   2006
           ISBN 1-904263-04-6
           pp 59

8   "The Golden Section"
           Scott Olsen
           Wooden Books Ltd of Glastonbury   2006AD
           ISBN 1-904263-47-X
           pp 59

Wikipedia References

Wikipedia contributors. (2019, March 14). Egyptian chronology. In *Wikipedia, The Free Encyclopedia*. Retrieved 10:21, May 21, 2019, from
https://en.wikipedia.org/w/index.php?title=Egyptian_chronology&oldid=887690333

Wikipedia contributors. (2019, April 22). Egyptian pyramid construction techniques. In *Wikipedia, The Free Encyclopedia*. Retrieved 10:22, May 21, 2019, from
https://en.wikipedia.org/w/index.php?title=Egyptian_pyramid_construction_techniques&oldid=893583551

Wikipedia contributors. (2019, May 17). Great Pyramid of Giza. In *Wikipedia, The Free Encyclopedia*. Retrieved 10:23, May 21, 2019, from
https://en.wikipedia.org/w/index.php?title=Great_Pyramid_of_Giza&oldid=897550957

Wikipedia contributors. (2019, May 4). Golden ratio. In *Wikipedia, The Free Encyclopedia*. Retrieved 10:24, May 21, 2019, from
https://en.wikipedia.org/w/index.php?title=Golden_ratio&oldid=895438780

Wikipedia contributors. (2019, May 2). List of Egyptian pyramids. In *Wikipedia, The Free Encyclopedia*. Retrieved 10:25, May 21, 2019, from
https://en.wikipedia.org/w/index.php?title=List_of_Egyptian_pyramids&oldid=895231692
(with downloadable spreadsheet)

Wikipedia contributors. (2019, May 8). Pyramid (geometry). In *Wikipedia, The Free Encyclopedia*. Retrieved 10:26, May 21, 2019, from
https://en.wikipedia.org/w/index.php?title=Pyramid_(geometry)&oldid=896164925

APPENDIX ONE
APPROXIMATION PAIRS

Paired Constants' Serial	Phidian Term (X) Symbol	or Exact Form Formula	Value	Ludolphine Term (Y) Symbol	or Approximate Form Formula	Value	Percentage Specific Defect
1	ϕ	$\dfrac{1+\sqrt{5}}{2}-1$	0.618033989		$\dfrac{\pi^2}{16}$	0.616850275	0.191528897
2	Φ	$\dfrac{1+\sqrt{5}}{2}$	1.618033989		$\dfrac{L}{b}$	1.618993187	-0.05928169
3	Φ	$\dfrac{1+\sqrt{5}}{2}$	1.618033989		$\dfrac{f}{b}$	1.618993187	-0.05928169
4	Φ	$\dfrac{1+\sqrt{5}}{2}$	1.618033989		$\dfrac{4}{\pi}\sqrt{1+\dfrac{\pi^2}{16}}$	1.618993187	-0.05928169
5	Φ	$\dfrac{1+\sqrt{5}}{2}$	1.618033989		$\sqrt{\left(\dfrac{4}{\pi}\right)^2+1}$	1.618993187	-0.05928169
6	Φ	$\dfrac{1+\sqrt{5}}{2}$	1.618033989		$\left(\dfrac{14}{11}\right)^2$	1.619834711	-0.11129074
7	C	$\sqrt{1+\dfrac{3\pi^2}{8}+\dfrac{\pi}{4}*\sqrt{16+\pi^2}}$	2.948865412	C_A	$\sqrt{5+\dfrac{3\pi^2}{8}}$	2.949762982	-0.0304378
8	C	$\sqrt{1+\dfrac{3\pi^2}{8}+\dfrac{\pi}{4}*\sqrt{16+\pi^2}}$	2.948865412	C_B	$\sqrt{1+6\phi+\pi f}$	2.950069406	-0.04082906
9	C	$\sqrt{1+\dfrac{3\pi^2}{8}+\dfrac{\pi}{4}\sqrt{16+\pi^2}}$	2.948865412	C_B	$\sqrt{1+6\phi+\pi\sqrt{1+\dfrac{\pi^2}{16}}}$	2.950069406	-0.04082906
10	C	$\sqrt{1+\dfrac{3\pi^2}{8}+\dfrac{\pi}{4}*\sqrt{16+\pi^2}}$	2.948865412	C_C	$\sqrt{5+6\phi}$	2.95096661	-0.07125444
11	f	$\sqrt{1+\dfrac{\pi^2}{16}}$	1.271554275		$b\sqrt{\dfrac{1}{b^2}+1}\quad\left(where\ b=\dfrac{\pi}{4}\right)$	1.271554275	0
12	4	2^2	4		$\sqrt{\pi^2+\dfrac{\pi^4}{16}}$	3.99470557	0.132360751

Table A1
A Tabular Comparison of Phidian and Ludolphine Near-Identities
with Some Additional Near-Identities

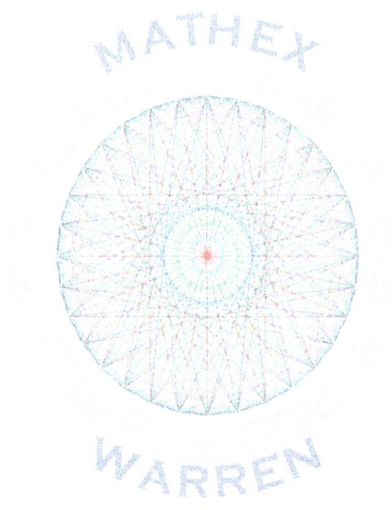

The Hermetic Defect ℎ as
A Difference between Ludolphine and Phidian Functions

by

James R Warren BSc MSc PhD PGCE

Introduction

The Minor Phidian Ratio, ϕ, is defined as:-

$$\phi = \frac{1 - \sqrt{5}}{2}$$

Equation 1.1

Where it is convenient to use a positive form, we may employ the Positive Minor Ratio of Phidias, ϕ', defined as:

$$\phi' = \frac{\sqrt{5} - 1}{2}$$

Equation 1.2

and the Major Phidian Ratio, Φ, as:-

$$\Phi = \frac{1 + \sqrt{5}}{2}$$

Equation 1.3

Like the Ludolphine Constant, the Ratios of Phidias (sometimes one or both are called "The Golden Section", "The Golden Mean" or "The Golden Ratio", etcetera) have many remarkable and elegant properties, which are well-rehearsed in the literature[1].

The Ratio of Phidias is frequently visualised as a "Golden Rectangle" whose aspect ratio is Φ:1. More formally, it is a lineal bi-product of the construction of a pentagram or star polygon. The ratio is a major metric in the classical arts and aesthetics. The Ludolphine Constant is the ratio of a circle's perimeter length to the length of its diameter. As such it is a metric of minimum-energy envelopes and is therefore of central importance to engineering, and the physical sciences.

Numerical Values of ϕ, ϕ' and Φ are presented in Table 1.1, together with the Ludolphine Constant, π. All four numbers are transcendental and perforce approximate. The numbers are stated to fifteen figures, as computed via EXCEL®. The numbers can be extended infinitely using a number of reliable processes. The extensions can never terminate in a strictly mathematical sense, nor do the successive digits repeat or form patterns. We can of course interrupt or abridge generative algorithms physically.

Constant Name	Constant Symbol	Value
Minor Phidian Ratio	ϕ	-0.618033988749895
Positive Minor Ratio of Phidias	ϕ'	0.618033988749895
Major Phidian Ratio	Φ	1.618033988749890
Ludolphine Constant	π	3.141592653589790

Table 1.1
Values of Selected Constants

In a previous study we observed that the Minor and Major Ratio of Phidias may accurately be approximated by sundry functions, especially some involving the Ludolphine Constant, π.

Table 1.2 presents a selection of such relationships together with error estimates in the shape of Percentage Specific Defects.

The Approximant, ξ

From Table 1.2, Paired Constant Serial 5 offers the following Simple Approximant, ξ:-

$$\xi = \sqrt{\left(\frac{4}{\pi}\right)^2 + 1}$$

Equation 1.4

ξ has a numeric value of 1.618993186606230. This is remarkably near the value of Φ, which is 1.618033988749890. There are of course estimates that are worse and better than ξ. But this estimate is a good starting-point for further discussions.

So we may now define the difference between the ideal Φ and the estimate ξ as the Hermetic Defect, $\mathfrak{h}$. We call this shortfall the hermetic defect because we need to call it something and the word hermetic always sounds cool.

$$\mathfrak{h} = \Phi - \xi = \left(\frac{1 + \sqrt{5}}{2}\right) - \sqrt{\left(\frac{4}{\pi}\right)^2 + 1} \approx -0.000959197856338$$

Equation 1.5

The absolute fractional difference is less than one thousandth. We can develop this and perhaps improve upon it.

Paired Constants' Serial	Phidian Term (X) — or Exact Form		Value	Ludolphine Term (Y) — or Approximate Form		Value	Percentage Specific Defect
	Symbol	Formula		Symbol	Formula		
1	ϕ	$\dfrac{1+\sqrt5}{2}-1$	0.618033989		$\dfrac{\pi^2}{16}$	0.616850275	0.191528897
2	Φ	$\dfrac{1+\sqrt5}{2}$	1.618033989		$\dfrac{L}{b}$	1.618993187	-0.05928169
3	Φ	$\dfrac{1+\sqrt5}{2}$	1.618033989		$\dfrac{f}{b}$	1.618993187	-0.05928169
4	Φ	$\dfrac{1+\sqrt5}{2}$	1.618033989		$\dfrac{4}{\pi}\sqrt{1+\dfrac{\pi^2}{16}}$	1.618993187	-0.05928169
5	Φ	$\dfrac{1+\sqrt5}{2}$	1.618033989		$\sqrt{\left(\dfrac{4}{\pi}\right)^2+1}$	1.618993187	-0.05928169
6	Φ	$\dfrac{1+\sqrt5}{2}$	1.618033989		$\left(\dfrac{14}{11}\right)^2$	1.619834711	-0.11129074
7	C	$\sqrt{1+\dfrac{3\pi^2}{8}+\dfrac{\pi}{4}*\sqrt{16+\pi^2}}$	2.948865412	C_A	$\sqrt{5+\dfrac{3\pi^2}{8}}$	2.949762982	-0.0304378
8	C	$\sqrt{1+\dfrac{3\pi^2}{8}+\dfrac{\pi}{4}*\sqrt{16+\pi^2}}$	2.948865412	C_B	$\sqrt{1+6\phi+\pi f}$	2.950069406	-0.04082906
9	C	$\sqrt{1+\dfrac{3\pi^2}{8}+\dfrac{\pi}{4}*\sqrt{16+\pi^2}}$	2.948865412	C_B	$\sqrt{1+6\phi+\pi\sqrt{1+\dfrac{\pi^2}{16}}}$	2.950069406	-0.04082906
10	C	$\sqrt{1+\dfrac{3\pi^2}{8}+\dfrac{\pi}{4}*\sqrt{16+\pi^2}}$	2.948865412	C_C	$\sqrt{5+6\phi}$	2.95096661	-0.07125444
11	f	$\sqrt{1+\dfrac{\pi^2}{16}}$	1.271554275		$b\sqrt{\dfrac{1}{b^2}+1}\quad\left(where\ b=\dfrac{\pi}{4}\right)$	1.271554275	0
12	4	2^2	4		$\sqrt{\pi^2+\dfrac{\pi^4}{16}}$	3.99470557	0.132360751

Table 1.2
Comparisons of the Ratio of Phidias and some of its Associated Functions
with
The Ludolphine Constant and Some of its Associated Functions

<u>Some Useful Functions</u>

Firstly, as per usual, we shall define our measure of the agreement between two compared numbers, Percentage Specific Defect, as:-

$$PSD(x, y) = 100\frac{x - y}{x}$$
Equation 1.6

Where x is the Resulting Number according to Canonical Theory, and y is the Estimate to be Compared. PSD(x,y) preserves the sign and sense of the discrepancy, and is to be preferred to Standard Deviation and Root Mean Square to which however it is mathematically related.

The Number of Combinations of r Objects drawn from a Population of n Objects, $^{n}C_{r}$, is given by:-

$$^{n}C_{r} = BE(n, r) = \frac{n!}{r!\,(n - r)!}$$
Equation 1.7

It is important to remember that $^{n}C_{r}$ expresses the *count* of combinations; not their *value*.

In connection with consideration of The Ratio of Phidias it is often important to discuss the Fibonacci Series, which is a succession of integers generated by summing the preceding two in the sequence such that (for $j_1 = 1$ and $j_2 = 1$):-

$$j_i = j_{i-1} + j_{i-2}$$
Equation 1.8

The Fibonacci Number Fia(i) is:-

$$Fia(i) = \sum_{i=1}^{\frac{i-1}{2}} \frac{(i - k - 1)!}{k!\,((i - k - 1) - k)!} = \sum_{i=1}^{\frac{i-1}{2}} \frac{(i - k - 1)!}{k!\,(1 - 2.k - 1)!}$$
Equation 1.9

This yields the Ordinary Fibonacci Series Fia(1) = 1, Fia(2) = 1, Fia(3) = 2, Fia(4) = 3, Fia(5) = 5 , Fia(6) = 8...

Allied to Equation 1.9 is the Alternating Fibonacci Series Fib(i) given by:-

$$Fib(i) = \frac{(\phi')^i - \left(-(\phi')\right)^{-i}}{\sqrt{5}}$$
Equation 1.10

or for Fib(j):-

$$Fib(j) = \frac{(-(\phi'))^{-j} - (\phi')^j}{\sqrt{5}}$$

Equation 1.11

which yields the Alternating Fibonacci Series Fib(1) = -1, Fib(2) = 1, Fib(3) = -2, Fib(4) = 3, Fib(5) = -5 , Fib(6) = 8...

Interestingly, it follows that the positive integral powers of ϕ' are given by:-

$$(\phi')^j = \phi.Fib(j) - Fib(j-1)$$

Equation 1.12

The Incomplete Gamma Function

The Incomplete Gamma Function, $\Gamma(s,t)$ is of potential use in the treatment of systems of factorials, especially (false) factorials that involve non-integers.

It is defined by:-

$$\Gamma(s,t) = \int_t^{INF} e^{-x}.x^{s-1}dx$$

Equation 1.13

My platform, and in particular Mathcad® Prime® 3.1, has no problem driving this to beyond twelve figures when INF is 18.

We must hold the IGF in reserve until we encounter a clinch that justifies its deployment.

Literature Convergent Series for Φ

The Wikipedia article upon the Golden Ratio[1] offers:-

$$\Phi = \frac{13}{8} + \sum_{i=0}^{INF} \frac{(-1)^{i+1}.(2.i+1)!}{4^{2.i+3}.i!.(i+2)!}$$

Equation 1.14

from which it appears to me that:-

$$\phi' = \frac{5}{8} + \sum_{i=0}^{INF} \frac{(-1)^{i+1}.(2.i+1)!}{4^{2.i+3}.i!.(i+2)!}$$

Equation 1.15

Mathcad computes Equation 1.15 as 0.618033988749894 when INF is 18.

Some Ways of Accurately Estimating the Ludolphine Constant

Like The Ratio of Phidias, the Ludolphine Constant π is transcendental and no exact value can be computed. Notwithstanding that, however, there are algorithms that can calculate it to trillions of decimal places.

For practical people like us, for whom fifteen-digit precisions are more than adequate, it is reasonable to select such estimation formulae as might assist theoretical developments, in the present context the product $\phi.\pi$

In such a spirit, we shall review a small selection of potentially-interesting equations from the literature. And Wikipedia is as reliable a source as most!

Ramajuhan's Formula[2]

During Late Victorian times, or possibly in Edwardian, the great Indian genius Srinivasa Ramanujan derived many original algebraic equations, including several relative to the efficient establishment of high-precision Ludolphine Constant values.

We will focus on one of those highlighted by R William Gosper Jr. in a 2009 memorial paper[3]:-

$$\frac{1}{\pi} = \frac{2\sqrt{2}}{9801} \sum_{k=0}^{INF2} \frac{(4k)!.\,(1103 + 26390k)}{(k!)^4 \times 396^{4k}}$$

Equation 1.16

On my platform, definition of INF2 = 1 drove to exhaustion at fourteen places of decimals.

The Ramanujan-Sato Form

Even with INF3 = 64 this form could not achieve three-figure accuracy on my HP 500400 Pavilion running Mathcad® Prime® 3.1. In my opinion, this is likely to be due to numerical errors arising from the application of the combinatorial function.

$$\frac{1}{\pi} = \frac{1}{4} \cdot \sum_{k=0}^{INF3} \frac{\binom{2k}{k}^2}{k+1} \cdot \frac{1}{2^{4k}}$$

Equation 1.17

The FC Størmer Form

$$\pi = 4\left[44.\,Atan\left(\frac{1}{57}\right) + 7.\,Atan\left(\frac{1}{239}\right) - 12.\,Atan\left(\frac{1}{682}\right) + 24.\,Atan\left(\frac{1}{12943}\right)\right]$$

Equation 1.18

The FC Størmer Form is said to be fantastically accurate (presumably depending upon whatever is used to determine arctangents) and it gave fifteen-digit accuracy upon my platform.

The Chudnovsky Algorithm[4]

This algorithm is brief enough to be consolidated into a single succinct formula, but I shall specify it as Wikipedia does in five digestible parts:-

$$C = 426880\sqrt{10005}$$

Equation 1.19

$$M_q = \frac{(6q)!}{(3q)!\,(q!)^2}$$

Equation 1.20

$$L_q = 545140134q + 13591409$$

Equation 1.21

$$X_q = (-262537412640768000)^q$$

Equation 1.22

Using which:-

$$\pi = C\left(\sum_{q=0}^{INF4}\frac{M_q L_q}{X_q}\right)^{-1}$$

Equation 1.23

With INF4 = 1 this algorithm drove to the full fifteen-figure precision. With INF4 = 0 fourteen figures were achieved.

The Failure of Simple Simplification

I hope you will forgive the pleonasm (if that is what it is).
Recollect that:-

$$\mathfrak{h} = \Phi - \xi = \left(\frac{1+\sqrt{5}}{2}\right) - \sqrt{\left(\frac{4}{\pi}\right)^2 + 1} \approx -0.000959197856338$$

Equation 1.5

If $\mathfrak{h} = 0$, then a naive transposition might yield:-

$$\Phi = \sqrt{\left(\frac{4}{\pi}\right)^2 + 1}$$

Equation 1.24

Or:-

$$\Phi^2 = \frac{16}{\pi^2} + 1$$

Equation 1.25

Or:-

$$\sqrt{\Phi} = \frac{\Phi\pi}{4}$$

Equation 1.26

Unfortunately, this cheap resolution will not do: The relevant Percentage Specific Defects demonstrate that Equation 1.26 is a worse estimator than Equation 1.24:-

$$PSD\left(\sqrt{\Phi}, \frac{\Phi\pi}{4}\right) = 0.095810346618442$$

Equation 1.27

$$PSD\left(\sqrt{\Phi}, \frac{4}{\pi}\right) = -0.095902230878257$$

Equation 1.28

$$PSD\left(\Phi, \sqrt{\left(\frac{4}{\pi}\right)^2 + 1}\right) = -0.059281687715287$$

Equation 1.29

although both bad forms still give an estimate of root Φ better than a one-in-one-thousandth of the true value.

The Binomial Expansion of $\mathfrak{h}$

The binomial expansion of the Hermetic Defect $\mathfrak{h}$ is given by:

$$\mathfrak{h} = (\Phi - \xi)^n$$
Equation 1.30

where n is a positive integer, the Degree of the Polynomial Expansion.

Determining the Desired Precision of $\mathfrak{h}$

The Absolute Number of Decimal Places, D, that is required in the difference is approximated by:-

$$D = -n\log_{10}(-\mathfrak{h})$$
Equation 1.31

We can aim for fifteen-figure precision, in which case n = 5 (i.e. a degree-5 Quintic Equation must be sought), because when n = 5, Equation 1.31 resolves as 15.0904590024529.

The Quintic

Each of the five (identical) ξ values represents an r-value determining the coefficients of the quintic equation such that:-

$$r_1 = r_2 = r_3 = r_4 = r_5 = \xi = -\sqrt{\left(\frac{4}{\pi}\right)^2 + 1}$$

Equation 1.32

The Quintic may be represented as:-

$$P(5) = \sum_{k=0}^{n} c_{n,k} \cdot \Phi^k$$

$$= \prod_{i=1}^{n} r_i + \prod_{i=1}^{n} r_i \cdot \left(\sum_{i=1}^{n} \frac{1}{r_i}\right) \cdot \Phi$$

$$+ \prod_{i=1}^{n} r_i \cdot \left(\sum_{j=1}^{n-1}\sum_{i=1}^{n-j} \frac{1}{r_j \cdot r_{i+j}}\right) \cdot \Phi^2 + \prod_{i=1}^{n} r_i \cdot \left(\sum_{i=1}^{n-1}\sum_{j=i+1}^{n} r_i \cdot r_j\right) \cdot \Phi^3 + \left(\sum_{i=1}^{n} r_i\right) \cdot \Phi^4$$

$$+ \Phi^5$$

Equation 1.33

whose five polynomial coefficients resolve to:-

$$c_{5,0} = \prod_{i=1}^{n} r_i = (-1)^n \cdot \left(\sqrt{\left(\frac{4}{\pi}\right)^2 + 1}\right)^n = -11.1230811408161$$

Equation 1.34a

$$c_{5,1} = \prod_{i=1}^{n} r_i \cdot \left(\sum_{i=1}^{n} \frac{1}{r_i}\right) = (-1)^{n-1} \cdot n \cdot \left(\sqrt{\left(\frac{4}{\pi}\right)^2 + 1}\right)^{n-1} = 34.35184666877$$

Equation 1.34b

$$c_{5,2} = \prod_{i=1}^{n} r_i \cdot \left(\sum_{j=1}^{n-1} \sum_{i=1}^{n-j} \frac{1}{r_j \cdot r_{i+j}}\right) = (-1)^n \cdot \left(\sqrt{\left(\frac{4}{\pi}\right)^2 + 1}\right)^{n-2} \cdot 10 = -42.4360608221941$$

Equation 1.34c

$$c_{5,3} = \prod_{i=1}^{n} r_i \cdot \left(\sum_{i=1}^{n-1} \sum_{j=i+1}^{n} r_i \cdot r_j\right) = 10\left(\left(\frac{4}{\pi}\right)^2 + 1\right) = 26.211389382774$$

Equation 1.34d

$$c_{5,4} = \left(\sum_{i=1}^{n} r_i\right) = -n \cdot \left(\sqrt{\left(\frac{4}{\pi}\right)^2 + 1}\right) = -8.09496593303116$$

Equation 1.34e

$$c_{5,5} = 1$$

Equation 1.34f

from which:-

$$P(5) = \sum_{k=0}^{n} c_{n,k} \cdot \Phi^k = 0.000000000000036$$

Equation 1.35

The numerical value of P(5) is either 0 exactly or some very small positive number like that of Equation 1.35, depending upon the precise symbolic formulation.
For reference:-

$$\mathfrak{h} = (\Phi - \xi)^5 = -8.11971895741518 \times 10^{-16}$$

Equation 1.36

The expansion of P(5) gives:-

$$P(5) = \sum_{k=0}^{n} c_{n,k} \cdot \Phi^k = c_{5,0} \cdot \Phi^0 + c_{5,1} \cdot \Phi^1 + c_{5,2} \cdot \Phi^2 + c_{5,3} \cdot \Phi^3 + c_{5,4} \cdot \Phi^4 + c_{5,5} \cdot \Phi^5$$

Equation 1.37

The relevant substitutions from Equations 1.34* permit:-

$$P(5) = \sum_{k=0}^{n} c_{n,k} \cdot \Phi^k$$

$$= (-1)^n \cdot \left(\sqrt{\left(\frac{4}{\pi}\right)^2 + 1} \right)^n \cdot \Phi^0 + (-1)^{n-1} \cdot n \cdot \left(\sqrt{\left(\frac{4}{\pi}\right)^2 + 1} \right)^{n-1} \cdot \Phi^1$$

$$+ (-1)^n \cdot \left(\sqrt{\left(\frac{4}{\pi}\right)^2 + 1} \right)^{n-2} \cdot 10 \cdot \Phi^2 + 10 \cdot \left(\left(\frac{4}{\pi}\right)^2 + 1 \right) \cdot \Phi^3$$

$$+ -n \cdot \left(\sqrt{\left(\frac{4}{\pi}\right)^2 + 1} \right) \cdot \Phi^4 + 1 \cdot \Phi^5$$

Equation 1.38

substitution of 5 for n gives:-

$$P(5) = \sum_{k=0}^{5} c_{5,k} \cdot \Phi^k$$

$$= -\left(\sqrt{\left(\frac{4}{\pi}\right)^2 + 1} \right)^5 \cdot \Phi^0 + n \cdot \left(\sqrt{\left(\frac{4}{\pi}\right)^2 + 1} \right)^4 \cdot \Phi^1 - \left(\sqrt{\left(\frac{4}{\pi}\right)^2 + 1} \right)^3 \cdot 10 \cdot \Phi^2$$

$$+ 10 \left(\left(\frac{4}{\pi}\right)^2 + 1 \right) \cdot \Phi^3 + -5 \cdot \left(\sqrt{\left(\frac{4}{\pi}\right)^2 + 1} \right) \cdot \Phi^4 + 1 \cdot \Phi^5$$

Equation 1.39

from which by organising roots it follows that:-

$$P(5) = \sum_{k=0}^{5} c_{5,k} \cdot \Phi^k$$

$$= -\left(\left(\frac{4}{\pi}\right)^2 + 1\right)^{\frac{5}{2}} \cdot \Phi^0 + n \cdot \left(\left(\frac{4}{\pi}\right)^2 + 1\right)^{\frac{4}{2}} \cdot \Phi^1 - \left(\left(\frac{4}{\pi}\right)^2 + 1\right)^{\frac{3}{2}} \cdot 10 \cdot \Phi^2$$

$$+ 10\left(\left(\frac{4}{\pi}\right)^2 + 1\right)^{\frac{2}{2}} \cdot \Phi^3 + -5 \cdot \left(\left(\frac{4}{\pi}\right)^2 + 1\right)^{\frac{1}{2}} \cdot \Phi^4 + 1^{\frac{2}{2}} \cdot \Phi^5$$

Equation 1.40

from which it follows that:-

$$P(5) = \sum_{k=0}^{5} c_{5,k} \cdot \Phi^k = \sum_{i=0}^{5} (-1)^{i+1} \cdot \left(\sqrt{\left(\frac{4}{\pi}\right)^2 + 1}\right)^{n-i} \cdot \Phi^i$$

Equation 1.41

Equation 1.41 discloses our Hermetic Defect $\mathfrak{h}$ naked and unmasked as nothing other than a classical binomial expansion.

After substituting $\mathfrak{h}^5$ for P(5) I was able to economise Equation 1.41 by deriving:-

$$P(5) = \mathfrak{h}^5 = n! \sum_{i=0}^{n} \frac{(-1)^i}{i! \, (n-i)!} \cdot \left(\sqrt{1 + \frac{16}{\pi^2}}\right)^i \cdot \Phi^{n-i}$$

Equation 1.42

from which:-

$$\mathfrak{h} = \sqrt[n]{n! \sum_{i=0}^{n} \frac{(-1)^i}{i! \, (n-i)!} \cdot \left(\sqrt{1 + \frac{16}{\pi^2}}\right)^i \cdot \Phi^{n-i}}$$

Equation 1.43

Equation 1.43 estimates $\mathfrak{h}$ as 0.001678488210535 when n = 5, which strikes me as a considerable vitiation compared with $\mathfrak{h} = \Phi - \xi = $ -0.000959197856338.

But of course a creature born in a stable is quite likely a horse. And even the oldest and most senatorial of those noble beings is more useful than my equation.

Criticism

Binomial expansion describes a description: It adds *no new information.*

Students may be forgiven for thinking that by using the tools I use I can prove that virtually anything resembles virtually anything else. I must take care to avoid the legerdemain of the magus. And yet mathematics is wholly human and describes this world as it appears to us: Not to the intelligent Superagent which made it by His inscrutable intent and technique.

So it is not pertinent for us to complain that this universe is full of holes and interstices. These lacunae give liberty and latitude to all who might venture to fill them, if only to worship the Father who multiplied such playgrounds, to honor the Roman slaves who sought to free us with truth, to build a mill or to pass an idle hour.

Kronecker said that God invented the natural numbers and that everything else was the work of man.

Kronecker was wrong. I do not presume to postscribe God's province, but I ask you what needs God with number when He can engender as much as is good?

Mathematics is not a science. Any science, or science generally, presents a series of problems each of which has *one true and unique solution* even though the human animal can never know that solution *for certain.* True science is privy only to its Creator.

Mathematics is quite different. Like the other fine arts it has infinite expressions, both known and latent. It is expressed with a view to elegance, economy and sensibility, too often like all arts with a precious or fastidious connoisseurship.

Of course, both science and mathematics are indispensable to any medicine, engineering or other technology. But then again so is lingual literature, and so is draftsmanship.

Speaking of the arts, mathematics is somewhat like the Law but more like Politics or Collecting. All four avocations appeal to a certain kind of mind, obsessional obviously, but also stricken with a sort of punctilious malice, or sublimated peccadillo, idle and patrician.

As with politics the issues addressed by mathematics are wholly adventitious. There is an interminate quest for a solution that is satisfying or economic or aesthetically-pleasing, and more so than the rest of the possible proposals, in a field in which, like Politics, many or most ideas are merely stupid.

If we compare the valuable work of two heroes we will salute two remarkable men, comparable only in industry.

Leonhard Euler was a wealthy and leisured scion of an educated family that conferred the finest, and most expensive, education. Leonhard grew to be a prudent and continent man, devoutly religious. Leonhard, an elector in a free democracy could demand, and get, frequent audiences with kings and empresses, who contended with each other to endow him the richest emoluments. Frederick, King of Prussia, invited Euler to confute the atheist Diderot in person, and he did so hilariously. Euler was still working at his desk in Petersburg when he suffered a stroke and died at the age of seventy-six.

Srinivasa Ramanujan was an impoverished Tamil born in the backstreets of Erode, an obscure inland town in the Madras Presidency of the then British India. Ramanujan enjoyed no franchise and little other social advantage. Ramanujan had none of the infantile familiarity with Arabic numerals or Roman script that so advantaged the thinkers of the western world. Srinivasa grew to be a prudent and continent man, devoutly religious. By chance, his inexplicable unpublished formulations, expressed however in the European idiom, came to the notice of the

English mathematician and lecturer Godfrey Hardy. Hardy and his employer, The University of Cambridge, sponsored Ramanujan to take up a research fellowship at Trinity College, and published much of his archive, perhaps twenty years of work. It would never have crossed anyone's mind to invite Ramanujan to debate with fellow Trinity man, The Earl Russell, or indeed to introduce the men. Ramanujan ascribed his abilities to his gods, and especially the goddess tutelary of his family. Sadly, Ramanujan took ill with a long-standing liver infection and returned to India to die. He was only thirty-two.

Mathematics is a language, and as language inherits the limitations of its mortal authors, plain to see in this starlight.

Euler's Identity[5,6]

Much in the present discussion is due to the prolific Swiss genius Leonhard Euler (1707-1783) (pronounced "oiler" as in the bloke with the lubricant).

Euler produced enough mathematics to fill, it is said, sixty octavo volumes, but many of his most celebrated contributions were not truly original, because, knowingly or not, they had been anticipated.

This is in no manner to disparage Euler. He was not a plagiarist. He was a man who celebrated his delight in a grand comedy and contributed his portion to the Gaiety of Nations. He may be said to be the father of Russian mathematics, a tradition second only to that of France.

Roger Cotes (1682-1716) and Johann Bernoulli (1667-1748) may have anticipated Euler's Identity, notionally invented in 1748, but the equation, said to be the most beautiful ever devised, was kind of floating around in the intellectual atmosphere of the day, inchoate and incondite, implied but unstated, until the moment of its explication arrived. And Euler never published it. Euler was nine when Cotes died at Cambridge, England, but Bernoulli and Euler were personal friends as well as colleagues.

This is that equation:-

$$e^{i\pi} + 1 = 0$$
Euler's Identity

e is the Base of Napierian Logarithms, otherwise known as Euler's Constant about 2.718281828459045; i is the Square Root of Minus One; π is the Ludolphine Constant, otherwise known as Archimedes Constant or Archimedes Number, the ratio of a circle's perimeter to its radius, and a central parameter in mathematics and physics (roughly 3.141592653589793).

Euler showed that the Euler Identity is a corollary of:-

$$e^z = \lim_{n \to \infty} \left(1 + \frac{z}{n}\right)^n$$
Euler's Limit

which you may think most suggestive in the context of our present discussion.

But men of lesser intellect such as me can only really stand back and admire, and make our own poor efforts. How can it be, I ask in wonder, that the exponent of one transcendental number by another can be the same as an integer, never mind a negative integer?

Today The Wikipedia Foundation is the greatest contributor to scholarship but it self-consciously, you might say ostentatiously, repudiates all pretension to originality, let alone priority.

I am minded of the Words of Ecclesiastes, that there is "no new thing under the sun"[7].

Developments of Ramanujan's Formula for π

We saw earlier that Ramanujan offered an estimate of π as:-

$$\frac{1}{\pi} = \frac{2\sqrt{2}}{9801} \sum_{k=0}^{INF2} \frac{(4k)!.(1103 + 26390k)}{(k!)^4 \times 396^{4k}}$$

Equation 1.16

If we perform zero iterations of this formula (i.e. INF2 = 0) we discover that the Percentage Specific Defect relative to the fifteen-figure fiducial Mathcad value for π is a mere -0.000002432635954, or in other words we get better than seven-figure precision.

Let us call this INF2 = 0 estimate π_{Rama0}.

Then simplification yields:

$$\pi_{Rama0} = \frac{9801}{1103.2.\sqrt{2}} = 3.14159273001331$$

Equation 1.44

All three of the integers in Equation 1.44 are prime.

On the other hand, if we inch forward a little to set INF2 = 1 we get Ramanujan π accurate to the system precision of fifteen figures in which case PSD is of course zero.

Let us call this INF2 = 1 estimate π_{Rama1}.

Then simplification yields:

$$\pi_{Rama1} = \frac{9801}{2.\sqrt{2} \sum_{k=0}^{INF2} \frac{(4k)!(1103 + 26390k)}{(k!)^4.396^{4k}}} = 3.14159265358979$$

Equation 1.45

Further consolidation, including elimination of the redundant summation, gives us a developed intermediate:-

$$\pi_{Rama1} = \frac{9801}{2.\sqrt{2}\left[1103 + \frac{24(1103 + 26390)}{396^4}\right]} = 3.14159265358979$$

Equation 1.46

Though clearly not in its simplest terms Equation 1.46 may be adequate to computational economy, certainly if a pocket calculator is used.

Prime Factorisation of Equation 1.46[11]

Several of the integers in Equation 1.46 are of course prime, but that is not to say that further analysis or condensation may not be suggested by a full factorisation. In that spirit:-

$$\pi_{Rama1} = \frac{3^4.11^2}{2.\sqrt{2}\left[1103 + \frac{2^3 3^1 (1103 + 2.5.7.13.29)}{(2^2.3^2.11)^4}\right]} = 3.14159265358979$$

Equation 1.47

Further manipulation enables the Simplest Prime Factorisation as:-

$$\pi_{Rama1} = \frac{3^4.11^2}{2.\sqrt{2}\left[1103 + \frac{19 \cdot 1447}{2^5.3^7.11^4}\right]} = 3.14159265358979$$

Equation 1.48

If it is preferable to employ a Ramanujan π expressed in low-value primes then I suggest the following form:-

$$\pi_{Rama1} = \frac{3^4.11^2}{\sqrt{2}\left[2(19 \cdot 29 + 2^3 \cdot 3 \cdot 23) + \frac{19 \cdot (3 \cdot 241 + 2^2 \cdot 181)}{2^5 \cdot 3^7 \cdot 11^4}\right]} = 3.14159265358979$$

Equation 1.49

I re-iterate that π_{Rama1} (nor anything else) is an exact value of π, and no-one suggested it was. But it is a good enough estimate for most scientific or industrial applications.

The Exact Relation of The Ludolphine Constant and The Ratio of Phidias

The Ludolphine Constant, π, and The Phidian Ratio, Φ, are exactly related by[8]:-

$$\Phi = 2\cos\frac{\pi}{5}$$

Equation 1.50

Neither does this provide an exact articulation between Φ and π, because the value of the Cosine is itself defined by a series summation or a series product, and is itself transcendental.

But Equation 1.50 is a simple and most elegant expression, very convenient of study.

So to examine Equation 1.50, and ultimately the behavior of the Hermetic Defect $\hbar$, we must now focus upon the precise and efficient treatment of The Cosine Function.

Let us look back in history to the seventeenth-century in Western Europe and the first organised attempts systematically to compute cosine values. Much of this effort was motivated by the demands of intercontinental navigation, not an economically-insurable activity as the technology of the day stood.

In 1655 Wallis discovered that:-

$$\pi = 2 \prod_{i=1}^{\infty} \frac{(2i)^2}{(2i-1).(2i+1)}$$

Equation 1.51

This equation is delightfully elegant, but converges with extreme sloth, such that when INF = 500, π_{est} = 3.1400238186006: Essentially three-figure accuracy.

In 1671 Gregory, noted that:-

$$\frac{\pi}{4} = \tan^{-1} 1$$

Equation 1.52

from which:-

$$\pi = 4 \sum_{i=0}^{\infty} (-1)^i \frac{1}{2i+1}$$

Equation 1.53

Unfortunately, whilst simple, this summation converges very slowly, such that when INF = 500, π_{est} = 3.14358865958579, essentially three-figure precision.

Naturally, there was widespread desire to find formulae for cosines of angles that human calculators could compute both with maximum speed and minimum probability of error. To achieve this it is often useful to have increasing factorials in the denominator of summated or multiplied fractions, because this usually betokens speedy convergence.

Wolfram[13] presents alternative summations and products defined by:-

$$\cos x = \sum_{j=0}^{\infty} (-1)^j \frac{x^{2j}}{(2j)!}$$

Equation 1.54

and alternatively:-

$$\cos x = \prod_{j=1}^{\infty} \left[1 - \frac{4x^2}{\pi^2(2j-1)^2} \right]$$

Equation 1.55

Both these formulae are as beautiful as the two previous, and the summation is much more expeditious.

Resorting to the obvious substitution of π/5 for x we may firstly estimate the value of π as summation π_{CS} according to Equation 1.54:-

$$\cos x = \sum_{j=0}^{\infty} (\ 1)^i \frac{x^{2j}}{(2j)!} = \sum_{j=0}^{INF} (-1)^j \frac{\left(\frac{\pi}{5}\right)^{2j}}{(2j)!} = \sum_{j=0}^{INF} (-1)^j \frac{\pi^{2j}}{5^{2j}.(2j)!} = 0.0809016994374947$$

Equation 1.56

When INF = 7, we achieve full fifteen-figure accuracy with this.
Using the product formula we may write:-

$$\cos x = \prod_{j=1}^{\infty} \left[1 - \frac{4x^2}{\pi^2(2j-1)^2}\right] = \prod_{j=1}^{INF} \left[1 - \frac{4\left(\frac{\pi}{5}\right)^2}{\pi^2(2j-1)^2}\right] = \prod_{j=1}^{INF} \left[1 - \left(\frac{2}{5(2j-1)}\right)^2\right]$$
$$= 0.809048597216218$$

Equation 1.57

Note that π cancels out of the product. Equation 1.57 is not in its simplest form, but when INF = 1024 the PSD of the estimate is only -0.003906326009215, a mere four-figure accuracy in the estimation of π.

The Behavior of the Harmonic Defect ɧ

We have seen that the behavior of ɧ can vary greatly with the nature of the approximant, and we further suspected that numerical issues such perhaps as underflow or other low-level machine shortcomings would affect it.

We sought good-quality estimators both for Φ and π, as well as efficient functions for cosines and other ancillaries, and we tested these for our object system.

Accepting meantime the Φ and π provided by our systems are fiducial, being confident that they give fifteen-figure accuracy, we will move forward to study the Hermetic Defect Φ-ξ for Equation 1.50 as serviced by the Cosine Sum Series Equation 1.56.

That is to say:-

$$s_j = 2 \sum_{j=0}^{n} (-1)^j \cdot \frac{\pi^{2j}}{5^{2j}.(2j)!}$$

Equation 1.58

where n is the Limit of Series Summation defined between zero and eight.
In such context:-

$$\xi_n = \sum_{i=1}^{n} s_i$$
Equation 1.59

where ξ_n is the Estimate of Φ for Summation Limit n.

We wish ξ_n to approach Φ quickly as n increases and to do so promptly so that $\mathfrak{h}$ vanishes:-

$$\mathfrak{h} \approx \Phi - \xi_n \approx 0$$
Equation 1.60

Tabulation of Successive Hermetic Defects

Table 1.3 displays the progress of the Hermetic Defect $\mathfrak{h}$ computed according to the cosine iterated according to the progress of Equation 1.57.

Table 1.4 is the same, except that a Graphical Function f(Hermetic Defect) $\equiv \mathfrak{h}*_{j-1}$ is shown, because it is plotted in Figure 1.1 to give an impression of the way Hermetic Defect simultaneously oscillates and decays with cosine iterations.

The Graphical Hermetic Defect h*$_{j-1}$ is given by:-

$$\mathfrak{h}^*_{j-1} = \frac{Sign(\mathfrak{h}_{j-1})}{\left| log_n |\mathfrak{h}_{j-1}| \right|}$$
Equation 1.61

I emphasise that $\mathfrak{h}*$ has *no theoretical status*: It is only computed to service the artwork presentation.

j	j-1	s_{j-1}	ξ_{j-1}	$\hbar_{j-1}$
1	0	2.000000000000000	2.000000000000000	-0.381966011250105
2	1	-0.394784176043574	1.605215823956430	0.012818164793469
3	2	0.012987878804534	1.618203702760960	-0.000169714011064
4	3	-0.000170913634413	1.618032789126550	0.000001199623349
5	4	0.000001204892827	1.618033994019370	-0.000000005269478
6	5	-0.000000005285251	1.618033988734120	0.000000000015773
7	6	0.000000000015807	1.618033988749930	-0.000000000000034
8	7	-0.000000000000034	1.618033988749890	0.000000000000000

Table 1.3
Hermetic Defect Behavior
conforming with the Rapidly-Convergent Cosine of
Equation 1.57

Hermetic Defect
Metric

j	j-1	$\mathfrak{h}^{*}_{j-1} =$ SGN/abs($\log_n$(abs($\mathfrak{h}_{j-1}$)))
1	0	-1.039043460617510
2	1	0.229521411683692
3	2	-0.115188849824531
4	3	0.073348720825015
5	4	-0.052462224115101
6	5	0.040204681394835
7	6	-0.032251093300092

Table 1.4
The Artwork Hermetic Defect $\mathfrak{h}^{*}_{j-1}$
conforming with the Rapidly-Convergent Cosine of
Equation 1.57

Scaled Illustration of the Behavior of the Hermetic Defect

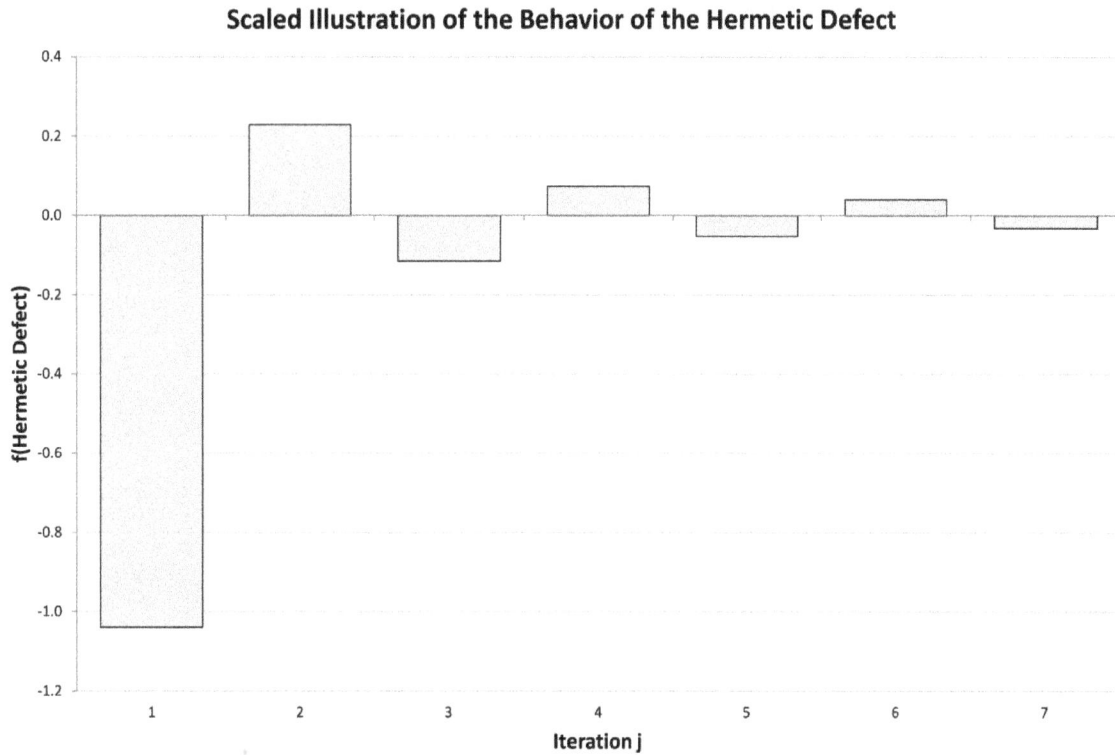

Figure 1.1
Illustrative Column Chart of $\mathfrak{h}$ ($\mathfrak{h}^{*}_{j-1}$) Convergence

<u>References</u>

1   Wikipedia contributors. (2020, September 15).
        Golden ratio.
        In Wikipedia, The Free Encyclopedia. Retrieved 19:17, September 19, 2020,
        from https://en.wikipedia.org/w/index.php?title=Golden_ratio&oldid=978554987

2   Wikipedia contributors. (2020, September 4).
        Approximations of π.
        In Wikipedia, The Free Encyclopedia. Retrieved 10:34, September 21, 2020,
        from
https://en.wikipedia.org/w/index.php?title=Approximations_of_%CF%80&oldid=976728569

3   "Ramanujan's Series for $1/\pi$: A Survey∗"
        Nayandeep Deka Baruah, Bruce C. Berndt, and Heng Huat Chan
        In Memory of V. Ramaswamy Aiyer,
        Founder of the Indian Mathematical Society in 1907
        The Mathematical Association of America
        August-September 2009
        p 568 Eqn. (2.1)

4   Wikipedia contributors. (2020, August 23).
        Chudnovsky algorithm.
        In Wikipedia, The Free Encyclopedia. Retrieved 10:26, September 21, 2020,
        from
https://en.wikipedia.org/w/index.php?title=Chudnovsky_algorithm&oldid=974574247

5   Wikipedia contributors. (2020, September 17).
        Leonhard Euler.
        In Wikipedia, The Free Encyclopedia. Retrieved 08:56, September 22, 2020,
        from
https://en.wikipedia.org/w/index.php?title=Leonhard_Euler&oldid=978838654

6   Wikipedia contributors. (2020, September 21).
        Euler's identity.
        In Wikipedia, The Free Encyclopedia. Retrieved 09:21, September 22, 2020,
        from
https://en.wikipedia.org/w/index.php?title=Euler%27s_identity&oldid=979584320

7   **KGV Ecclesiastes 1:9**
        "The wind goeth toward the south, and turneth about unto the north; it whirleth about continually, and the wind returneth again according to his circuits. All the rivers run into the sea; yet the sea is not full; unto the place from whence the rivers come, thither they return again. All things are full of labour; man cannot utter it: the eye is not satisfied with seeing, nor the ear filled with hearing. The thing that hath been, it is that which shall be; and that which is done is that

which shall be done: and there is no new thing under the sun. Is there any thing whereof it may be said, See, this is new? it hath been already of old time, which was before us. There is no remembrance of former things; neither shall there be any remembrance of things that are to come with those that shall come after. I the Preacher was king over Israel in Jerusalem."

8   Azimuth
     Pi and the Golden Ratio
     John Carlos Baez
     https://johncarlosbaez.wordpress.com/2017/03/07/pi-and-the-golden-ratio/

9   Pi, Pho and Fibonacci – The Golden Ratio, 1.618
     https://www.goldennumber.net/pi-phi-fibonacci/

10   Phi's Fascinating Figures
     R Knott
     University of Surrey
     http://www.maths.surrey.ac.uk/hosted-sites/R.Knott/Fibonacci/propsOfPhi.html

11   MathPapa (Prime) Factorisation Calculator
     https://www.mathpapa.com/factoring-calculator/

12   Wikipedia contributors. (2020, August 29).
     Trigonometric functions.
     In Wikipedia, The Free Encyclopedia. Retrieved 13:48, September 24, 2020, from
     https://en.wikipedia.org/w/index.php?title=Trigonometric_functions&oldid=975677206

13   WolframMathWorld™
     Cosine
     https://mathworld.wolfram.com/Cosine.html

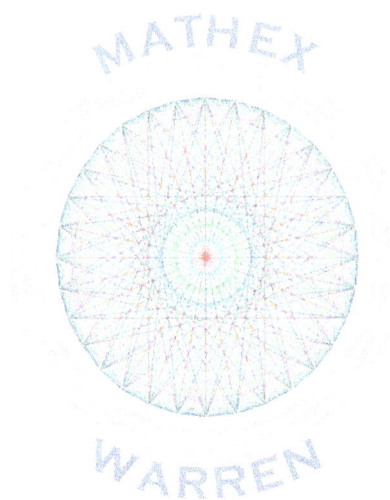

CHAPTER THREE

Computer Efficiencies

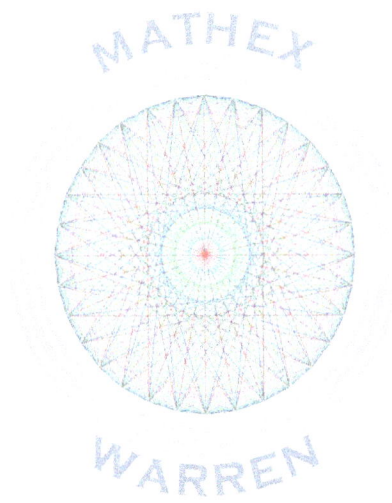

A Program for Computing the
Relative and Absolute Time Costs of
Certain Common Procedures

by

James R Warren BSc MSc PhD PGCE

In theory, it is not only possible exactly to define the precisions of a computed number, by reference to register sizes and so forth, but also exactly to define the execution time duration of any particular operator or mathematical function. This is supposed to be because the circuit architectures involved in multiplication relative to addition are designedly-established and the latencies of the various electronic components known.

For example, it is often thought that multiplication and division of real numbers takes exactly the same time, and both take exactly three times as long as an addition or a subtraction. This is naive.

In practice, of course, the world is not ideal and many vagaries impinge upon the actual calculation of a given number: Electrical transients and other power-supply instabilities; interruptions from the Internet or peripheral interrogation and polling devices; physical perturbants such as temperature fluctuations; not to mention inefficiencies or instabilities arising from the (mis)design or implementation of software.

Accordingly, if we are to enjoy a tenable knowledge of the absolute and relative execution times of the mathematical operations that we use, we must experiment upon the particular computational configuration that we employ. In my case today this is MS-DOS operated through the MicroSoft® Windows 7 Home Premium GUI, supporting compiled QB64 operating upon a 64-bit Hewlett-Packard Pavilion 500400 microcomputer. The machine was fitted with an AMD A10-6700 APU operating at 3.70 GHz. When operated at high-level, this system is only good to fifteen places of decimals at best and some softwares are markedly worse.

To effect the necessary rating of procedures mediated by this assembly I adapted TARIFF.bas, an (interpreted) Visual Basic program that I wrote in 2004AD so that QB64 object code could elaborate on my current system. The resulting source code TARIFF (Version u) is presented in Appendix A.

In high-level programming a practical difficulty is that most elementary operators complete in a matter of ten or a hundred nanoseconds, and that accordingly reliable timing depends upon the operation being repeated on many million separate occasions. In the case of TARIFF this could vary from 5000 times for the long and very slow subroutine for Riemann's Zeta Function, to as many as five hundred million for the B*C Multiplication Operator. It was arranged that the iterative elaboration of each of the thirty (later reduced to twenty-nine) functions approximated ten seconds. The total time was of course very precisely measured as described below.

Whilst the arguments of the function should randomly be varied on each iteration in order to emulate the use of the tool during applications, it should otherwise be ensured that the iterations are invariant.

I made no attempt to interrogate the low-level status of the CPU clock circuitry. We should also note that that "tick" function readily-accessible through embedded code will only yield

$\pm 1/18^{th}$ of a second accuracy. Rather than these, I preferred the QB64 command TIMER(.001), which is nominally accurate to ± 0.001 seconds.

A third issue of interest is that most of the tested operators and functions required argument randomisations and all required assignment. It was therefore essential that these scientific and syntactic imperatives were discounted from the substantive test timings. This was done by embedding all the repeated ancillaries in a Carcase Routine elaborated immediately after the Composite (pay) Routine, and to the same number of repetitions.

These general principles are illustrated by the TARIFF code segments reproduced below for the case of the MUL Multiplication Operation.

Segment OPxMUL()

This is the Composite Routine that actually executes the multiplication operation for timing.

The segments ADD, SUB, MUL and DIV that ran the four cardinal operations Addition, Subtraction, Multiplication and Division were too brief reliably to measure the appropriate durations and therefore were configured to perform fifteen operations to each assignment.

With only one operation they were found often to return negative times. That of course is absurd. My aging Pavilion is very good but it not a time machine!

The BRT (parenthesis) operator reported negative times under almost all conditions, no matter how protracted iteratively, and accordingly I abandoned it: Hence the reported results include twenty-nine rather than thirty mathematical operators.

RANDARG generates eight double-precision real numbers, each randomly composed within a suitable range. In the case of OPxMUL only B and C are actually employed, but in both OPxMUL and its carcass RANDARG is called to furnish a different octet of random numbers, of which the other twelve are redundant in this particular operation.

```
     Sub OPxMUL()
' A Subroutine to Compute a Product of Two Real Numbers
' ( Arrays IRG() and DRG() are
'   Common Shared by static Public Declaration )
' FIFTEEN INTERNAL MULTIPLICATIONS
'
     Call RANDARG(I, J, A, B, C, D, E, F, G, H)
     X = B * C * B * C * B * C * B * C * B * C * B * C * B * C * B * C
     End Sub
```

Subroutine OPxMUL()
Code Sequence One

Segment OPxMULcarc()

This routine includes RANDARG and one only assignment in order exactly to mimic OPxMUL, *except for the fifteen multiplications*.

The carcass can therefore be run to determine how much "dead time" should be deducted from OPxMUL().

```
      Sub OPxMULcarc()
' A Subroutine to Compute a Product of Two Real Numbers
' ( Arrays IRG() and DRG() are
'   Common Shared by static Public Declaration )
'
      Call RANDARG(I, J, A, B, C, D, E, F, G, H )
      X = B
      End Sub
```

Subroutine OPxMULcarc()
Code Sequence Two

Segment OPHOLDERxMUL(arguments)

OPHOLDERxMUL is designed to run the MUL composite and carcase subroutines one after the other, each LI times. I programed LI of necessity to be a long integer: LI can be as large as five hundred million in this rating application.

LI is chosen from a roster of such suitably-selected large numbers housed in the input array LNC(KN), where KN is the number of Operations or functions to be tested (i.e. twenty-nine).

The resulting double-precision timings, usually around ten seconds, are output into DLTM(KN).

```
      Sub OPHOLDERxMUL(KN, LNC(), KO(), SOC(), SOP(), DLTM())
' A Subroutine to Invoke and Time an Operation or Function Elaboration
'     Arguments:-
'         KN       The Operation or Function Serial Number
'         LNC()    The Number of Cycles
'         NOP      The Number of Operations or Functions
'         KO()     The Array of Operations Serial Numbers
'         SOC()    The Array of Operation Codes
'         SOP()    The Array of Operation Descriptions
'         DLTM()   The Array of Operation or Function Timings
'
' Do Composite
      DLT1 = TIMER(.001)
      For LI = 1 To LNC(KN)
         Call OPxMUL
      Next LI
      DLT2 = TIMER(.001)
' Do Carcass
      DLT3 = TIMER(.001)
      For LI = 1 To LNC(KN)
         Call OPxMULcarc
      Next LI
      DLT4 = TIMER(.001)
' Calculate Mill Time
      DLTA = DLT2 - DLT1: DLTB = DLT4 - DLT3
      DLTM(KN) = DLTA - DLTB
' Terminate
      End Sub
```

Subroutine OPHOLDERxMUL()
Code Sequence Three

The Experiments

Five identical experiments were completed using the identical code of TariffQB64u.bas in its compiled .exe object form.

The identical input text sequential file was TARIFF.INP.

Summary data of the five identical experiments (which varied slightly in terms of execution time) is listed in Table B1 of Appendix B, whilst Table B2 presents the results of the fifth and last experiment, including the Performed Cycles and Number of Operations for each tested mathematical operator or function.

An EXCEL[®] tabulation of the full results averages and selected derived data is presented in Appendix C, and an abstract is offered in Table One below:-

Operation Number	Operation Code	Operation Description	Mean Mill Time (μMT)	Pop.SD Mill Time $\sigma(\mu$MT)	Coefficient of Variation Mill Time $C_v(\mu$MT)= σ/μ	Negative Log10 of Mean Mill Time (μMT) $-Log_{10}(\mu$MT)	Relative Mill Time against μMT$_{MUL}$ as Unity μMT/μMT$_{MUL}$	Relative Mill Time Metric ρ_{OP} $1+Log_{10}$ (μMT/ μMT$_{MUL}$)
1	PWR	B^C	2.0582E-07	2.1765E-09	0.0106	6.686521	336.0078	3.526349
2	SQR	B^2	1.0904E-07	1.2098E-09	0.0111	6.962414	178.0148	3.250456
3	MUL	B*C	6.1253E-10	1.5555E-10	0.2539	9.212870	1.0000	1.000000
4	DIV	B/C	9.5133E-09	2.1328E-10	0.0224	8.021667	15.5311	2.191203
5	ADD	B+C	7.5432E-10	3.2078E-11	0.0425	9.122444	1.2315	1.090426
6	SUB	B-C	7.1795E-10	1.8803E-11	0.0262	9.143908	1.1721	1.068962
7	ASS	B:=C	3.4880E-09	8.9117E-10	0.2555	8.457424	5.6944	1.755447
8	EXP	EXP(B)	7.4724E-08	2.6407E-09	0.0353	7.126540	121.9917	3.086330
9	DLGN	LOGN(C)	1.1203E-07	3.7545E-09	0.0335	6.950673	182.8929	3.262197
10	DL10	LOG10(C)	1.6034E-07	3.8902E-09	0.0243	6.794969	261.7588	3.417901
11	SIN	SIN(A)	4.0144E-08	1.1139E-09	0.0277	7.396379	65.5377	2.816491
12	COS	COS(A)	4.4064E-08	1.3105E-09	0.0297	7.355916	71.9373	2.856954
13	TAN	TAN(E)	9.2428E-08	4.2157E-09	0.0456	7.034196	150.8946	3.178674
14	ASIN	ASIN(A)	1.5448E-07	8.9204E-10	0.0058	6.811116	252.2051	3.401754
15	ACOS	ACOS(A)	1.5149E-07	8.6465E-10	0.0057	6.819610	247.3204	3.393260
16	ATAN	ATAN(B)	1.3848E-07	3.3135E-09	0.0239	6.858600	226.0840	3.354270
17	ATN	ATAN(B)	1.4240E-07	3.1889E-09	0.0224	6.846478	232.4837	3.366392
18	SNH	SINH(B)	1.7573E-07	2.1521E-09	0.0122	6.755149	286.8938	3.457721
19	COH	COSH(B)	1.7732E-07	3.6455E-09	0.0206	6.751242	289.4863	3.461628
20	TAH	TANH(B)	3.5212E-07	9.2244E-09	0.0262	6.453314	574.8520	3.759556
21	ASNH	ASINH(A)	1.1967E-07	2.6896E-09	0.0225	6.922007	195.3722	3.290863
22	ACOH	ACOSH(A)	2.4084E-08	4.8383E-09	0.2009	7.618271	39.3187	2.594599
23	ATAH	ATANH(C)	2.0452E-08	2.0426E-09	0.0999	7.689264	33.3892	2.523606
24	SRT	SQRT(B)	4.4524E-08	1.5961E-09	0.0358	7.351406	72.6883	2.861464
25	LAC	Loop Access	5.5360E-09	4.0525E-10	0.0732	8.256804	9.0379	1.956066
26	AR1	1D Real Array Access	1.0832E-08	1.5662E-09	0.1446	7.965291	17.6839	2.247579
27	AR2	2D Real Array Access	1.3936E-08	8.0616E-09	0.5785	7.855862	22.7514	2.357008
28	ZET	ZETA(B)	4.4536E-04	7.4189E-07	0.0017	3.351289	727078.7984	6.861581
29	FAC	FACT(JX)	2.8456E-08	4.1558E-10	0.0146	7.545826	46.4562	2.667044
		Total	4.4777E-04	8.0841E-07	2.1069E+00	2.1212E+02	7.3102E+05	8.4056E+01
		Mean	1.5440E-05	2.7876E-08	7.2651E-02	7.3144E+00	2.5208E+04	2.8985E+00
		Population SD	8.1247E-05	1.3495E-07	0.1177756	1.06865518	132641.2683	1.0686552

Table One
Selected Results of the TARIFF Rating Experiments

<u>Plot Diagrams</u>

Plot One below is of Absolute Average Mill Time in seconds as a column for each of the twenty-nine mathematical operations.

You can see how the protracted Zeta Function overwhelms everything rendering, as noted above, such absolute timings quite useless for a personal appreciation of costs:-

**Mill Time Averages for
Twenty-Nine Operators and Functions programmed in QB64**

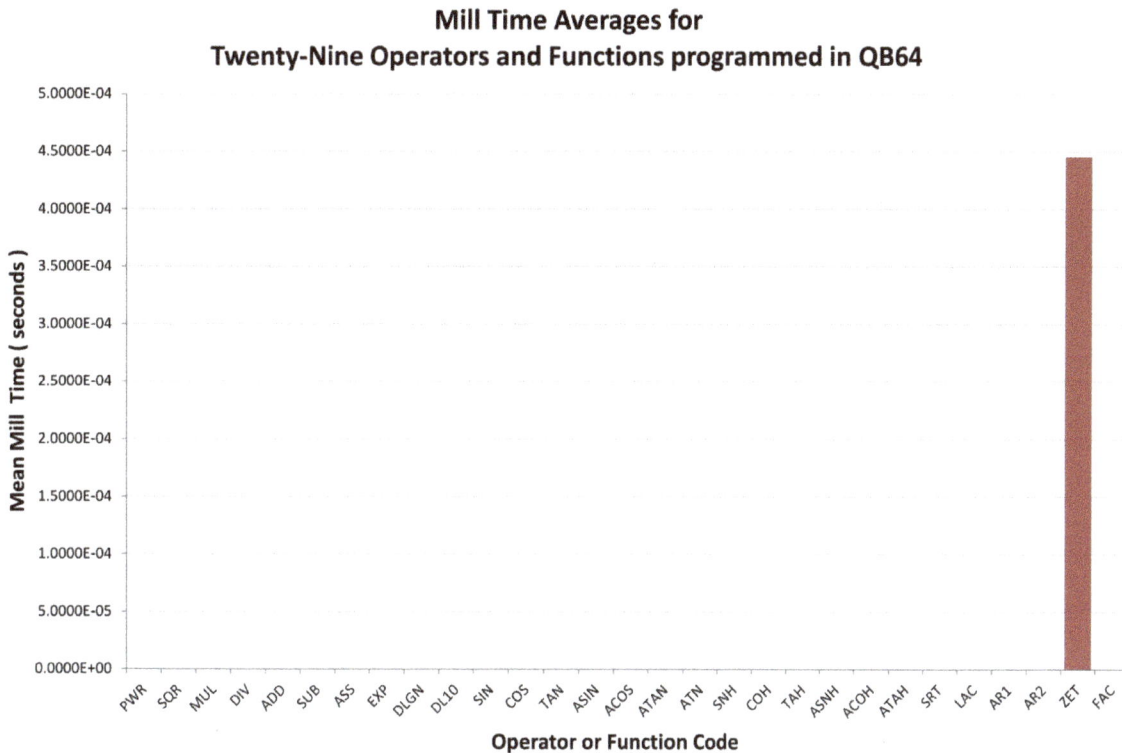

**Plot One
Absolute Mean Execution Times Drowned Out By Zeta**

The non-Zeta operations do in fact express as little maroon lines in screen renditions of this chart, but not apparently in this textual context.

Plot Two is a logarithmic expression of the same data recast as relative to MT_{MUL}. This is ρ_{OP} defined as:-:-

$$\rho_{OP} = 1 + \log_{10}\left(\frac{\mu MT}{\mu MT_{MUL}}\right)$$
Equation 1

where μMT is the Mean Operational Mill Time across all five experiments and μMT_{MUL} is the Mean Operational Mill Time for the MUL Multiplication Operator, chosen only for being the swiftest.

Relative Mill Time for MUL and ρ_{MUL} are both by definition unity.

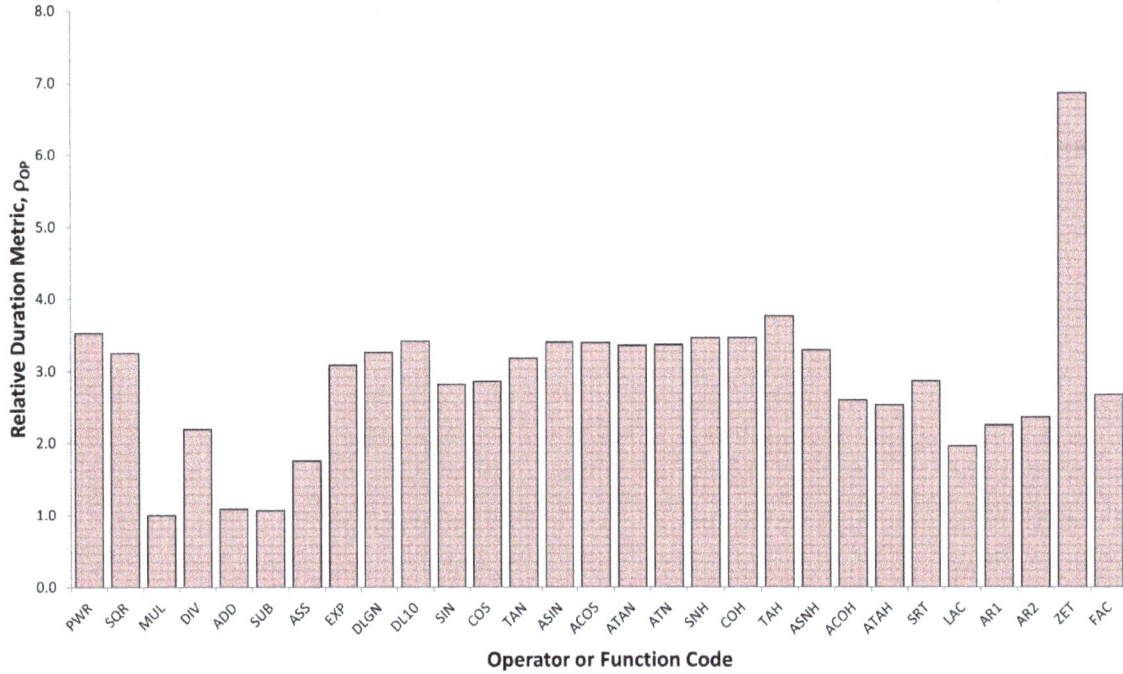

Plot Two
Relative Mill Time Metric ρOP
for the Twenty-Nine Operators or Functions

The scale of Zeta is now much better controlled and we can see that DIV is about 15.5 times as costly in time as MUL, and even ADD and SUB are significantly more expensive than MUL. Whole classes of logarithmic, circular and hyperbolic standard functions, most of them intrinsic to QB64, have ρ values around 3.5 indicating approximate relative durations of $330 \times \mu MT_{MUL}$.

Coefficient of Variation, C_v, is defined as:-

$$C_v(\mu MT) = \frac{\sigma(\mu MT)}{\mu MT}$$

Equation 2

where $\sigma(\mu MT)$ is the Population Standard Deviation of the Operator Mill Time across the five experiments, whilst μMT is the Mean of those durations.

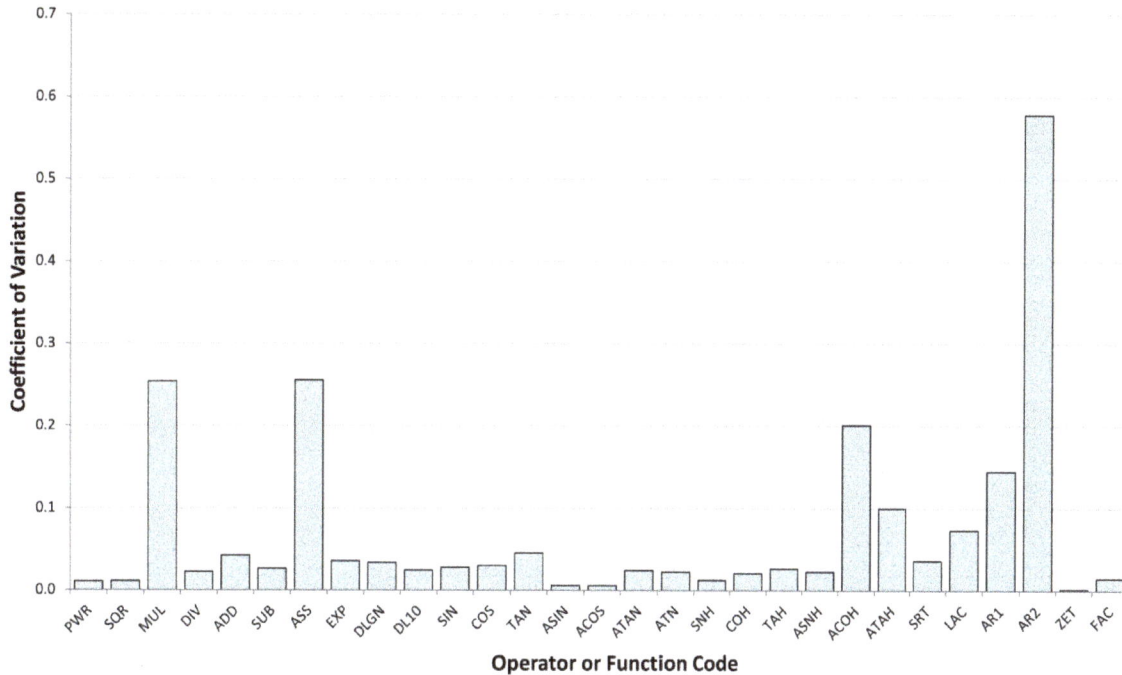

**Mill Time Coefficient of Variation for
Twenty-Nine Operators and Functions programmed in QB64**

**Plot Three
Coefficient of Variation C_v
for the Twenty-Nine Operators or Functions**

We can see straight away that Multiplication MUL and Assignment ASS show a roughly 25% variation about the mean whilst Hyperbolic Arccosine is also unstable and my naively-programed Zeta Function ZET extremely variable, fluctuating by over fifty percent.

A fourth chart illustrates the minimum, maximum and mean Mill Times for each of the twenty-nine operators across the five available experiment episodes. Note that the metric is $-\log_{10}(MT)$ and that the best agents are towards the top of the diagram.

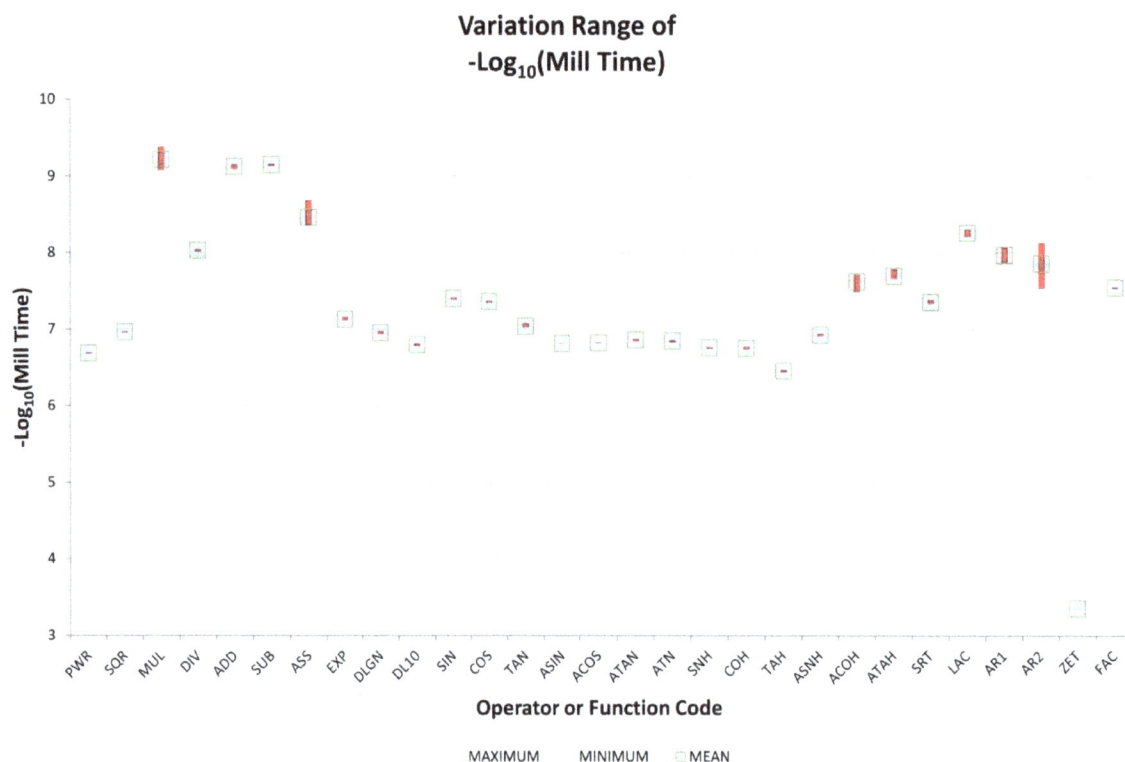

Plot Four
Minimum, Maximum and Mean –Log$_{10}$(Mill Time)
for the Twenty-Nine Operators or Functions
across the Five Experiment Episodes

The red lines express variation and we can see that this format mirrors the implication of Coefficient of Variation. Only MUL, AS, ACOH, AR2 and especially ZET are visibly variable.

Review of the Summary Table

Perusal of Table One disclosed some interesting facts:-

(a)   The briefest operation is MUL (multiplication) which executed in some 6.1253×10^{-10} seconds, about 0.61253 nanoseconds. Unexpectedly, this is the fastest operation. Addition and subtraction take respectively 23% and 17% longer, whilst division takes more than fifteen and one half times as long as multiplication.

(b)   It is very difficult for the human mind to interpret the list of Absolute Mean Mill Times, µ(MT).

(c)   Notwithstanding, process standard deviations are about two orders of magnitude less than the mean, proving that the experiments are tightly repeatable and therefore reliable despite of any adventitious microevents. This further discredits any lingering prejudices we may indulge regarding the superiority of low-level, theoretical rating formulae.

The standard deviation findings are borne-out by the Coefficients of Variation, the dividend of standard deviation over the mean. These are typically around 0.02 to 0.04, except for ASS (assignment) which is 25.55% and AR2 (two-dimensional array access) which is 57.85%. The very heavy functions such as ACOS and ZET have coefficients of variation in the thousandths.

(d)   The negative logarithm of the average Mill Time, $-\log_{10}(\mu MT)$, has the virtue of controlling the extreme dispersion, but unfortunately inverts the relative mill times so that MUL is 9.212870, whereas the very protracted Zeta Function ZET is 3.351289.

(e)   MUL is the briefest operation in our system, and accordingly it is convenient to define a Relative Mill Time $\mu MT/\mu MT_{MUL}$ with the execution time of the multiplication operator MUL taken to be unity.

It is then clear that (e.g.) the General Power Operator PWR is more than 336 times as expensive as MUL, and even more interestingly, the Square Operator SQR is more than 178 times as expensive as MUL multiplication. This confirms the old programmers' adage that it is much cheaper to multiply a number by itself than to square it.

All circular and hyperbolic functions are seen to be what they are: Expensive.

Meanwhile, Riemann's Zeta Function, programmed naïvely by me, is nearly three-quarters of a million times more time-costly than multiplication.

(f)   A more satisfactory Mill Time metric is ρ_{OP}, defined as $1+\log_{10}(\mu MT/\mu MT_{MUL})$. This also sets the Mill Time of MUL to unity, but has the added virtue of representing more protracted operations with larger, but nevertheless manageable, figures.

Therefore, whilst a relatively cheap operation like DIV is 2.191203, the very time-consuming ZET is still "only" 6.861581.

(g)   The population standard deviations of $-\log_{10}(\mu MT)$ and $1+\log_{10}(\mu MT/\mu MT_{MUL})$ are equivalent. That is to say:-

$$\sqrt{\left\{\sum_{i=1}^{n}\left[-log_{10}(\mu MT) - \frac{\sum_{i=1}^{n}-log_{10}(\mu MT)}{n}\right]\right\}^2}$$
$$= \sqrt{\left\{\sum_{i=1}^{n}\left[1 + log_{10}\left(\frac{\mu MT}{\mu MT_{MUL}}\right) - \frac{\sum_{i=1}^{n}1 + log_{10}\left(\frac{\mu MT}{\mu MT_{MUL}}\right)}{n}\right]\right\}^2} = 0$$

Equation 3

Note that Equation Three holds good for any base, and even more remarkably, for any arbitrarily-chosen Operator Mean Mill Time, any of the twenty-nine available values of μMT_{OP}. (It is only that μMT_{MUL} is convenient for us).

A brief simplification exercise confirms the identity of these two complex quadratic equations, and their mutual equivalence to zero.

<u>References</u>

1   Abramowitz and Stegun
        "Handbook of Mathematical Functions"
        "With Formulas, Graphs and Mathematical Tables"
        Milton Abramowitz and Irene A Stegun
        Dover Publications, Inc., New York   1965
        SBN 486-61272-4
        pp 1046

2   QB64 Main Page with Downloadable Manual
        https://www.qb64.org/wiki/Main_Page

APPENDIX A

RATING PROGRAM
TariffQB64u.bas
SOURCE CODE

```
'           Program TARIFF
' A Program
' To Compute a Set of Empirical Operation and Function Mill Times
' Based upon Data Located from data text file TARIFF.INP
'           Written by:-
'
'                James R Warren BSc MSc PhD PGCE
'                "Southgate"
'                31 Victoria Avenue
'                Bloxwich
'                Walsall
'                WS3 3HS
'
'                5 September 2004
'
'      This Program is Written in QBasic for elaboration as a compiled
'      QB64 .exe object
'      This source version composited and adapted: 12 July 2019
'
'      Version u
'      14 July 2019
'      Compiled QB64 on a 64-bit Hewlett-Packard Pavilion 500400:
'      Total Execution Time = 1 Hour 45 Minutes
'
'      REM $DYNAMIC

' Initialise the Random Number Generator
      Randomize (Timer)
' VARIABLE TYPE DEFAULTS
      DefDbl A-H, O-R, T-Z
      DefStr S
      DefInt I-K, M-N
      DefLng L
' Declare Dynamic Array Dimensions
      COMMON SHARED NUnits
      COMMON SHARED NRX, NRY, NAR, NP
' Assign the Dynamic Arrays' Maxima and Dimension the Dynamic Arrays
'      ( None )
' Pseudo-Static Array Declarations
      COMMON SHARED IRG(), DRG(), AR1(), AR2()
      COMMON SHARED KO(), SOC(), SOP(), LNC(), DLTM()
' Assign the Pseudo-Static Array Dimensions Maxima and Dimension the Pseudo-
Static Arrays
      NRX = 8: NRY = 2
      ReDim IRG(NRX, NRY), DRG(NRX, NRY)
      NAR = 30
      ReDim AR1(NAR), AR2(NAR, NAR)
      NP = 30
      ReDim KO(NP), SOC(NP), SOP(NP), LNC(NP), DLTM(NP)
' Object Definitions
'      ( None )
' Publicise Object Instantiations
'      ( None )
' Static Array Declarations
'      ( None )
' Dynamic Array Declarations
'      ( None )
' Constant Definitions
      Const PI = 3.14159265358979
      Const PI2 = 6.28318530717959
      Const HP = 1.5707963267949
      Const RC = 2.506628274631
      Const EMC = 0.577215664901533
```

```
              Const ENAP = 2.71828182845905
              Const DL10 = 2.30258509299405
' Declare Logical Unit Number Holders
      COMMON SHARED IU, IV, IW, IX
' Declare String Constant Data ( for Common Shared emulation )
      COMMON SHARED SC, SM, SCR
' Declare Formats ( for Common Shared emulation )
      COMMON SHARED SI4, SF10P8, SF11P6
' Define Random Number Range Parameters
      IRG(1, 1) = 1: IRG(1, 2) = 30
      IRG(2, 1) = 1: IRG(2, 2) = 30
      DRG(1, 1) = 0: DRG(1, 2) = 1
      DRG(2, 1) = 0.1: DRG(2, 2) = 10
      DRG(3, 1) = 0.1: DRG(3, 2) = 100
      DRG(4, 1) = 0: DRG(4, 2) = -1000
      DRG(5, 1) = 0: DRG(5, 2) = HP-0.00001
      DRG(6, 1) = -10: DRG(6, 2) = 10
      DRG(7, 1) = 1: DRG(7, 2) = 10
      DRG(8, 1) = -1: DRG(8, 2) = 1
' String Constant Definitions
      SC = ":": SM = ",": SCR = Chr$(13) + Chr$(10)
' Format Definitions
      SI4 = "####": SF10P8 = "#.00000000": SF11P6 = "####.000000"
' Perform Actions
      DoCompute
' Terminate
      End


      Sub DoCompute()
' Load Operations Descriptions
      Call OPSIN(NOP, KO(), SOC(), SOP(), LNC())
' Do Timings
' **** Start of Job Blocks
      PRINT "Entering Sequence:-" :PRINT
      DStart = Timer(.001)
' ** Job 1 Block ( The Real Power of a Real Number )
      Call OPHOLDERxPWR(1, LNC(), KO(), SOC(), SOP(), DLTM())
'     Call METAtrailing(1, LNC(), KO(), SOC(), SOP(), DLTM())
' ** Job 2 Block ( The Real Square of a Real Number )
      Call OPHOLDERxSQR(2, LNC(), KO(), SOC(), SOP(), DLTM())
'     Call METAtrailing(2, LNC(), KO(), SOC(), SOP(), DLTM())
' ** Job 3 Block ( The Product of Two Real Numbers )
      Call OPHOLDERxMUL(3, LNC(), KO(), SOC(), SOP(), DLTM())
'     Call METAtrailing(3, LNC(), KO(), SOC(), SOP(), DLTM())
' ** Job 4 Block ( The Dividend of Two Real Numbers )
      Call OPHOLDERxDIV(4, LNC(), KO(), SOC(), SOP(), DLTM())
'     Call METAtrailing(4, LNC(), KO(), SOC(), SOP(), DLTM())
' ** Job 5 Block ( The Sum of Two Real Numbers )
      Call OPHOLDERxADD(5, LNC(), KO(), SOC(), SOP(), DLTM())
'     Call METAtrailing(5, LNC(), KO(), SOC(), SOP(), DLTM())
' ** Job 6 Block ( The Difference of Two Real Numbers )
      Call OPHOLDERxSUB(6, LNC(), KO(), SOC(), SOP(), DLTM())
'     Call METAtrailing(6, LNC(), KO(), SOC(), SOP(), DLTM())
' ** Job 7 Block ( The Assignment of a Real Number )
      Call OPHOLDERxASS(7, LNC(), KO(), SOC(), SOP(), DLTM())
'     Call METAtrailing(7, LNC(), KO(), SOC(), SOP(), DLTM())
' ** Job 8 Block ( The Bracketing of a Real Number )
      Call OPHOLDERxBRT(8, LNC(), KO(), SOC(), SOP(), DLTM())
'     Call METAtrailing(8, LNC(), KO(), SOC(), SOP(), DLTM())
' ** Job 9 Block ( The Napierian Exponent of a Real Number )
      Call OPHOLDERxEXP(9, LNC(), KO(), SOC(), SOP(), DLTM())
'     Call METAtrailing(9, LNC(), KO(), SOC(), SOP(), DLTM())
' ** Job 10 Block ( The Napierian Logarithm of a Real Number )
```

```
          Call OPHOLDERxDLGN(10, LNC(), KO(), SOC(), SOP(), DLTM())
            Call METAtrailing(10, LNC(), KO(), SOC(), SOP(), DLTM())
' ** Job 11 Block ( The Radix Ten Logarithm of a Real Number )
          Call OPHOLDERxDL10(11, LNC(), KO(), SOC(), SOP(), DLTM())
            Call METAtrailing(11, LNC(), KO(), SOC(), SOP(), DLTM())
' ** Job 12 Block ( The Sine of a Real Number )
          Call OPHOLDERxSIN(12, LNC(), KO(), SOC(), SOP(), DLTM())
            Call METAtrailing(12, LNC(), KO(), SOC(), SOP(), DLTM())
' ** Job 13 Block ( The Cosine of a Real Number )
          Call OPHOLDERxCOS(13, LNC(), KO(), SOC(), SOP(), DLTM())
            Call METAtrailing(13, LNC(), KO(), SOC(), SOP(), DLTM())
' ** Job 14 Block ( The Tangent of a Real Number )
          Call OPHOLDERxTAN(14, LNC(), KO(), SOC(), SOP(), DLTM())
            Call METAtrailing(14, LNC(), KO(), SOC(), SOP(), DLTM())
' ** Job 15 Block ( The Inverse Sine of a Real Number )
          Call OPHOLDERxASIN(15, LNC(), KO(), SOC(), SOP(), DLTM())
            Call METAtrailing(15, LNC(), KO(), SOC(), SOP(), DLTM())
' ** Job 16 Block ( The Inverse Cosine of a Real Number )
          Call OPHOLDERxACOS(16, LNC(), KO(), SOC(), SOP(), DLTM())
            Call METAtrailing(16, LNC(), KO(), SOC(), SOP(), DLTM())
' ** Job 17 Block ( The Inverse Tangent of a Real Number )
          Call OPHOLDERxATAN(17, LNC(), KO(), SOC(), SOP(), DLTM())
            Call METAtrailing(17, LNC(), KO(), SOC(), SOP(), DLTM())
' ** Job 18 Block ( The Inverse Tangent of a Real Number )
          Call OPHOLDERxATN(18, LNC(), KO(), SOC(), SOP(), DLTM())
            Call METAtrailing(18, LNC(), KO(), SOC(), SOP(), DLTM())
' ** Job 19 Block ( The Hyperbolic Sine of a Real Number )
          Call OPHOLDERxSINH(19, LNC(), KO(), SOC(), SOP(), DLTM())
            Call METAtrailing(19, LNC(), KO(), SOC(), SOP(), DLTM())
' ** Job 20 Block ( The Hyperbolic Cosine of a Real Number )
          Call OPHOLDERxCOSH(20, LNC(), KO(), SOC(), SOP(), DLTM())
            Call METAtrailing(20, LNC(), KO(), SOC(), SOP(), DLTM())
' ** Job 21 Block ( The Hyperbolic Tangent of a Real Number )
          Call OPHOLDERxTANH(21, LNC(), KO(), SOC(), SOP(), DLTM())
            Call METAtrailing(21, LNC(), KO(), SOC(), SOP(), DLTM())
' ** Job 22 Block ( The Inverse Hyperbolic Sine of a Real Number )
          Call OPHOLDERxASINH(22, LNC(), KO(), SOC(), SOP(), DLTM())
            Call METAtrailing(22, LNC(), KO(), SOC(), SOP(), DLTM())
' ** Job 23 Block ( The Inverse Hyperbolic Cosine of a Real Number )
          Call OPHOLDERxACOSH(23, LNC(), KO(), SOC(), SOP(), DLTM())
            Call METAtrailing(23, LNC(), KO(), SOC(), SOP(), DLTM())
' ** Job 24 Block ( The Inverse Hyperbolic Tangent of a Real Number )
          Call OPHOLDERxATANH(24, LNC(), KO(), SOC(), SOP(), DLTM())
            Call METAtrailing(24, LNC(), KO(), SOC(), SOP(), DLTM())
' ** Job 25 Block ( The Square Root of a Positive Real Number )
          Call OPHOLDERxSRT(25, LNC(), KO(), SOC(), SOP(), DLTM())
            Call METAtrailing(25, LNC(), KO(), SOC(), SOP(), DLTM())
' ** Job 26 Block ( FOR-NEXT Loop Access )
          Call OPHOLDERxLAC(26, LNC(), KO(), SOC(), SOP(), DLTM())
            Call METAtrailing(26, LNC(), KO(), SOC(), SOP(), DLTM())
' ** Job 27 Block ( One-Dimensional Real Array Access )
          Call OPHOLDERxAR1(27, LNC(), KO(), SOC(), SOP(), DLTM())
            Call METAtrailing(27, LNC(), KO(), SOC(), SOP(), DLTM())
' ** Job 28 Block ( Two-Dimensional Real Array Access )
          Call OPHOLDERxAR2(28, LNC(), KO(), SOC(), SOP(), DLTM())
            Call METAtrailing(28, LNC(), KO(), SOC(), SOP(), DLTM())
' ** Job 29 Block ( The Zeta Function of a Real Number )
          Call OPHOLDERxZET(29, LNC(), KO(), SOC(), SOP(), DLTM())
            Call METAtrailing(29, LNC(), KO(), SOC(), SOP(), DLTM())
' ** Job 30 Block ( The Double Precision Factorial of a Short Integer )
          Call OPHOLDERxFAC(30, LNC(), KO(), SOC(), SOP(), DLTM())
            Call METAtrailing(30, LNC(), KO(), SOC(), SOP(), DLTM())
' **** End of Job Blocks
```

```
' Output the Results
      Call OPSOUT( NOP, KO(), SOC(), SOP(), LNC(), DLTM())
      DFinish = Timer(.001)
      PMT = DFinish - DStart
      PRINT "Total Execution Time = ";PMT
' Terminate
      End Sub

      Function DSINH(X)
' A Function to Return the Double-Precision Hyperbolic Sine for Argument X
'     Argument:-
'         X       The Double-Precision Argument of DSINH
'

      DSINH = (Exp(X) - Exp(-X)) / 2
      End Function


      Function DCOSH(X)
' A Function to Return the Double-Precision Hyperbolic Cosine for Argument X
'     Argument:-
'         X       The Double-Precision Argument of DCOSH
'

      DCOSH = (Exp(X) + Exp(-X)) / 2
      End Function

      Function DTANH(X)
' A Function to Return the Double-Precision Hyperbolic Tangent for Argument X
'     Argument:-
'         X       The Double-Precision Argument of DTANH
'

      DTANH = (Exp(X) - Exp(-X)) / (Exp(X) + Exp(-X))
      End Function

      Function ZETA(ZK, IZ)
' A Function to Return the Double-Precision Riemann's Zeta Function for Argument
ZK.
' The Definitional Order(1/n) Algorithm is Employed.
'     Arguments:-
'         ZK      The Double-Precision Argument of ZETA
'         IZ      The Number of Iteration Cycles
'

      ZETA = 0#
      For K = 1 To IZ
         ZETA = ZETA + 1 / K ^ ZK
      Next K
      End Function

      Function DLOG10(X)
' A Function to Return the Double-Precision Base Ten Logarithm of Argument X
'     Arguments:-
'         X       The Argument whose Base Ten Logarithm is to be Determined
'

      DLOG10 = Log(X) / DL10
      End Function

      Function JARG(JLIM)
' A Function to Return A Random Short Integer between 0 and 120 as
' An Argument of the Factorial Function FACT(JK)
'     Arguments:-
'         JX      The Argument whose Factorial is to be Determined
'

      JARG=INT(JLIM*RND+0.5)
      End Function
```

```
      Function FACT(JX)
' A Function to Return A Double-Precision Factorial of the Short Integer JX
'     Arguments:-
'         JX       The Argument whose Factorial is to be Determined
'
      FACT=1
      IF JX>0 THEN
          FOR I=1 TO JX
              FACT=FACT*I
          NEXT I
      END IF
      End Function

      Sub OPSIN(NOP, KO(), SOC(), SOP(), LNC())
' A Subroutine to Input Operations' Descriptors Data
'     Arguments:-
'         NOP      The Number of Operations or Functions
'         KO()     The Array of Operations Serial Numbers
'         SOC()    The Array of Operation Codes
'         SOP()    The Array of Operation Descriptions
'         LNC()    The Array of Required Cycles
'
' Open the File
      IU=1
      SPath="C:\BYJRW\WORKHAND\ELLIPERI\TARIFF\"
      SFile="TARIFF"
      SExt=".INP"
      OPEN SPath+SFile+SExt FOR INPUT AS IU
' Input the Data
      Input #IU, NOP
      For I = 1 To NOP
          Input #IU, KO(I), SOC(I), SOP(I), LNC(I)
      Next I
' Terminate
      Close IU
      End Sub

      Sub OPSOUT( NOP, KO(), SOC(), SOP(), LNC(), DLTM())
' A Subroutine to Output Operations' Descriptors Data with
' Operation or Function Timings
'     Arguments:-
'         NOP      The Number of Operations or Functions
'         KO()     The Array of Operations Serial Numbers
'         SOC()    The Array of Operation Codes
'         SOP()    The Array of Operation Descriptions
'         LNC()    The Array of Required Cycles
'         DLTM()   The Array of Operation or Function Timings
'
' Open the File
      IV=2
      SPath="C:\BYJRW\WORKHAND\ELLIPERI\TARIFF\"
      SFile="TARIFF"
      SExt=".OUT"
      OPEN SPath+SFile+SExt FOR OUTPUT AS IV
' Output the Data
      Write #IV,
      Write  #IV,  "Operation  Number",  "Operation  Code",  "Operation
Description", "Performed Cycles", "Execution Time ( seconds )"
      For I = 1 To NOP
          Write #IV, KO(I), SOC(I), SOP(I), LNC(I), DLTM(I)
      Next I
' Terminate
```

```
        Close IV
        End Sub

        Sub RANDARG(I, J, A, B, C, D, E, F, G, H)
' A Subroutine to Generate a Selection of Random Numbers based
' upon the Ranges specified by Arrays IRG() and DRG()
'       Arguments:-
'           I       A Positive Short Integer
'           J       A Negative Short Integer
'           A       A Positive Double-Precision Real Number
'           B       A Positive Double-Precision Real Number
'           C       A Positive Double-Precision Real Number
'           D       A Negative Double-Precision Real Number
'           E       An Angle Between Zero and Half-Pi Radians
'           F       The Range of Validity of a Hyperbolic Sine
'           G       The Range of Validity of a Hyperbolic Cosine
'           H       The Range of Validity of a Hyperbolic Tangent
' ( Arrays IRG() and DRG() are
'   Common Shared by static Public Declaration )
'
' ** Under Standard Configuration the Random Number Ranges
' ** are as Follows:-
' **
' **         Argument        Lower Bound         Upper Bound
' **         I               1                   30
' **         J               1                   30
' **         A               0                   1
' **         B               0.1                 10
' **         C               0.1                 100
' **         D               0                   -1000
' **         E               0                   PI/2-0.001
' **         F               -10                 10
' **         G               1                   10
' **         H               -1                  1
' **
'
' Compute the Random Integers
        I = IRG(1, 1) + Int((IRG(1, 2) - IRG(1, 1)) * RND)
        J = IRG(2, 1) + Int((IRG(2, 2) - IRG(2, 1)) * RND)
' Compute the Random Reals
        A = DRG(1, 1) + (DRG(1, 2) - DRG(1, 1)) * RND
        B = DRG(2, 1) + (DRG(2, 2) - DRG(2, 1)) * RND
        C = DRG(3, 1) + (DRG(3, 2) - DRG(3, 1)) * RND
        D = DRG(4, 1) + (DRG(4, 2) - DRG(4, 1)) * RND
' Compute the Random Angles
        E = DRG(5, 1) + (DRG(5, 2) - DRG(5, 1)) * RND
' Compute the Random Hyperbolic Arguments
        F = DRG(6, 1) + (DRG(6, 2) - DRG(6, 1)) * RND
        G = DRG(7, 1) + (DRG(7, 2) - DRG(7, 1)) * RND
        H = DRG(8, 1) + (DRG(8, 2) - DRG(8, 1)) * RND
' Terminate
        End Sub

        Sub METAtrailing(KN, LNC(), KO(), SOC(), SOP(), DLTM())
' A Subroutine to Display Selected Output Values
'       Arguments:-
'           KN      The Operation or Function Serial Number
'           LNC()   The Number of Cycles
'           NOP     The Number of Operations or Functions
'           KO()    The Array of Operations Serial Numbers
'           SOC()   The Array of Operation Codes
'           SOP()   The Array of Operation Descriptions
'           DLTM()  The Array of Operation or Function Timings
```

```
'
        PRINT
        PRINT "Serial = ";KN;" Description = "; SOP(KN)
        PRINT "Required Iterations Executed = ";LNC(KN)
        PRINT "Operation Execution Time = ";DLTM(KN)
        SLEEP
        PRINT
' Terminate
        End Sub


' ****** START OF HOLDER SECTION

        Sub OPHOLDERxPWR(KN, LNC(), KO(), SOC(), SOP(), DLTM())
' A Subroutine to Invoke and Time an Operation or Function Elaboration
'     Arguments:-
'         KN        The Operation or Function Serial Number
'         LNC()     The Number of Cycles
'         NOP       The Number of Operations or Functions
'         KO()      The Array of Operations Serial Numbers
'         SOC()     The Array of Operation Codes
'         SOP()     The Array of Operation Descriptions
'         DLTM()    The Array of Operation or Function Timings
'
' Do Composite
        DLT1=TIMER(.001)
        For LI = 1 To LNC(KN)
            Call OPxPWR
        Next LI
        DLT2 = TIMER(.001)
' Do Carcass
        DLT3 = TIMER(.001)
        For LI = 1 To LNC(KN)
            Call OPxPWRcarc
        Next LI
        DLT4 = TIMER(.001)
' Calculate Mill Time
        DLTA = DLT2 - DLT1: DLTB = DLT4 - DLT3
        DLTM(KN) = DLTA - DLTB
' Terminate
        End Sub

        Sub OPHOLDERxSQR(KN, LNC(), KO(), SOC(), SOP(), DLTM())
' A Subroutine to Invoke and Time an Operation or Function Elaboration
'     Arguments:-
'         KN        The Operation or Function Serial Number
'         LNC()     The Number of Cycles
'         NOP       The Number of Operations or Functions
'         KO()      The Array of Operations Serial Numbers
'         SOC()     The Array of Operation Codes
'         SOP()     The Array of Operation Descriptions
'         DLTM()    The Array of Operation or Function Timings
'
' Do Composite
        DLT1=TIMER(.001)
        For LI = 1 To LNC(KN)
            Call OPxSQR
        Next LI
        DLT2 = TIMER(.001)
' Do Carcass
        DLT3 = TIMER(.001)
        For LI = 1 To LNC(KN)
            Call OPxSQRcarc
        Next LI
```

```
        DLT4 = TIMER(.001)
' Calculate Mill Time
        DLTA = DLT2 - DLT1: DLTB = DLT4 - DLT3
        DLTM(KN) = DLTA - DLTB
' Terminate
        End Sub

        Sub OPHOLDERxMUL(KN, LNC(), KO(), SOC(), SOP(), DLTM())
' A Subroutine to Invoke and Time an Operation or Function Elaboration
'       Arguments:-
'           KN          The Operation or Function Serial Number
'           LNC()       The Number of Cycles
'           NOP         The Number of Operations or Functions
'           KO()        The Array of Operations Serial Numbers
'           SOC()       The Array of Operation Codes
'           SOP()       The Array of Operation Descriptions
'           DLTM()      The Array of Operation or Function Timings
'
' Do Composite
        DLT1 = TIMER(.001)
        For LI = 1 To LNC(KN)
            Call OPxMUL
        Next LI
        DLT2 = TIMER(.001)
' Do Carcass
        DLT3 = TIMER(.001)
        For LI = 1 To LNC(KN)
            Call OPxMULcarc
        Next LI
        DLT4 = TIMER(.001)
' Calculate Mill Time
        DLTA = DLT2 - DLT1: DLTB = DLT4 - DLT3
        DLTM(KN) = DLTA - DLTB
' Terminate
        End Sub

        Sub OPHOLDERxDIV(KN, LNC(), KO(), SOC(), SOP(), DLTM())
' A Subroutine to Invoke and Time an Operation or Function Elaboration
'       Arguments:-
'           KN          The Operation or Function Serial Number
'           LNC()       The Number of Cycles
'           NOP         The Number of Operations or Functions
'           KO()        The Array of Operations Serial Numbers
'           SOC()       The Array of Operation Codes
'           SOP()       The Array of Operation Descriptions
'           DLTM()      The Array of Operation or Function Timings
'
' Do Composite
        DLT1 = TIMER(.001)
        For LI = 1 To LNC(KN)
            Call OPxDIV
        Next LI
        DLT2 = TIMER(.001)
' Do Carcass
        DLT3 = TIMER(.001)
        For LI = 1 To LNC(KN)
            Call OPxDIVcarc
        Next LI
        DLT4 = TIMER(.001)
' Calculate Mill Time
        DLTA = DLT2 - DLT1: DLTB = DLT4 - DLT3
        DLTM(KN) = DLTA - DLTB
' Terminate
```

```
        End Sub

        Sub OPHOLDERxADD(KN, LNC(), KO(), SOC(), SOP(), DLTM())
' A Subroutine to Invoke and Time an Operation or Function Elaboration
'     Arguments:-
'         KN        The Operation or Function Serial Number
'         LNC()     The Number of Cycles
'         NOP       The Number of Operations or Functions
'         KO()      The Array of Operations Serial Numbers
'         SOC()     The Array of Operation Codes
'         SOP()     The Array of Operation Descriptions
'         DLTM()    The Array of Operation or Function Timings
'
' Do Composite
        DLT1 = TIMER(.001)
        For LI = 1 To LNC(KN)
            Call OPxADD
        Next LI
        DLT2 = TIMER(.001)
' Do Carcass
        DLT3 = TIMER(.001)
        For LI = 1 To LNC(KN)
            Call OPxADDcarc
        Next LI
        DLT4 = TIMER(.001)
' Calculate Mill Time
        DLTA = DLT2 - DLT1: DLTB = DLT4 - DLT3
        DLTM(KN) = DLTA - DLTB
' Terminate
        End Sub

        Sub OPHOLDERxSUB(KN, LNC(), KO(), SOC(), SOP(), DLTM())
' A Subroutine to Invoke and Time an Operation or Function Elaboration
'     Arguments:-
'         KN        The Operation or Function Serial Number
'         LNC()     The Number of Cycles
'         NOP       The Number of Operations or Functions
'         KO()      The Array of Operations Serial Numbers
'         SOC()     The Array of Operation Codes
'         SOP()     The Array of Operation Descriptions
'         DLTM()    The Array of Operation or Function Timings
'
' Do Composite
        DLT1 = TIMER(.001)
        For LI = 1 To LNC(KN)
            Call OPxSUB
        Next LI
        DLT2 = TIMER(.001)
' Do Carcass
        DLT3 = TIMER(.001)
        For LI = 1 To LNC(KN)
            Call OPxSUBcarc
        Next LI
        DLT4 = TIMER(.001)
' Calculate Mill Time
        DLTA = DLT2 - DLT1: DLTB = DLT4 - DLT3
        DLTM(KN) = DLTA - DLTB
' Terminate
        End Sub

        Sub OPHOLDERxASS(KN, LNC(), KO(), SOC(), SOP(), DLTM())
' A Subroutine to Invoke and Time an Operation or Function Elaboration
'     Arguments:-
```

```
'        KN        The Operation or Function Serial Number
'        LNC()     The Number of Cycles
'        NOP       The Number of Operations or Functions
'        KO()      The Array of Operations Serial Numbers
'        SOC()     The Array of Operation Codes
'        SOP()     The Array of Operation Descriptions
'        DLTM()    The Array of Operation or Function Timings
'
' Do Composite
        DLT1 = TIMER(.001)
        For LI = 1 To LNC(KN)
            Call OPxASS
        Next LI
        DLT2 = TIMER(.001)
' Do Carcass
        DLT3 = TIMER(.001)
        For LI = 1 To LNC(KN)
            Call OPxASScarc
        Next LI
        DLT4 = TIMER(.001)
' Calculate Mill Time
        DLTA = DLT2 - DLT1: DLTB = DLT4 - DLT3
        DLTM(KN) = DLTA - DLTB
' Terminate
        End Sub


        Sub OPHOLDERxBRT(KN, LNC(), KO(), SOC(), SOP(), DLTM())
' A Subroutine to Invoke and Time an Operation or Function Elaboration
'     Arguments:-
'        KN        The Operation or Function Serial Number
'        LNC()     The Number of Cycles
'        NOP       The Number of Operations or Functions
'        KO()      The Array of Operations Serial Numbers
'        SOC()     The Array of Operation Codes
'        SOP()     The Array of Operation Descriptions
'        DLTM()    The Array of Operation or Function Timings
'
' Do Composite
        DLT1 = TIMER(.001)
        For LI = 1 To LNC(KN)
            Call OPxBRT
        Next LI
        DLT2 = TIMER(.001)
' Do Carcass
        DLT3 = TIMER(.001)
        For LI = 1 To LNC(KN)
            Call OPxBRTcarc
        Next LI
        DLT4 = TIMER(.001)
' Calculate Mill Time
        DLTA = DLT2 - DLT1: DLTB = DLT4 - DLT3
        DLTM(KN) = DLTA - DLTB
' Terminate
        End Sub


        Sub OPHOLDERxEXP(KN, LNC(), KO(), SOC(), SOP(), DLTM())
' A Subroutine to Invoke and Time an Operation or Function Elaboration
'     Arguments:-
'        KN        The Operation or Function Serial Number
'        LNC()     The Number of Cycles
'        NOP       The Number of Operations or Functions
'        KO()      The Array of Operations Serial Numbers
'        SOC()     The Array of Operation Codes
```

```
'          SOP()   The Array of Operation Descriptions
'          DLTM()  The Array of Operation or Function Timings
'
' Do Composite
          DLT1 = TIMER(.001)
          For LI = 1 To LNC(KN)
              Call OPxEXP
          Next LI
          DLT2 = TIMER(.001)
' Do Carcass
          DLT3 = TIMER(.001)
          For LI = 1 To LNC(KN)
              Call OPxEXPcarc
          Next LI
          DLT4 = TIMER(.001)
' Calculate Mill Time
          DLTA = DLT2 - DLT1: DLTB = DLT4 - DLT3
          DLTM(KN) = DLTA - DLTB
' Terminate
          End Sub


          Sub OPHOLDERxDLGN(KN, LNC(), KO(), SOC(), SOP(), DLTM())
' A Subroutine to Invoke and Time an Operation or Function Elaboration
'      Arguments:-
'          KN      The Operation or Function Serial Number
'          LNC()   The Number of Cycles
'          NOP     The Number of Operations or Functions
'          KO()    The Array of Operations Serial Numbers
'          SOC()   The Array of Operation Codes
'          SOP()   The Array of Operation Descriptions
'          DLTM()  The Array of Operation or Function Timings
'
' Do Composite
          DLT1 = TIMER(.001)
          For LI = 1 To LNC(KN)
              Call OPxDLGN
          Next LI
          DLT2 = TIMER(.001)
' Do Carcass
          DLT3 = TIMER(.001)
          For LI = 1 To LNC(KN)
              Call OPxDLGNcarc
          Next LI
          DLT4 = TIMER(.001)
' Calculate Mill Time
          DLTA = DLT2 - DLT1: DLTB = DLT4 - DLT3
          DLTM(KN) = DLTA - DLTB
' Terminate
          End Sub


          Sub OPHOLDERxDL10(KN, LNC(), KO(), SOC(), SOP(), DLTM())
' A Subroutine to Invoke and Time an Operation or Function Elaboration
'      Arguments:-
'          KN      The Operation or Function Serial Number
'          LNC()   The Number of Cycles
'          NOP     The Number of Operations or Functions
'          KO()    The Array of Operations Serial Numbers
'          SOC()   The Array of Operation Codes
'          SOP()   The Array of Operation Descriptions
'          DLTM()  The Array of Operation or Function Timings
'
' Do Composite
          DLT1 = TIMER(.001)
```

```
        For LI = 1 To LNC(KN)
            Call OPxDL10
        Next LI
        DLT2 = TIMER(.001)
' Do Carcass
        DLT3 = TIMER(.001)
        For LI = 1 To LNC(KN)
            Call OPxDL10carc
        Next LI
        DLT4 = TIMER(.001)
' Calculate Mill Time
        DLTA = DLT2 - DLT1: DLTB = DLT4 - DLT3
        DLTM(KN) = DLTA - DLTB
' Terminate
        End Sub

        Sub OPHOLDERxSRT(KN, LNC(), KO(), SOC(), SOP(), DLTM())
' A Subroutine to Invoke and Time an Operation or Function Elaboration
'       Arguments:-
'           KN      The Operation or Function Serial Number
'           LNC()   The Number of Cycles
'           NOP     The Number of Operations or Functions
'           KO()    The Array of Operations Serial Numbers
'           SOC()   The Array of Operation Codes
'           SOP()   The Array of Operation Descriptions
'           DLTM()  The Array of Operation or Function Timings
'
' Do Composite
        DLT1 = TIMER(.001)
        For LI = 1 To LNC(KN)
            Call OPxSRT
        Next LI
        DLT2 = TIMER(.001)
' Do Carcass
        DLT3 = TIMER(.001)
        For LI = 1 To LNC(KN)
            Call OPxSRTcarc
        Next LI
        DLT4 = TIMER(.001)
' Calculate Mill Time
        DLTA = DLT2 - DLT1: DLTB = DLT4 - DLT3
        DLTM(KN) = DLTA - DLTB
' Terminate
        End Sub

        Sub OPHOLDERxSIN(KN, LNC(), KO(), SOC(), SOP(), DLTM())
' A Subroutine to Invoke and Time an Operation or Function Elaboration
'       Arguments:-
'           KN      The Operation or Function Serial Number
'           LNC()   The Number of Cycles
'           NOP     The Number of Operations or Functions
'           KO()    The Array of Operations Serial Numbers
'           SOC()   The Array of Operation Codes
'           SOP()   The Array of Operation Descriptions
'           DLTM()  The Array of Operation or Function Timings
'
' Do Composite
        DLT1 = TIMER(.001)
        For LI = 1 To LNC(KN)
            Call OPxSIN
        Next LI
        DLT2 = TIMER(.001)
' Do Carcass
```

```
      DLT3 = TIMER(.001)
      For LI = 1 To LNC(KN)
         Call OPxSINcarc
      Next LI
      DLT4 = TIMER(.001)
' Calculate Mill Time
      DLTA = DLT2 - DLT1: DLTB = DLT4 - DLT3
      DLTM(KN) = DLTA - DLTB
' Terminate
      End Sub

      Sub OPHOLDERxCOS(KN, LNC(), KO(), SOC(), SOP(), DLTM())
' A Subroutine to Invoke and Time an Operation or Function Elaboration
'     Arguments:-
'        KN       The Operation or Function Serial Number
'        LNC()    The Number of Cycles
'        NOP      The Number of Operations or Functions
'        KO()     The Array of Operations Serial Numbers
'        SOC()    The Array of Operation Codes
'        SOP()    The Array of Operation Descriptions
'        DLTM()   The Array of Operation or Function Timings
'
' Do Composite
      DLT1 = TIMER(.001)
      For LI = 1 To LNC(KN)
         Call OPxCOS
      Next LI
      DLT2 = TIMER(.001)
' Do Carcass
      DLT3 = TIMER(.001)
      For LI = 1 To LNC(KN)
         Call OPxCOScarc
      Next LI
      DLT4 = TIMER(.001)
' Calculate Mill Time
      DLTA = DLT2 - DLT1: DLTB = DLT4 - DLT3
      DLTM(KN) = DLTA - DLTB
' Terminate
      End Sub

      Sub OPHOLDERxTAN(KN, LNC(), KO(), SOC(), SOP(), DLTM())
' A Subroutine to Invoke and Time an Operation or Function Elaboration
'     Arguments:-
'        KN       The Operation or Function Serial Number
'        LNC()    The Number of Cycles
'        NOP      The Number of Operations or Functions
'        KO()     The Array of Operations Serial Numbers
'        SOC()    The Array of Operation Codes
'        SOP()    The Array of Operation Descriptions
'        DLTM()   The Array of Operation or Function Timings
'
' Do Composite
      DLT1 = TIMER(.001)
      For LI = 1 To LNC(KN)
         Call OPxTAN
      Next LI
      DLT2 = TIMER(.001)
' Do Carcass
      DLT3 = TIMER(.001)
      For LI = 1 To LNC(KN)
         Call OPxTANcarc
      Next LI
      DLT4 = TIMER(.001)
```

```
' Calculate Mill Time
      DLTA = DLT2 - DLT1: DLTB = DLT4 - DLT3
      DLTM(KN) = DLTA - DLTB
' Terminate
      End Sub


      Sub OPHOLDERxASIN(KN, LNC(), KO(), SOC(), SOP(), DLTM())
' A Subroutine to Invoke and Time an Operation or Function Elaboration
'      Arguments:-
'          KN        The Operation or Function Serial Number
'          LNC()     The Number of Cycles
'          NOP       The Number of Operations or Functions
'          KO()      The Array of Operations Serial Numbers
'          SOC()     The Array of Operation Codes
'          SOP()     The Array of Operation Descriptions
'          DLTM()    The Array of Operation or Function Timings
'
' Do Composite
      DLT1 = TIMER(.001)
      For LI = 1 To LNC(KN)
          Call OPxASIN
      Next LI
      DLT2 = TIMER(.001)
' Do Carcass
      DLT3 = TIMER(.001)
      For LI = 1 To LNC(KN)
          Call OPxASINcarc
      Next LI
      DLT4 = TIMER(.001)
' Calculate Mill Time
      DLTA = DLT2 - DLT1: DLTB = DLT4 - DLT3
      DLTM(KN) = DLTA - DLTB
' Terminate
      End Sub

      Sub OPHOLDERxACOS(KN, LNC(), KO(), SOC(), SOP(), DLTM())
' A Subroutine to Invoke and Time an Operation or Function Elaboration
'      Arguments:-
'          KN        The Operation or Function Serial Number
'          LNC()     The Number of Cycles
'          NOP       The Number of Operations or Functions
'          KO()      The Array of Operations Serial Numbers
'          SOC()     The Array of Operation Codes
'          SOP()     The Array of Operation Descriptions
'          DLTM()    The Array of Operation or Function Timings
'
' Do Composite
      DLT1 = TIMER(.001)
      For LI = 1 To LNC(KN)
          Call OPxACOS
      Next LI
      DLT2 = TIMER(.001)
' Do Carcass
      DLT3 = TIMER(.001)
      For LI = 1 To LNC(KN)
          Call OPxACOScarc
      Next LI
      DLT4 = TIMER(.001)
' Calculate Mill Time
      DLTA = DLT2 - DLT1: DLTB = DLT4 - DLT3
      DLTM(KN) = DLTA - DLTB
' Terminate
      End Sub
```

```
        Sub OPHOLDERxATAN(KN, LNC(), KO(), SOC(), SOP(), DLTM())
' A Subroutine to Invoke and Time an Operation or Function Elaboration
'     Arguments:-
'         KN        The Operation or Function Serial Number
'         LNC()     The Number of Cycles
'         NOP       The Number of Operations or Functions
'         KO()      The Array of Operations Serial Numbers
'         SOC()     The Array of Operation Codes
'         SOP()     The Array of Operation Descriptions
'         DLTM()    The Array of Operation or Function Timings
'
' Do Composite
        DLT1 = TIMER(.001)
        For LI = 1 To LNC(KN)
            Call OPxATAN
        Next LI
        DLT2 = TIMER(.001)
' Do Carcass
        DLT3 = TIMER(.001)
        For LI = 1 To LNC(KN)
            Call OPxATANcarc
        Next LI
        DLT4 = TIMER(.001)
' Calculate Mill Time
        DLTA = DLT2 - DLT1: DLTB = DLT4 - DLT3
        DLTM(KN) = DLTA - DLTB
' Terminate
        End Sub

        Sub OPHOLDERxATN(KN, LNC(), KO(), SOC(), SOP(), DLTM())
' A Subroutine to Invoke and Time an Operation or Function Elaboration
'     Arguments:-
'         KN        The Operation or Function Serial Number
'         LNC()     The Number of Cycles
'         NOP       The Number of Operations or Functions
'         KO()      The Array of Operations Serial Numbers
'         SOC()     The Array of Operation Codes
'         SOP()     The Array of Operation Descriptions
'         DLTM()    The Array of Operation or Function Timings
'
' Do Composite
        DLT1 = TIMER(.001)
        For LI = 1 To LNC(KN)
            Call OPxATN
        Next LI
        DLT2 = TIMER(.001)
' Do Carcass
        DLT3 = TIMER(.001)
        For LI = 1 To LNC(KN)
            Call OPxATNcarc
        Next LI
        DLT4 = TIMER(.001)
' Calculate Mill Time
        DLTA = DLT2 - DLT1: DLTB = DLT4 - DLT3
        DLTM(KN) = DLTA - DLTB
' Terminate
        End Sub

        Sub OPHOLDERxSINH(KN, LNC(), KO(), SOC(), SOP(), DLTM())
' A Subroutine to Invoke and Time an Operation or Function Elaboration
'     Arguments:-
'         KN        The Operation or Function Serial Number
```

```
'           LNC()    The Number of Cycles
'           NOP      The Number of Operations or Functions
'           KO()     The Array of Operations Serial Numbers
'           SOC()    The Array of Operation Codes
'           SOP()    The Array of Operation Descriptions
'           DLTM()   The Array of Operation or Function Timings
'
' Do Composite
      DLT1 = TIMER(.001)
      For LI = 1 To LNC(KN)
          Call OPxSINH
      Next LI
      DLT2 = TIMER(.001)
' Do Carcass
      DLT3 = TIMER(.001)
      For LI = 1 To LNC(KN)
          Call OPxSINHcarc
      Next LI
      DLT4 = TIMER(.001)
' Calculate Mill Time
      DLTA = DLT2 - DLT1: DLTB = DLT4 - DLT3
      DLTM(KN) = DLTA - DLTB
' Terminate
      End Sub

      Sub OPHOLDERxCOSH(KN, LNC(), KO(), SOC(), SOP(), DLTM())
' A Subroutine to Invoke and Time an Operation or Function Elaboration
'     Arguments:-
'           KN       The Operation or Function Serial Number
'           LNC()    The Number of Cycles
'           NOP      The Number of Operations or Functions
'           KO()     The Array of Operations Serial Numbers
'           SOC()    The Array of Operation Codes
'           SOP()    The Array of Operation Descriptions
'           DLTM()   The Array of Operation or Function Timings
'
' Do Composite
      DLT1 = TIMER(.001)
      For LI = 1 To LNC(KN)
          Call OPxCOSH
      Next LI
      DLT2 = TIMER(.001)
' Do Carcass
      DLT3 = TIMER(.001)
      For LI = 1 To LNC(KN)
          Call OPxCOSHcarc
      Next LI
      DLT4 = TIMER(.001)
' Calculate Mill Time
      DLTA = DLT2 - DLT1: DLTB = DLT4 - DLT3
      DLTM(KN) = DLTA - DLTB
' Terminate
      End Sub

      Sub OPHOLDERxTANH(KN, LNC(), KO(), SOC(), SOP(), DLTM())
' A Subroutine to Invoke and Time an Operation or Function Elaboration
'     Arguments:-
'           KN       The Operation or Function Serial Number
'           LNC()    The Number of Cycles
'           NOP      The Number of Operations or Functions
'           KO()     The Array of Operations Serial Numbers
'           SOC()    The Array of Operation Codes
'           SOP()    The Array of Operation Descriptions
```

```
'         DLTM()  The Array of Operation or Function Timings
'
' Do Composite
      DLT1 = TIMER(.001)
      For LI = 1 To LNC(KN)
          Call OPxTANH
      Next LI
      DLT2 = TIMER(.001)
' Do Carcass
      DLT3 = TIMER(.001)
      For LI = 1 To LNC(KN)
          Call OPxTANHcarc
      Next LI
      DLT4 = TIMER(.001)
' Calculate Mill Time
      DLTA = DLT2 - DLT1: DLTB = DLT4 - DLT3
      DLTM(KN) = DLTA - DLTB
' Terminate
      End Sub

      Sub OPHOLDERxASINH(KN, LNC(), KO(), SOC(), SOP(), DLTM())
' A Subroutine to Invoke and Time an Operation or Function Elaboration
'     Arguments:-
'         KN      The Operation or Function Serial Number
'         LNC()   The Number of Cycles
'         NOP     The Number of Operations or Functions
'         KO()    The Array of Operations Serial Numbers
'         SOC()   The Array of Operation Codes
'         SOP()   The Array of Operation Descriptions
'         DLTM()  The Array of Operation or Function Timings
'
' Do Composite
      DLT1 = TIMER(.001)
      For LI = 1 To LNC(KN)
          Call OPxASINH
      Next LI
      DLT2 = TIMER(.001)
' Do Carcass
      DLT3 = TIMER(.001)
      For LI = 1 To LNC(KN)
          Call OPxASINHcarc
      Next LI
      DLT4 = TIMER(.001)
' Calculate Mill Time
      DLTA = DLT2 - DLT1: DLTB = DLT4 - DLT3
      DLTM(KN) = DLTA - DLTB
' Terminate
      End Sub

      Sub OPHOLDERxACOSH(KN, LNC(), KO(), SOC(), SOP(), DLTM())
' A Subroutine to Invoke and Time an Operation or Function Elaboration
'     Arguments:-
'         KN      The Operation or Function Serial Number
'         LNC()   The Number of Cycles
'         NOP     The Number of Operations or Functions
'         KO()    The Array of Operations Serial Numbers
'         SOC()   The Array of Operation Codes
'         SOP()   The Array of Operation Descriptions
'         DLTM()  The Array of Operation or Function Timings
'
' Do Composite
      DLT1 = TIMER(.001)
      For LI = 1 To LNC(KN)
```

```
                Call OPxACOSH
            Next LI
            DLT2 = TIMER(.001)
' Do Carcass
            DLT3 = TIMER(.001)
            For LI = 1 To LNC(KN)
                Call OPxACOSHcarc
            Next LI
            DLT4 = TIMER(.001)
' Calculate Mill Time
            DLTA = DLT2 - DLT1: DLTB = DLT4 - DLT3
            DLTM(KN) = DLTA - DLTB
' Terminate
            End Sub


            Sub OPHOLDERxATANH(KN, LNC(), KO(), SOC(), SOP(), DLTM())
' A Subroutine to Invoke and Time an Operation or Function Elaboration
'       Arguments:-
'           KN          The Operation or Function Serial Number
'           LNC()       The Number of Cycles
'           NOP         The Number of Operations or Functions
'           KO()        The Array of Operations Serial Numbers
'           SOC()       The Array of Operation Codes
'           SOP()       The Array of Operation Descriptions
'           DLTM()      The Array of Operation or Function Timings
'
' Do Composite
            DLT1 = TIMER(.001)
            For LI = 1 To LNC(KN)
                Call OPxATANH
            Next LI
            DLT2 = TIMER(.001)
' Do Carcass
            DLT3 = TIMER(.001)
            For LI = 1 To LNC(KN)
                Call OPxATANHcarc
            Next LI
            DLT4 = TIMER(.001)
' Calculate Mill Time
            DLTA = DLT2 - DLT1: DLTB = DLT4 - DLT3
            DLTM(KN) = DLTA - DLTB
' Terminate
            End Sub


            Sub OPHOLDERxLAC(KN, LNC(), KO(), SOC(), SOP(), DLTM())
' A Subroutine to Invoke and Time an Operation or Function Elaboration
'       Arguments:-
'           KN          The Operation or Function Serial Number
'           LNC()       The Number of Cycles
'           NOP         The Number of Operations or Functions
'           KO()        The Array of Operations Serial Numbers
'           SOC()       The Array of Operation Codes
'           SOP()       The Array of Operation Descriptions
'           DLTM()      The Array of Operation or Function Timings
'
' Do Composite
            DLT1 = TIMER(.001)
            For LI = 1 To LNC(KN)
                Call OPxLAC
            Next LI
            DLT2 = TIMER(.001)
' Do Carcass
            DLT3 = TIMER(.001)
```

```
        For LI = 1 To LNC(KN)
            Call OPxLACcarc
        Next LI
        DLT4 = TIMER(.001)
' Calculate Mill Time
        DLTA = DLT2 - DLT1: DLTB = DLT4 - DLT3
        DLTM(KN) = DLTA - DLTB
' Terminate
        End Sub


        Sub OPHOLDERxAR1(KN, LNC(), KO(), SOC(), SOP(), DLTM())
' A Subroutine to Invoke and Time an Operation or Function Elaboration
'       Arguments:-
'           KN          The Operation or Function Serial Number
'           LNC()       The Number of Cycles
'           NOP         The Number of Operations or Functions
'           KO()        The Array of Operations Serial Numbers
'           SOC()       The Array of Operation Codes
'           SOP()       The Array of Operation Descriptions
'           DLTM()      The Array of Operation or Function Timings
'
' Do Composite
        DLT1 = TIMER(.001)
        For LI = 1 To LNC(KN)
            Call OPxAR1
        Next LI
        DLT2 = TIMER(.001)
' Do Carcass
        DLT3 = TIMER(.001)
        For LI = 1 To LNC(KN)
            Call OPxAR1carc
        Next LI
        DLT4 = TIMER(.001)
' Calculate Mill Time
        DLTA = DLT2 - DLT1: DLTB = DLT4 - DLT3
        DLTM(KN) = DLTA - DLTB
' Terminate
        End Sub


        Sub OPHOLDERxAR2(KN, LNC(), KO(), SOC(), SOP(), DLTM())
' A Subroutine to Invoke and Time an Operation or Function Elaboration
'       Arguments:-
'           KN          The Operation or Function Serial Number
'           LNC()       The Number of Cycles
'           NOP         The Number of Operations or Functions
'           KO()        The Array of Operations Serial Numbers
'           SOC()       The Array of Operation Codes
'           SOP()       The Array of Operation Descriptions
'           DLTM()      The Array of Operation or Function Timings
'
' Do Composite
        DLT1 = TIMER(.001)
        For LI = 1 To LNC(KN)
            Call OPxAR2
        Next LI
        DLT2 = TIMER(.001)
' Do Carcass
        DLT3 = TIMER(.001)
        For LI = 1 To LNC(KN)
            Call OPxAR2carc
        Next LI
        DLT4 = TIMER(.001)
' Calculate Mill Time
```

```
      DLTA = DLT2 - DLT1: DLTB = DLT4 - DLT3
      DLTM(KN) = DLTA - DLTB
' Terminate
      End Sub

      Sub OPHOLDERxZET(KN, LNC(), KO(), SOC(), SOP(), DLTM())
' A Subroutine to Invoke and Time an Operation or Function Elaboration
'     Arguments:-
'        KN       The Operation or Function Serial Number
'        LNC()    The Number of Cycles
'        NOP      The Number of Operations or Functions
'        KO()     The Array of Operations Serial Numbers
'        SOC()    The Array of Operation Codes
'        SOP()    The Array of Operation Descriptions
'        DLTM()   The Array of Operation or Function Timings
'
' Do Composite
      DLT1 = TIMER(.001)
      For LI = 1 To LNC(KN)
         Call OPxZET
      Next LI
      DLT2 = TIMER(.001)
' Do Carcass
      DLT3 = TIMER(.001)
      For LI = 1 To LNC(KN)
         Call OPxZETcarc
      Next LI
      DLT4 = TIMER(.001)
' Calculate Mill Time
      DLTA = DLT2 - DLT1: DLTB = DLT4 - DLT3
      DLTM(KN) = DLTA - DLTB
' Terminate
      End Sub

      Sub OPHOLDERxFAC(KN, LNC(), KO(), SOC(), SOP(), DLTM())
' A Subroutine to Invoke and Time an Operation or Function Elaboration
'     Arguments:-
'        KN       The Operation or Function Serial Number
'        LNC()    The Number of Cycles
'        NOP      The Number of Operations or Functions
'        KO()     The Array of Operations Serial Numbers
'        SOC()    The Array of Operation Codes
'        SOP()    The Array of Operation Descriptions
'        DLTM()   The Array of Operation or Function Timings
'
' Do Composite
      DLT1 = TIMER(.001)
      For LI = 1 To LNC(KN)
         Call OPxFAC
      Next LI
      DLT2 = TIMER(.001)
' Do Carcass
      DLT3 = TIMER(.001)
      For LI = 1 To LNC(KN)
         Call OPxFACcarc
      Next LI
      DLT4 = TIMER(.001)
' Calculate Mill Time
      DLTA = DLT2 - DLT1: DLTB = DLT4 - DLT3
      DLTM(KN) = DLTA - DLTB
' Terminate
      End Sub
```

```
' ****** START OF COMPOSITE SECTION

        Sub OPxPWR()
' A Subroutine to Compute a Real Power of a Real
'       Arguments:-
' ( Arrays IRG() and DRG() are
'   Common Shared by static Public Declaration )
'
        Call RANDARG(I, J, A, B, C, D, E, F, G, H)
        X = B ^ C
        End Sub

        Sub OPxSQR()
' A Subroutine to Compute a Real Square of a Real Number
'       Arguments:-
' ( Arrays IRG() and DRG() are
'   Common Shared by static Public Declaration )
'
        Call RANDARG(I, J, A, B, C, D, E, F, G, H)
        X = B ^ 2
        End Sub

        Sub OPxMUL()
' A Subroutine to Compute a Product of Two Real Numbers
' ( Arrays IRG() and DRG() are
'   Common Shared by static Public Declaration )
' FIFTEEN INTERNAL MULTIPLICATIONS
'
        Call RANDARG(I, J, A, B, C, D, E, F, G, H)
        X = B * C * B * C * B * C * B * C * B * C * B * C * B * C * B * C
        End Sub

        Sub OPxDIV()
' A Subroutine to Compute a Dividend of Two Real Numbers
' ( Arrays IRG() and DRG() are
'   Common Shared by static Public Declaration )
' FIFTEEN INTERNAL DIVISIONS
'
        Call RANDARG(I, J, A, B, C, D, E, F, G, H)
        X = C / B / C / B / C / B / C / B / C / B / C / B / C / B / C / B
        End Sub

        Sub OPxADD()
' A Subroutine to Compute a Sum of Two Real Numbers
' ( Arrays IRG() and DRG() are
'   Common Shared by static Public Declaration )
' FIFTEEN INTERNAL ADDITIONS
'
        Call RANDARG(I, J, A, B, C, D, E, F, G, H)
        X = B + C + B + C + B + C + B + C + B + C + B + C + B + C + B + C
        End Sub

        Sub OPxSUB()
' A Subroutine to Compute a Difference of Two Real Numbers
' ( Arrays IRG() and DRG() are
'   Common Shared by static Public Declaration )
' FIFTEEN INTERNAL SUBTRACTIONS
'
        Call RANDARG(I, J, A, B, C, D, E, F, G, H)
        X = B - C - B - C - B - C - B - C - B - C - B - C - B - C - B - C
        End Sub

        Sub OPxASS()
```

```
' A Subroutine to Assign a Real Number
' ( Arrays IRG() and DRG() are
'   Common Shared by static Public Declaration )
'
        Call RANDARG(I, J, A, B, C, D, E, F, G, H)
        X = B
        End Sub

        Sub OPxBRT()
' A Subroutine to Bracket a Real Number
' ( Arrays IRG() and DRG() are
'   Common Shared by static Public Declaration )
'
        Call RANDARG(I, J, A, B, C, D, E, F, G, H)
        X = (B)
        End Sub

        Sub OPxEXP()
' A Subroutine to Compute a Napierian Exponent of a Real Number
' ( Arrays IRG() and DRG() are
'   Common Shared by static Public Declaration )
'
        Call RANDARG(I, J, A, B, C, D, E, F, G, H)
        X = Exp(B)
        End Sub

        Sub OPxDLGN()
' A Subroutine to Compute a Napierian Logarithm of a Real Number
' ( Arrays IRG() and DRG() are
'   Common Shared by static Public Declaration )
'
        Call RANDARG(I, J, A, B, C, D, E, F, G, H)
        X = Log(C)
        End Sub

        Sub OPxDL10()
' A Subroutine to Compute a Radix Ten Logarithm of a Real Number
' ( Arrays IRG() and DRG() are
'   Common Shared by static Public Declaration )
'
        Call RANDARG(I, J, A, B, C, D, E, F, G, H)
        X = DLOG10(C)
        End Sub

        Sub OPxSRT()
' A Subroutine to Compute a Square Root of a Positive Real Number
' ( Arrays IRG() and DRG() are
'   Common Shared by static Public Declaration )
'
        Call RANDARG(I, J, A, B, C, D, E, F, G, H)
        X = Sqr(B)
        End Sub

        Sub OPxSIN()
' A Subroutine to Compute a Sine of a Real Number
' ( Arrays IRG() and DRG() are
'   Common Shared by static Public Declaration )
'
        Call RANDARG(I, J, A, B, C, D, E, F, G, H)
        X = Sin(A)
        End Sub

        Sub OPxCOS()
```

```
' A Subroutine to Compute a Cosine of a Real Number
' ( Arrays IRG() and DRG() are
'   Common Shared by static Public Declaration )

        Call RANDARG(I, J, A, B, C, D, E, F, G, H)
        X = Cos(A)
        End Sub

        Sub OPxTAN()
' A Subroutine to Compute a Tangent of a Real Number
' ( Arrays IRG() and DRG() are
'   Common Shared by static Public Declaration )

        Call RANDARG(I, J, A, B, C, D, E, F, G, H)
        X = TAN(E)
        End Sub

        Sub OPxASIN()
' A Subroutine to Compute an Inverse Sine of Two Real Numbers
' ( Arrays IRG() and DRG() are
'   Common Shared by static Public Declaration )

        Call RANDARG(I, J, A, B, C, D, E, F, G, H)
        X = _ASIN(A)
        End Sub

        Sub OPxACOS()
' A Subroutine to Compute an Inverse Cosine of Two Real Numbers
' ( Arrays IRG() and DRG() are
'   Common Shared by static Public Declaration )

        Call RANDARG(I, J, A, B, C, D, E, F, G, H)
        X = _ACOS(A)
        End Sub

        Sub OPxATAN()
' A Subroutine to Compute an Inverse Tangent of a Real Number
' ( Arrays IRG() and DRG() are
'   Common Shared by static Public Declaration )

        Call RANDARG(I, J, A, B, C, D, E, F, G, H)
        X = ATN(B)
        End Sub

        Sub OPxATN()
' A Subroutine to Compute an Inverse Tangent of a Real Number
' ( Arrays IRG() and DRG() are
'   Common Shared by static Public Declaration )

        Call RANDARG(I, J, A, B, C, D, E, F, G, H)
        X = Atn(B)
        End Sub

        Sub OPxSINH()
' A Subroutine to Compute a Hyperbolic Sine of a Real Number
' ( Arrays IRG() and DRG() are
'   Common Shared by static Public Declaration )

        Call RANDARG(I, J, A, B, C, D, E, F, G, H)
        X = DSINH(F)
        End Sub

        Sub OPxCOSH()
```

```
' A Subroutine to Compute a Hyperbolic Cosine of a Real Number
' ( Arrays IRG() and DRG() are
'   Common Shared by static Public Declaration )
'
        Call RANDARG(I, J, A, B, C, D, E, F, G, H)
        X = DCOSH(G)
        End Sub

        Sub OPxTANH()
' A Subroutine to Compute a Hyperbolic Tangent of a Real Number
' ( Arrays IRG() and DRG() are
'   Common Shared by static Public Declaration )
'
        Call RANDARG(I, J, A, B, C, D, E, F, G, H)
        X = DTANH(H)
        End Sub

        Sub OPxASINH()
' A Subroutine to Compute an Inverse Hyperbolic Sine of a Real Number
' ( Arrays IRG() and DRG() are
'   Common Shared by static Public Declaration )
'
        Call RANDARG(I, J, A, B, C, D, E, F, G, H)
        X = _ASINH(A)
        End Sub

        Sub OPxACOSH()
' A Subroutine to Compute an Inverse Hyperbolic Cosine of a Real Number
' ( Arrays IRG() and DRG() are
'   Common Shared by static Public Declaration )
'
        Call RANDARG(I, J, A, B, C, D, E, F, G, H)
        X = _ACOSH(A)
        End Sub

        Sub OPxATANH()
' A Subroutine to Compute an Inverse Hyperbolic Tangent of a Real Number
' ( Arrays IRG() and DRG() are
'   Common Shared by static Public Declaration )
'
        Call RANDARG(I, J, A, B, C, D, E, F, G, H)
        X = _ATANH(C)
        End Sub

        Sub OPxLAC()
' A Subroutine to Execute a FOR-NEXT Loop Access
' ( Arrays IRG() and DRG() are
'   Common Shared by static Public Declaration )
'
        Call RANDARG(I, J, A, B, C, D, E, F, G, H)
        For I = 1 To 1
           II = I
        Next I
        End Sub

        Sub OPxAR1()
' A Subroutine to Execute a One-Dimensional Real Array Access
' ( Arrays IR1(), IR2(), IRG() and DRG() are
'   Common Shared by static Public Declaration )
'
        Call RANDARG(I, J, A, B, C, D, E, F, G, H)
        AR1(I) = B
        X = AR1(I)
```

```
        End Sub

        Sub OPxAR2()
' A Subroutine to Execute a Two-Dimensional Real Array Access
' ( Arrays AR1(), AR2(), IRG() and DRG() are
'   Common Shared by static Public Declaration )
'
        Call RANDARG(I, J, A, B, C, D, E, F, G, H)
        AR2(I, J) = B
        X = AR2(I, J)
        End Sub

        Sub OPxZET()
' A Subroutine to Compute a Zeta Function of a Real Number
' ( Arrays IRG() and DRG() are
'   Common Shared by static Public Declaration )
'
        Call RANDARG(I, J, A, B, C, D, E, F, G, H)
        X = ZETA(B, 2000)
        End Sub

        Sub OPxFAC()
' A Subroutine to Compute a Factorial of a Short Integer
' ( Arrays IRG() and DRG() are
'   Common Shared by static Public Declaration )
'
        JLIM=11
        JX=JARG(JLIM)
        DD=FACT(JX)
        End Sub

' ****** START OF CARCASE SECTION

        Sub OPxPWRcarc()
' A Subroutine to Compute a Real Power of a Real
' ( Arrays IRG() and DRG() are
'   Common Shared by static Public Declaration )
'
        Call RANDARG(I, J, A, B, C, D, E, F, G, H)
        X = B
        End Sub

        Sub OPxSQRcarc()
' A Subroutine to Compute a Real Square of a Real Number
' ( Arrays IRG() and DRG() are
'   Common Shared by static Public Declaration )
'
        Call RANDARG(I, J, A, B, C, D, E, F, G, H)
        X = B
        End Sub

        Sub OPxMULcarc()
' A Subroutine to Compute a Product of Two Real Numbers
' ( Arrays IRG() and DRG() are
'   Common Shared by static Public Declaration )
'
        Call RANDARG(I, J, A, B, C, D, E, F, G, H )
        X = B
        End Sub

        Sub OPxDIVcarc()
' A Subroutine to Compute a Dividend of Two Real Numbers
' ( Arrays IRG() and DRG() are
```

```
'    Common Shared by static Public Declaration )

         Call RANDARG(I, J, A, B, C, D, E, F, G, H)
         X = B
         End Sub

         Sub OPxADDcarc()
'  A Subroutine to Compute a Sum of Two Real Numbers
'  ( Arrays IRG() and DRG() are
'    Common Shared by static Public Declaration )

         Call RANDARG(I, J, A, B, C, D, E, F, G, H)
         X = B
         End Sub

         Sub OPxSUBcarc()
'  A Subroutine to Compute a Difference of Two Real Numbers
'  ( Arrays IRG() and DRG() are
'    Common Shared by static Public Declaration )

         Call RANDARG(I, J, A, B, C, D, E, F, G, H)
         X = B
         End Sub

         Sub OPxASScarc()
'  A Subroutine to Assign a Real Number
'  ( Arrays IRG() and DRG() are
'    Common Shared by static Public Declaration )

         Call RANDARG(I, J, A, B, C, D, E, F, G, H)
         End Sub

         Sub OPxBRTcarc()
'  A Subroutine to Bracket a Real Number
'  ( Arrays IRG() and DRG() are
'    Common Shared by static Public Declaration )

         Call RANDARG(I, J, A, B, C, D, E, F, G, H)
         X = B
         End Sub

         Sub OPxEXPcarc()
'  A Subroutine to Compute a Napierian Exponent of a Real Number
'  ( Arrays IRG() and DRG() are
'    Common Shared by static Public Declaration )

         Call RANDARG(I, J, A, B, C, D, E, F, G, H)
         X = B
         End Sub

         Sub OPxDLGNcarc()
'  A Subroutine to Compute a Napierian Logarithm of a Real Number
'  ( Arrays IRG() and DRG() are
'    Common Shared by static Public Declaration )

         Call RANDARG(I, J, A, B, C, D, E, F, G, H)
         X = C
         End Sub

         Sub OPxDL10carc()
'  A Subroutine to Compute a Radix Ten Logarithm of a Real Number
'  ( Arrays IRG() and DRG() are
'    Common Shared by static Public Declaration )
```

```
        Call RANDARG(I, J, A, B, C, D, E, F, G, H)
        X = C
        End Sub

        Sub OPxSRTcarc()
' A Subroutine to Compute a Square Root of a Positive Real Number
' ( Arrays IRG() and DRG() are
'   Common Shared by static Public Declaration )

        Call RANDARG(I, J, A, B, C, D, E, F, G, H)
        X = B
        End Sub

        Sub OPxSINcarc()
' A Subroutine to Compute a Sine of a Real Number
' ( Arrays IRG() and DRG() are
'   Common Shared by static Public Declaration )

        Call RANDARG(I, J, A, B, C, D, E, F, G, H)
        X = A
        End Sub

        Sub OPxCOScarc()
' A Subroutine to Compute a Cosine of a Real Number
' ( Arrays IRG() and DRG() are
'   Common Shared by static Public Declaration )

        Call RANDARG(I, J, A, B, C, D, E, F, G, H)
        X = A
        End Sub

        Sub OPxTANcarc()
' A Subroutine to Compute a Tangent of a Real Number
' ( Arrays IRG() and DRG() are
'   Common Shared by static Public Declaration )

        Call RANDARG(I, J, A, B, C, D, E, F, G, H)
        X = E
        End Sub

        Sub OPxASINcarc()
' A Subroutine to Compute an Inverse Sine of a Real Number
' ( Arrays IRG() and DRG() are
'   Common Shared by static Public Declaration )

        Call RANDARG(I, J, A, B, C, D, E, F, G, H)
        X = A
        End Sub

        Sub OPxACOScarc()
' A Subroutine to Compute an Inverse Cosine of a Real Number
' ( Arrays IRG() and DRG() are
'   Common Shared by static Public Declaration )

        Call RANDARG(I, J, A, B, C, D, E, F, G, H)
        X = A
        End Sub

        Sub OPxATANcarc()
' A Subroutine to Compute an Inverse Tangent of a Real Number
' ( Arrays IRG() and DRG() are
'   Common Shared by static Public Declaration )
```

```
        Call RANDARG(I, J, A, B, C, D, E, F, G, H)
        X = C
        End Sub

        Sub OPxATNcarc()
' A Subroutine to Compute an Inverse Tangent of a Real Number
' ( Arrays IRG() and DRG() are
'   Common Shared by static Public Declaration )
'
        Call RANDARG(I, J, A, B, C, D, E, F, G, H)
        X = B
        End Sub

        Sub OPxSINHcarc()
' A Subroutine to Compute a Hyperbolic Sine of a Real Number
' ( Arrays IRG() and DRG() are
'   Common Shared by static Public Declaration )
'
        Call RANDARG(I, J, A, B, C, D, E, F, G, H)
        X = A
        End Sub

        Sub OPxCOSHcarc()
' A Subroutine to Compute a Hyperbolic Cosine of a Real Number
' ( Arrays IRG() and DRG() are
'   Common Shared by static Public Declaration )
'
        Call RANDARG(I, J, A, B, C, D, E, F, G, H)
        X = A
        End Sub

        Sub OPxTANHcarc()
' A Subroutine to Compute a Hyperbolic Tangent of a Real Number
' ( Arrays IRG() and DRG() are
'   Common Shared by static Public Declaration )
'
        Call RANDARG(I, J, A, B, C, D, E, F, G, H)
        X = C
        End Sub

        Sub OPxASINHcarc()
' A Subroutine to Compute an Inverse Hyperbolic Sine of a Real Number
' ( Arrays IRG() and DRG() are
'   Common Shared by static Public Declaration )
'
        Call RANDARG(I, J, A, B, C, D, E, F, G, H)
        X = A
        End Sub

        Sub OPxACOSHcarc()
' A Subroutine to Compute an Inverse Hyperbolic Cosine of a Real Number
' ( Arrays IRG() and DRG() are
'   Common Shared by static Public Declaration )
'
        Call RANDARG(I, J, A, B, C, D, E, F, G, H)
        X = A
        End Sub

        Sub OPxATANHcarc()
' A Subroutine to Compute an Inverse Hyperbolic Tangent of a Real Number
' ( Arrays IRG() and DRG() are
'   Common Shared by static Public Declaration )
```

```
    Call RANDARG(I, J, A, B, C, D, E, F, G, H)
    X = C
    End Sub

    Sub OPxLACcarc()
' A Subroutine to Execute a FOR-NEXT Loop Access
' ( Arrays IRG() and DRG() are
'   Common Shared by static Public Declaration )
'
'    Call RANDARG(I, J, A, B, C, D, E, F, G, H)
'    For I = 1 To 1
        II = I
'    Next I
    End Sub

    Sub OPxAR1carc()
' A Subroutine to Execute a One-Dimensional Real Array Access
'   Common Shared by static Public Declaration )
'
    Call RANDARG(I, J, A, B, C, D, E, F, G, H)
    AR1(I) = B
    X = B
    End Sub

    Sub OPxAR2carc()
' A Subroutine to Execute a Two-Dimensional Real Array Access
' ( Arrays AR1(), AR2(), IRG() and DRG() are
'   Common Shared by static Public Declaration )
'
    Call RANDARG(I, J, A, B, C, D, E, F, G, H)
    AR2(I, J) = B
    X = B
    End Sub

    Sub OPxZETcarc()
' A Subroutine to Compute a Zeta Function of a Real Number
' ( Arrays IRG() and DRG() are
'   Common Shared by static Public Declaration )
'
    Call RANDARG(I, J, A, B, C, D, E, F, G, H)
    X = B
    End Sub

    Sub OPxFACcarc()
' A Subroutine to Compute a Real Square of a Real Number
' ( Arrays IRG() and DRG() are
'   Common Shared by static Public Declaration )
'
    JX=JARG(11)
    DD=39916800
    End Sub
```

APPENDIX B

RATING EXPERIMENTS
DESCRIPTIVE TABLES

Episode	1	2	3	4	5
Manual Logged Start Time	15:18	17:28	19:27	13:54	15:42
Time (.OUT file)	17:03	19:13	21:12	15:39	17:27
Date	13/07/2019	13/07/2019	13/07/2019	14/07/2019	14/07/2019
Program	TARIFFQB64u.bas	TARIFFQB64u.bas	TARIFFQB64u.bas	TARIFFQB64u.bas	TARIFFQB64u.bas
Input File	TARIFF.INP	TARIFF.INP	TARIFF.INP	TARIFF.INP	TARIFF.INP
Output File	TARIFFexTARIFFu1.OUT	TARIFFexTARIFFu2.OUT	TARIFFexTARIFFu3.OUT	TARIFFexTARIFFu4.OUT	TARIFFexTARIFFu5.OUT
Total Execution Time	6319.903000000000	6318.552999999990	6305.501999999990	6318.489999999990	6321.712000000000
Total Execution Time (minutes)	105.3317167	105.3092167	105.0917	105.3081667	105.3618667
Apparent Total Execution Time	01:45	01:45	01:45	01:47	01:45
n-Fold Internal Operations					
ADD	15	15	15	15	15
SUB	15	15	15	15	15
MUL	15	15	15	15	15
DIV	15	15	15	15	15
All Other Operations	1	1	1	1	1

Table B1
Execution History Summary Sheet
for the Five Rating Experiments

TARIFFQB64u

Operation Number	Operation Code	Operation Description	Performed Cycles	Execution Time (seconds)	Module Operation Repetition	Number of Operations	Operation Mill Time MT (Seconds)	-Log_{10}(MT)	MT/MT_{MUL}	1+Log_{10} (MT/ MT_{MUL})
1	PWR	B^C	50000000	10.378000000115000	1	50000000	2.0756E-07	6.682856	306.3767	3.486256
2	SQR	B^2	50000000	5.529000000022700	1	50000000	1.1058E-07	6.956323	163.2257	3.212789
3	MUL	B*C	500000000	5.080999999983100	15	7500000000	6.7747E-10	9.169112	1.0000	1.000000
4	DIV	B/C	50000000	7.164000000043000	15	750000000	9.5520E-09	8.019906	14.0996	2.149206
5	ADD	B+C	50000000	5.601999999989500	15	7500000000	7.4693E-10	9.126718	1.1025	1.042394
6	SUB	B-C	50000000	5.460000000064000	15	7500000000	7.2800E-10	9.137869	1.0746	1.031243
7	ASS	B:=C	50000000	0.201000000000310	1	50000000	4.0200E-09	8.395774	5.9339	1.773338
8	EXP	EXP(B)	50000000	3.790000000081500	1	50000000	7.5800E-08	7.120331	111.8874	3.048781
9	DLGN	LOGN(C)	50000000	5.823000000039500	1	50000000	1.1646E-07	6.933823	171.9051	3.235289
10	DL10	LOG10(C)	50000000	8.036000000000500	1	50000000	1.6072E-07	6.793930	237.2368	3.375182
11	SIN	SIN(A)	50000000	2.059999999976700	1	50000000	4.1200E-08	7.385103	60.8148	2.784009
12	COS	COS(A)	50000000	2.275000000014500	1	50000000	4.5500E-08	7.341989	67.1620	2.827123
13	TAN	TAN(E)	50000000	4.674000000063400	1	50000000	9.3480E-08	7.029281	137.9846	3.139831
14	ASIN	ASIN(A)	50000000	7.712000000068100	1	50000000	1.5424E-07	6.811803	227.6717	3.357309
15	ACOS	ACOS(A)	50000000	7.497000000030200	1	50000000	1.4994E-07	6.824082	221.3245	3.345030
16	ATAN	ATAN(B)	50000000	7.231999999963300	1	50000000	1.4464E-07	6.839712	213.5013	3.329400
17	ATN	ATAN(B)	50000000	7.019999999968000	1	50000000	1.4040E-07	6.852633	207.2427	3.316479
18	SNH	SINH(B)	50000000	8.925000000029100	1	50000000	1.7850E-07	6.748362	263.4816	3.420750
19	COH	COSH(B)	50000000	8.908999999996500	1	50000000	1.7818E-07	6.749141	263.0093	3.419971
20	TAH	TANH(B)	50000000	17.244000000060000	1	50000000	3.4488E-07	6.462332	509.0730	3.706780
21	ASNH	ASINH(A)	50000000	6.068000000065700	1	50000000	1.2136E-07	6.915924	179.1380	3.253188
22	ACOH	ACOSH(A)	50000000	1.027999999984200	1	50000000	2.0560E-08	7.686977	30.3484	2.482135
23	ATAH	ATANH(C)	50000000	1.098000000126900	1	50000000	2.1960E-08	7.658368	32.4149	2.510744
24	SRT	SQRT(B)	50000000	2.072000000001100	1	50000000	4.1440E-08	7.382580	61.1691	2.786532
25	LAC	Loop Access	50000000	0.309000000011060	1	50000000	6.1800E-09	8.209012	9.1222	1.960101
26	AR1	1D Real Arra	50000000	0.495999999991850	1	50000000	9.9200E-09	8.003488	14.6428	2.165624
27	AR2	2D Real Arra	50000000	1.444000000031400	1	50000000	2.8880E-08	7.539403	42.6294	2.629709
28	ZET	ZETA(B)	5000	2.225999999951100	1	5000	4.4520E-04	3.351445	657154.1035	6.817667
29	FAC	FACT(JX)	50000000	1.438000000019200	1	50000000	2.8760E-08	7.541211	42.4523	2.627901
					85	2.445E+10	0.00044764	211.669488	660751.1283	83.2347623 Total
					2.93103448	843103621	1.5436E-05	7.29894785	22784.52167	2.87016422 Mean
					4.82758621	2264863643	8.1218E-05	1.05988791	119884.6418	1.05988791 Pop SD

Table B2
Specimen Table for the Input and Output of
Experiment EPISODE 5

APPENDIX C

MILL TIME RATING STATISTICS
FOR
THE TWENTY-NINE
OPERATIONS AND FUNCTIONS

Operation Number	Operation Code	Operation Description	Episode 1	Episode 2	Episode 3	Episode 4	Episode 5	Mean Mill Time (μMT)	Pop SD Mill Time σ(μMT)	Coefficient of Variation Mill Time $C_v(\mu MT)=\sigma/\mu$	Negative Log10 of Mean Mill Time (μMT) $-Log_{10}(\mu MT)$	Relative Mill Time against μMT_max as Unity μMT/μMT_max	Relative Mill Time Metric P_{or} $1+Log_{10}(\mu MT/\mu MT_{max})$	Minimum Mill Time	Mean Mill Time	Maximum Mill Time	$-Log_{10}$ Maximum Mill Time	$-Log_{10}$ Minimum Mill Time	$-Log_{10}$ Mean Mill Time
1	PWR	B^C	2.0546E-07	2.0492E-07	2.0244E-07	2.0870E-07	2.0756E-07	2.0582E-07	2.1765E-09	0.0106	6.686521	336.0078	3.526349	2.0244E-07	2.0582E-07	2.0870E-07	6.6805	6.6937	6.6865
2	SQR	B^2	1.0880E-07	1.0716E-07	1.1010E-07	1.0856E-07	1.1058E-07	1.0904E-07	1.2098E-09	0.0111	6.962414	178.0148	3.250456	1.0716E-07	1.0904E-07	1.1058E-07	6.9563	6.9700	6.9624
3	MUL	B*C	4.5800E-10	6.6667E-10	8.4160E-10	4.1893E-10	6.7747E-10	6.1253E-10	1.5555E-10	0.2539	9.212870	1.0000	1.000000	4.1893E-10	6.1253E-10	8.4160E-10	9.0749	9.3379	9.2129
4	DIV	B/C	9.4773E-09	9.5707E-09	9.1520E-09	9.8147E-09	9.5520E-09	9.5133E-09	2.1328E-10	0.0224	8.021667	15.5311	2.191203	9.1520E-09	9.5133E-09	9.8147E-09	8.0081	8.0385	8.0217
5	ADD	B+C	7.5400E-10	7.5533E-10	8.0800E-10	7.0733E-10	7.4693E-10	7.5432E-10	3.2078E-11	0.0425	9.122444	1.2315	1.090426	7.0733E-10	7.5432E-10	8.0800E-10	9.0926	9.1504	9.1224
6	SUB	B-C	6.9640E-10	6.9760E-10	7.4573E-10	7.2200E-10	7.2800E-10	7.1795E-10	1.8803E-11	0.0262	9.143908	1.1721	1.068962	6.9640E-10	7.1795E-10	7.4573E-10	9.1274	9.1571	9.1439
7	ASS	B:=C	3.9400E-09	2.0800E-09	4.5400E-09	2.8600E-09	4.0200E-09	3.4880E-09	8.9117E-10	0.2555	8.457424	5.6944	1.755447	2.0800E-09	3.4880E-09	4.5400E-09	8.3429	8.6819	8.4574
8	EXP	EXP(B)	7.6500E-08	6.9660E-08	7.4720E-08	7.6940E-08	7.5800E-08	7.4724E-08	2.6407E-09	0.0353	7.126540	121.9917	3.086330	6.9660E-08	7.4724E-08	7.6940E-08	7.1138	7.1570	7.1265
9	DLGN	LOGN(C)	1.1540E-07	1.1288E-07	1.0850E-07	1.0690E-07	1.1646E-07	1.1203E-07	3.7545E-09	0.0335	6.950673	182.8929	3.262197	1.0690E-07	1.1203E-07	1.1646E-07	6.9338	6.9710	6.9507
10	DL10	LOG10(C)	1.5456E-07	1.6062E-07	1.5928E-07	1.6670E-07	1.6072E-07	1.6034E-07	3.8902E-09	0.0243	6.794969	261.7588	3.417901	1.5456E-07	1.6034E-07	1.6670E-07	6.7781	6.8109	6.7950
11	SIN	SIN(A)	3.9040E-08	4.1520E-08	3.8740E-08	4.0220E-08	4.1200E-08	4.0144E-08	1.1139E-09	0.0277	7.396379	65.5377	2.816491	3.8740E-08	4.0144E-08	4.1520E-08	7.3817	7.4118	7.3964
12	COS	COS(A)	4.4060E-08	4.5560E-08	4.2600E-08	4.2600E-08	4.5500E-08	4.4064E-08	1.3105E-09	0.0297	7.355916	71.9373	2.856954	4.2600E-08	4.4064E-08	4.5560E-08	7.3414	7.3706	7.3559
13	TAN	TAN(E)	8.4080E-08	9.4600E-08	9.4660E-08	9.5320E-08	9.3480E-08	9.2428E-08	4.2157E-09	0.0456	7.034196	150.8946	3.178674	8.4080E-08	9.2428E-08	9.5320E-08	7.0208	7.0753	7.0342
14	ASIN	ASIN(A)	1.5374E-07	1.5454E-07	1.5616E-07	1.5374E-07	1.5424E-07	1.5448E-07	8.9204E-10	0.0058	6.811116	252.2051	3.401754	1.5374E-07	1.5448E-07	1.5616E-07	6.8064	6.8132	6.8111
15	ACOS	ACOS(A)	1.5168E-07	1.5258E-07	1.5144E-07	1.5182E-07	1.4994E-07	1.5149E-07	8.6465E-10	0.0057	6.819610	247.3204	3.393260	1.4994E-07	1.5149E-07	1.5258E-07	6.8165	6.8241	6.8196
16	ATAN	ATAN(B)	1.3470E-07	1.3778E-07	1.3828E-07	1.3702E-07	1.4464E-07	1.3848E-07	3.3135E-09	0.0239	6.858600	226.0840	3.354270	1.3470E-07	1.3848E-07	1.4464E-07	6.8397	6.8706	6.8586
17	ATN	ATAN(B)	1.4338E-07	1.4824E-07	1.4036E-07	1.3964E-07	1.4040E-07	1.4240E-07	3.1889E-09	0.0224	6.846478	232.4837	3.366392	1.3964E-07	1.4240E-07	1.4824E-07	6.8290	6.8550	6.8465
18	SNH	SINH(B)	1.7594E-07	1.7724E-07	1.7224E-07	1.7474E-07	1.7850E-07	1.7573E-07	2.1521E-09	0.0122	6.755149	286.8938	3.457721	1.7224E-07	1.7573E-07	1.7850E-07	6.7484	6.7639	6.7551
19	COH	COSH(B)	1.7926E-07	1.8054E-07	1.7022E-07	1.7840E-07	1.7818E-07	1.7732E-07	3.6455E-09	0.0206	6.751242	289.4863	3.461628	1.7022E-07	1.7732E-07	1.8054E-07	6.7434	6.7690	6.7512
20	TAH	TANH(B)	3.4654E-07	3.4584E-07	3.5390E-07	3.6942E-07	3.4488E-07	3.5212E-07	9.2244E-09	0.0262	6.453314	574.8520	3.759556	3.4488E-07	3.5212E-07	3.6942E-07	6.4325	6.4623	6.4533
21	ASNH	ASINH(A)	1.2000E-07	1.2008E-07	1.1458E-07	1.2234E-07	1.2136E-07	1.1967E-07	2.6896E-09	0.0225	6.922007	195.3722	3.290863	1.1458E-07	1.1967E-07	1.2234E-07	6.9124	6.9409	6.9220
22	ACOH	ACOSH(A)	3.2820E-08	2.5600E-08	1.9420E-08	2.2020E-08	2.0560E-08	2.4084E-08	4.8383E-09	0.2009	7.618271	39.3187	2.594599	1.9420E-08	2.4084E-08	3.2820E-08	7.4839	7.7118	7.6183
23	ATAH	ATANH(C)	2.1360E-08	2.1320E-08	2.1220E-08	1.6400E-08	2.1960E-08	2.0452E-08	2.0426E-09	0.0999	7.689264	33.3892	2.523606	1.6400E-08	2.0452E-08	2.1960E-08	7.6584	7.7852	7.6893
24	SRT	SQRT(B)	4.5100E-08	4.6080E-08	4.5100E-08	4.4900E-08	4.1440E-08	4.4524E-08	1.5961E-09	0.0358	7.351406	72.6883	2.861464	4.1440E-08	4.4524E-08	4.6080E-08	7.3365	7.3826	7.3514
25	LAC	Loop Access	5.5000E-09	4.9000E-09	5.5400E-09	5.5600E-09	6.1800E-09	5.5360E-09	4.0525E-10	0.0732	8.256804	9.0379	1.956066	4.9000E-09	5.5360E-09	6.1800E-09	8.2090	8.3098	8.2568
26	AR1	1D Real Array Access	1.1260E-08	8.5600E-09	1.1140E-08	1.3280E-08	9.9200E-09	1.0832E-08	1.5662E-09	0.1446	7.965291	17.6839	2.247579	8.5600E-09	1.0832E-08	1.3280E-08	7.8768	8.0675	7.9653
27	AR2	2D Real Array Access	9.0400E-09	1.5980E-08	8.3200E-09	7.4600E-09	2.8880E-08	1.3936E-08	8.0616E-09	0.5785	7.855862	22.7514	2.357008	7.4600E-09	1.3936E-08	2.8880E-08	7.5394	8.1273	7.8559
28	ZET	ZETA(B)	4.4600E-04	4.4400E-04	4.4560E-04	4.4600E-04	4.4520E-04	4.4536E-04	7.4189E-07	0.0017	3.351289	727078.7984	6.861581	4.4400E-04	4.4536E-04	4.4600E-04	3.3507	3.3526	3.3513
29	FAC	FACT(IX)	2.8940E-08	2.7920E-08	2.8000E-08	2.8660E-08	2.8760E-08	2.8456E-08	4.1558E-10	0.0146	7.545826	46.4562	2.667044	2.7920E-08	2.8456E-08	2.8940E-08	7.5385	7.5541	7.5458
		Total	4.8840E-04	4.6642E-04	4.7798E-04	4.8843E-04	4.4764E-04	4.4777E-04	8.0841E-07	2.1069E+00	2.1212E+02	7.3102E+05	8.4056E+01	4.4633E-04	4.4777E-04	4.4851E-04	2.1097E+02	2.1346E+02	2.1212E+02
		Mean	1.5462E-05	1.5394E-05	1.5448E-05	1.5463E-05	1.5436E-05	1.5440E-05	2.7876E-08	7.2651E-02	7.3144E+00	2.5208E+04	2.8985E+00	1.5391E-05	1.5440E-05	1.5466E-05	7.2750E+00	7.3605E+00	7.3144E+00
		Population SD	8.1364E-05	8.0999E-05	8.1291E-05	8.1364E-05	8.1218E-05	8.1247E-05	1.3495E-07	0.1177756	1.06865518	132641.2683	1.0686552	8.1E-05	8.1247E-05	8.1363E-05	1.04659433	1.09935266	1.06865518

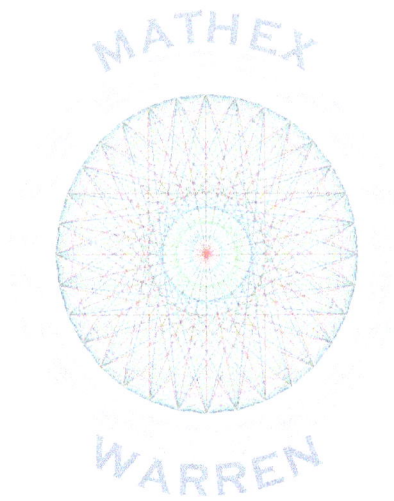

CHAPTER FOUR

Planetary Oblation

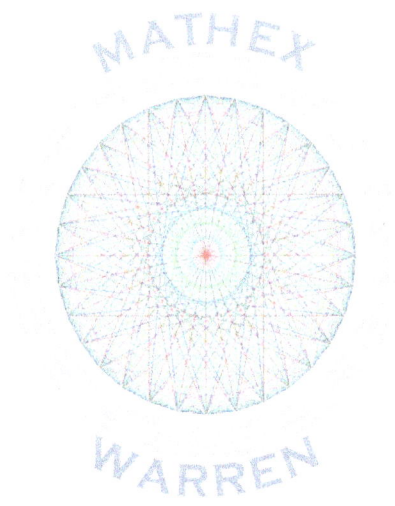

Observations upon the Estimation of
Ellipse Perimeter Length

by
James R Warren BSc MSc PhD PGCE

PART ONE
CRITICAL REVIEW OF SOME KEY ELLIPTIC INTEGRAL METHODS

Fundamental Variables

There is no "exact" value of physical ellipse perimeter which is determinable. There are only approximations of more or less quality achievable. A mathematical object called the Complete Elliptic Integral of the Second Kind, E(k), is intimately related to most high-accuracy approximators of ellipse perimeter.

A geometrical ellipse is a two-dimensional oval of a specific kind which has two principal radii or semi-axes, a and b. If the ellipse is oblate, like a polar meridian of the Earth, then the Horizontal Semi-Axis, a, is longer than the Vertical Semi-Axis, b.

A derivative property fundamental to the ellipse is its Eccentricity, ε, defined by:-

$$\varepsilon = \sqrt{1 - \frac{b^2}{a^2}}$$

Equation 1.1

Also, several algebraic expressions of ellipse mathematics are clarified by use of the Gauss-Kummer Parameter, h, which is defined by:-

$$h = \frac{(a-b)^2}{(a+b)^2}$$

Equation 1.2

Fiducial Values

On the whole, I rarely referred to Abramowitz and Stegun[1] or other canonical authorities to define numerical relations between a, b and Ellipse Perimeter Length P(a,b) or indeed the Elliptic Integral E(ε) and K(ε) (or K(ε^2)). That is because I needed to examine fifteen-figure results at the limits of accuracy of my chosen spreadsheet software Microsoft Excel® 2010 and my (64-bit) scratchpad PTC® Mathcad® Prime™ 3.1.

I applied my work and the chosen proprietary software on a 64-bit Hewlett-Packard Pavilion 500400 microcomputer.

Table 1.1 lists the EXCEL-derived fiducial values and selected computed outcomes defined for a "moderate" oblate ellipse of aspect ratio 4:3:-

Metric	Symbol	Value
Semi-Major Axis	a	1.000000000000000
Semi-Minor Axis	b	0.750000000000000
Right Focus	c	0.661437827766148
Left Focus	-c	-0.661437827766148
Eccentricity (Cbsd)	ε_{cbsd}	0.661437827766148
Eccentricity (Wikipedia)	ε_{wiki}	0.661437827766148
Origin Abscissal Co-Ordinate	h	0.000000000000000
Origin Ordinal Co-ordinate	k	0.000000000000000
Area	A	2.356194490192340
Radius of Equivalent Circle	r	0.866025403784439
Gauss-Kummer h	h_{gk}	0.020408163265306
Perimeter by Simple Summation of Σds	P_{sum}	5.525698505963820
Ramanujan Circumference h^3	P_{h3}	5.525872947729190
Ramanujan Circumference h^5	P_{h5}	5.525873040176910
Sykora (2005) Approximation	P_{syk}	5.525972946825880
Cantrell-Ramanujan (2004) Approximation	P_{CR}	5.525873040176910
Classical P Integral (64-interval Simpson)	$P_{Simpson}$	5.525873040177380
Classical P Integral (8-interval A&S Rule 18)	P_{Rule18}	5.525875337787820
Gauss-Kummer Series Integral	P_{GK}	5.525873040177380
Guillera Quadratic Process	P_{G2}	5.525873040177370
Guillera Quartic Process	P_{G4}	5.525873040177380
Ekwall (1973) Estimator	P_{Ek}	5.526127005789490
Complete Elliptic Integral of the First Kind	K	1.804461621553970
Complete Elliptic Integral of the Second Kind	E	1.381468260044340
Adlaj CEI of the First Kind	K_{adlaj}	1.804461621553970
Adlaj CEI of the Second Kind	E_{adlaj}	1.381468260044340
Adlaj Perimeter	P_{adlaj}	5.525873040177380

Table 1.1

We may abstract key target fiducial values as:-

P_{target}	22.1034921607095
ε_{target}	0.661437827766148
h_{target}	0.020408163265306
K_{target}	1.80446162155397
E_{target}	1.38146826004434

Candidate formulae for perimeter will be measured against these numbers.

<u>Functional Definitions</u>

Firstly, we may define our habitual quality metric, Percentage Specific Defect, in the usual manner:-

$$SDF(x, y) = 100 \times \left(\frac{x - y}{x} \right)$$

Equation 1.3

It should be noted that because this is a *percentage* discrepancy it underestimates the *fractional* relative accuracy by two orders of magnitude.

We also need to define the Base-ten Logarithm in the following terms:-

$$log10(x) = \frac{\ln(x)}{\ln(10)}$$

Equation 1.4

using which we may define the Number of Clear Decimal Positions (the more the better) as:-

$$dPlace(x) = if\left(x = 0, \text{"exact"}, floor\left(-log10(abs(x))\right)\right)$$

Equation 1.5

Abs() is the Mathcad Prime intrinsic function that effectively renders a positive of a negative number, used here to pre-empt attempts to find the logarithm of a negative number.

Mathcad Prime does not offer elliptic integrals as intrinsic functions, but it does offer definite integrals, which judiciously applied, and with sufficiently large bounds may be exploited to furnish $K(\varepsilon^2)$ and $E(\varepsilon)$ (Reference 2). Accordingly, the Elliptic Integral of the First Kind, $K(\varepsilon^2)$ can be given by:-

$$K(\varepsilon^2) = \int_0^{\pi/2} \frac{1}{\sqrt{1 - \varepsilon^2 . \sin(\theta)^2}} . d\theta$$

Equation 1.6

for which, with a 4 by 3 ellipse, $K(\varepsilon^2) = 1.80446162155397$ for a Percentage Specific Defect SDF(K_{target},$K(\varepsilon^2)$) = 0.00000000000062. This is fifteen-figure precision, or in dPlace terms +13.

With regard to the Elliptic Integral of the Second Kind, $E(\varepsilon)$, two integral formulae immediately suggest themselves:-

$$E(\varepsilon) = \int_0^{\pi/2} \sqrt{1 - \varepsilon^2 . \sin(\theta)^2} . d\theta$$

Equation 1.7

and:-

$$E1(\varepsilon) = \int_0^1 \frac{\sqrt{1 - \varepsilon^2 t^2}}{\sqrt{1 - t^2}} . dt$$

Equation 1.8

The Mathcad Prime 3.1 computed value using Equation 1.7 is $E(\varepsilon) = 1.38146826004434$ whilst the value using Equation 1.8 is 1.38146826004432. Since the figures differ in the fourteenth decimal position the relevant $SDF(E(\varepsilon),E1(\varepsilon))$ computes as 0.000000000001414 so that dPlace is $+11$.

Special Parameters

For the sake of brevity and later simplifications it is convenient to define several ancillary functions of the Semi-Axes a and b. These we will designate with selected lower-case Greek letters. In addition, we will define the Guillera Iteration Intermediates q_1 and q_2 in like terms.

$$\eta = \frac{b}{a} = \sqrt{1 - \varepsilon^2} = 0.75$$

Equation 1.9

$$\chi = \frac{b^2}{a^2} = \left(\frac{b}{a}\right)^2 = \eta^2 = 1 - \varepsilon^2 = 0.5625$$

Equation 1.10

$$\varepsilon = \sqrt{1 - \chi} = \sqrt{1 - \eta^2} = 0.661437827766148$$

Equation 1.11

$$\psi = \sqrt{\frac{b}{a}} = \sqrt[4]{1 - \varepsilon^2} = 0.866025403784439$$

Equation 1.12

$$\varepsilon = \sqrt{1 - \psi^4}$$

Equation 1.13

$$\lambda = 1 + \sqrt{\frac{b}{a}} = 1 + \psi = 1.86602540378444$$

Equation 1.14

$$\omega = \frac{1}{\chi} = \frac{a^2}{b^2} = \frac{16}{9} = 1.77777777777778$$

Equation 1.15

$$\zeta = \sqrt{ab} = 3.46410161513775$$

Equation 1.16

$$q_1 = \frac{1 - \sqrt{\frac{b}{a}}}{1 + \sqrt{\frac{b}{a}}} = \frac{1 - \psi}{1 + \psi} = 0.071796769724491$$

Equation 1.17

$$q_2 = \frac{1 - \sqrt[4]{1 - q_1{}^4}}{1 + \sqrt[4]{1 - q_1{}^4}} = \frac{1 - \gamma}{1 + \gamma} = 0.000003321508765$$

Equation 1.18

$$\gamma = \sqrt[4]{1 - q_1{}^4} = 0.999993357004535$$

Equation 1.19

$$\xi = \sqrt[4]{1 - ((a - b).(-2.\sqrt{ab} + a + b))^4} + 1 = 1 + \gamma = 1.99999335700454$$

Equation 1.20

$$\xi^5 = \left\{ \left[1 - ((a - b).(-2.\sqrt{ab} + a + b))^4 \right]^{\frac{1}{4}} + 1 \right\}^5 = 31.9994685638932$$

Equation 1.21

The Legendre Relation

The Legendre Relation is an important theorem that relates the Complete Elliptic Integrals E(), K() and the transcendental constant π known as The Ludolphine Constant, which has a value of about 3.14159265358979

Allowing that:-

$$t = h^2$$

Equation 1.22

and that:-

$$t_o = t^2 \qquad\qquad t_p = 1 - t^2$$

Equation 1.23a   **Equation 1.23b**

then:-

$$E_o K_p + E_p K_o - E_o K_p = \frac{\pi}{2}$$

Equation 1.24

or in an elaborated form of The Legendre Relation:-

$$E(t^2).K(1 - t^2) + E(1 - t^2).K(t^2) - E(t^2).K(1 - t^2) = \frac{\pi}{2}$$

Equation 1.25

Comparing the LHS of Equation 1.25 against $\pi/2$ for the a=4 b=3 ellipse within Mathcad Prime 3.1, the discrepancy SDF(LHS,π/2) was found to be 0.000108951797786. Accordingly, with the user-defined functions specified, the Legendre Relation scored some five-figure accuracy, demonstrating the operational value both of it and the antecedent methods.

The Adlaj Process for High-Precision Complete Elliptic Integral Computation

This process due to Semjon F Adlaj[3] of The Institution of the Russian Academy of Sciences at Moscow relies upon the theorem:-

$$\eta^2 + \varepsilon^2 = \left(\frac{b}{a}\right)^2 + \sqrt{1 - \left(\frac{b^2}{a^2}\right)}^{\,2} = 1$$

Equation 1.26

along with the facts that the Arithmetic-Geometric Means (AGMs) M() and N() can be employed to determine the Complete Elliptic Integrals K() and E().

η is the Prolation of the Ellipse (i.e. b/a). If η is unity, the ellipse is a circle and if η is zero it is a straight line. ε is as aforenoted the Eccentricity of the Ellipse.

The Determination of the
Complete Elliptic Integral of the First Kind, K

Firstly, we define the starting values of the iterates x_i and y_i as:-

$$x_0 = 1 \qquad\qquad y_0 = \eta$$

Equation 1.27a   **Equation 1.27b**

Then the first five successive iterates are:-

$$x_1 = \frac{x_0 + y_0}{2} \qquad y_1 = \sqrt{x_0 y_0} \qquad x_1 - y_1 = 0.09$$

$$x_2 = \frac{x_1 + y_1}{2} \qquad y_2 = \sqrt{x_1 y_1} \qquad x_2 - y_2 = 0.00001157$$

$$x_3 = \frac{x_2 + y_2}{2} \qquad y_3 = \sqrt{x_2 y_2} \qquad x_3 - y_3 = 1.821 \times 10^{-11}$$

$$x_4 = \frac{x_3 + y_3}{2} \qquad y_4 = \sqrt{x_3 y_3} \qquad x_4 - y_4 = 1.11 \times 10^{-16}$$

$$x_5 = \frac{x_4 + y_4}{2} \qquad y_5 = \sqrt{x_4 y_4} \qquad x_5 - y_5 = 1.11 \times 10^{-16}$$

Equations 1.28

The Adlaj Process converges quadratically and by the fifth iteration it is clear that the limits of system accuracy have been reached and any further iteration will hunt, spoiling the outcome.

In formal terms, the chosen convergent (y_4) is the Arithmetic-Geometric Mean, AGM(x,y) or M(x,y), in particular M(1, η) and since the first argument is unity this symbol is sometimes abridged to M(η).

So it is now possible to declare K as:-

$$K_{adlaj} = \frac{\pi}{2 y_4}$$

Equation 1.29

The assessed value of K_{adlaj} is 1.80446162155397 and SDF(K_{target},K_{adlaj}) is 8.6×10^{-14} which clearly demonstrates the efficacy of the method within the system limits of accuracy.

The Implicit Determination of
The Complete Elliptic Integral of the Second Kind, E

The Second phase of the Adlaj Process involves a determination of the Second Complete Elliptic Integral, but we will not examine that explicitly; rather proceed directly to the estimation of the relevant Ellipse Perimeter, P_{adlaj}.

Formally, we shall identify the value of a second species of arithmetic mean, N(1,η^2).

These initial values apply:-

$$u_0 = 1 \qquad\qquad v_0 = \eta^2$$
Equation 1.30a \qquad\qquad **Equation 1.30b**

$$w_0 = 0 \qquad\qquad z_0 = \sqrt{(u_0 - w_0) * (v_0 - w_0)}$$
Equation 1.30c \qquad\qquad **Equation 1.30d**

$$u_0 - v_0 = 0.4375$$
Equation 1.30e

The first four of these iterates are:-

$$u_1 = \frac{(u_0 + v_0)}{2} \qquad\qquad v_1 = w_0 + z_0$$

$$w_1 = w_0 - z_0 \qquad\qquad z_1 = \sqrt{(u_1 - w_1)*(v_1 - w_1)}$$

$$u_1 - v_1 = 0.03125$$

$$u_2 = \frac{(u_1 + v_1)}{2} \qquad\qquad v_2 = w_1 + z_1$$

$$w_2 = w_1 - z_1 \qquad\qquad z_2 = \sqrt{(u_2 - w_2)*(v_2 - w_2)}$$

$$u_2 - v_2 = 0.0000805434$$

$$u_3 = \frac{(u_2 + v_2)}{2} \qquad\qquad v_3 = w_2 + z_2$$

$$w_3 = w_2 - z_2 \qquad\qquad z_3 = \sqrt{(u_3 - w_3)*(v_3 - w_3)}$$

$$u_3 - v_3 = 0.000000000267526$$

$$u_4 = \frac{(u_3 + v_3)}{2} \qquad\qquad v_4 = w_3 + z_3$$

$$w_4 = w_3 - z_3 \qquad\qquad z_4 = \sqrt{(u_4 - w_4)*(v_4 - w_4)}$$

$$u_4 - v_4 = 0$$

Of course, the final tolerance factor u_4-v_4 is not exactly zero: It is some quantity too small for representation by Mathcad Prime 3.1. Like the AGM iteration for $M(\eta)$, this later iteration for $N(\eta^2)$ converges quadratically, approximately doubling the number of significant digits at each iteration.

It is now possible to move directly to Ellipse Perimeter, P_{adlaj}, as:-

$$P_{adlaj} = 4a.\frac{\pi}{2}.\frac{u_4}{y_4} = 2\pi a \frac{u_4}{y_4} = 22.1034921607095$$

Equation 1.31

The Percentage Specific Defect against the fiducial perimeter established by applying a 64-interval Simpson's Rule summation to the quarter-sector is given by $SDF(P_{target}, P_{adlaj}) = 0$. So the four-step Adlaj Process is "exact" within the capacity of my machine.

MacLaurin Series[4]

There are a number of "closed form" equations for the perimeter of an ellipse proposed for the approximation of that geometrical lineament. None are of course truly closed form, because most are truncations of one or another series expansion for one or more elliptic integrals, typically the MacLaurin Series. Those that are not series truncations are often inspired guesses, or purely empirical forms.

The MacLaurin Series for the Perimeter of the Ellipse, P_{mac}, was published by Colin MacLaurin in 1742AD and is the earliest precise estimator of the ellipse perimeter. It is given by:-

$$P_{mac} = 2\pi a \sum_{i=0}^{n} \left[\frac{-1}{2i-1} \left(\frac{(2i)!}{(2^i \cdot i!)^2} \right)^2 \cdot \varepsilon^{2i} \right] = 22.1034921607095$$
Equation 1.32

With my chosen a = 4, e = 0.661437827766148, and n = 35, the discrepancy SDF(P_{target},P_{mac}) was -0.000000000000048.

Abbott's Perimeter[5]

Abbott's Formula is:-

$$P_{abbott} = 4\sqrt{ab}\left(2E(c) - K(c)\right)$$
Equation 1.33

where:-

$$c = \frac{(a-b)^2}{4ab}$$
Equation 1.34

Michon's Approximation[9]

Michon's Approximation is given by:-

$$P_{michon} = 2\pi a \left(1 - \frac{1}{4}\varepsilon^2 - \frac{3}{64}\varepsilon^4 - \frac{5}{256}\varepsilon^6 - \frac{175}{16384}\varepsilon^8 - \frac{441}{65536}\varepsilon^{10} \right)$$
Equation 1.35

An interesting feature of Michon's Approximation is that it is a relentlessly subtractive series. This is possible because $2\pi a$ is equivalent to $2\pi r$ which latter is the circumference of the ellipse's excriptive circle, and that circumference is by definition longer than the ellipse perimeter.

Using the data a=4, b=3 and so ε=0.661437827766148, P_{michon} was computed to be 22.104700938046 and SDF(P_{target},P_{michon}) = -0.005468716561529. This is only five and a half thousanths of a percent but is nevertheless unimpressive when contrasted with some of the other methods we review.

The Elliptic Integral of the First Kind in its Relation to the Arithmetic Geometric Mean[6]

The following relation may sometimes be convenient:-

$$M(\eta) = \frac{\pi}{4} \cdot \frac{(1+\eta)}{K\left(\frac{1-\eta}{1+\eta}\right)}$$

Equation 1.36

Whilst not of central interest, I briefly tested this expression by setting M(η) to the output y_4 from the Adlaj Process. y_4 was 0.870506919089893 whilst the RHS of Equation 1.35 was 0.842233630773771. Therefore SDF(AGM$_{xy}$,RHS) was 3.2479 (%), and accordingly other methods should be sought to define K() when using small computers.

Estimators drawn from Wolfram Mathworld[7]

Wolfram Mathworld[TM] offers a number of resources for the estimation of ellipse perimeters.

The first class of these involve the Complete Gamma Function Γ(x), because certain estimators implicate non-integral binomial expansions. Unlike the 32-bit Mathcad versions offered by Mathsoft[TM] twenty years ago, modern PTC-sponsored free or cheap offerings do not offer Gamma functions as intrinsics, and as an amateur I find it difficult to justify £510 sterling for a full licence, especially if I have to pay the same every year. Therefore of course I had to program both gamma functions and binomial expansion functions myself.

Two flavors of Gamma Integral were tried with tmax set to 100:-

$$\Gamma_1(z) = \int_0^1 \left[ln\left(\frac{1}{t}\right) \right]^{z-1} . dt$$

Equation 1.37

and:-

$$\Gamma_2(z) = \int_0^{tmax} t^{z-1} . e^{-t} . dt$$

Equation 1.38

Obviously, the two outcomes should be identical for any given argument z. Experience proved very different when $\Gamma_1(z)$ and $\Gamma_2(z)$ were compared for:-

$$\Gamma\left(\frac{1}{2}\right) \equiv \sqrt{\pi}$$

Equation 1.39

The Ludolphine Constant π *is* a Mathcad Prime intrinsic and using that scratchpad program the following results arose:-

$$SDF\left(\Gamma_1\left(\frac{1}{2}\right), \sqrt{\pi}\right) = 0.000414149192793$$
Equation 1.40

$$SDF\left(\Gamma_2\left(\frac{1}{2}\right), \sqrt{\pi}\right) = 0.000000000051764$$
Equation 1.41

So the Γ_2 conformation gives seven orders of magnitude more precision, at least with the given argument.

I then turned my attention to composing a function for non-integral binomial expansions by employing the usual formula:-

$$BINOM(x, y) = \frac{\Gamma_2(x + 1)}{\Gamma_2(y + 1) \cdot \Gamma_2(x - y + 1)}$$
Equation 1.42

and incorporated this function into a suggested estimator:-

$$P_{wolf1} = \pi(a + b) \sum_{i=1}^{n} \left(BINOM\left(\frac{1}{2}, i\right)\right)^2 \cdot h^{2i}$$
Equation 1.43

where h is the Gauss-Kummer Parameter with a value for a=4 and b=3, and therefore h of 0.020408163265306. I strongly suspect that Equation 1.43 is itself defective but notwithstanding that I could not get Mathcad Prime 3.1 to work with the function defined in Equation 1.42.

Therefore, I turned my attention to elaborated expansions of Equation 1.43.
Firstly, I tried this:-

$$P_{wolf1a} = \pi(a + b)\left(1 + \frac{1}{4} \cdot h^2 + \frac{1}{64} \cdot h^4 + \frac{1}{2564} \cdot h^6\right)$$
Equation 1.44

This object yields a perimeter estimate P_{wolf1a} of 21.9934384253037 which implies a discrepancy of $SDF(P_{target}, P_{wolf1a}) = 0.497902026546732$ (i.e. nigh on a half a percent error).
Now, if we eliminate the double-powers of h by this simple expedient:-

$$P_{wolf1a} = \pi(a + b)\left(1 + \frac{1}{4} \cdot h + \frac{1}{64} \cdot h^2 + \frac{1}{2564} \cdot h^3\right)$$
Equation 1.45

the estimate P_{wolf1a} immediately improves to 22.1034921548298, with a percentage discrepancy of $SDF(P_{target}, P_{wolf1a}) = 0.000000026600767$, an improvement much better than seven orders of magnitude. Therefore, we are entitled to think that Equation 1.44 is a typesetting error.

Developments of the MacLaurin Series as a Perimeter Estimator[7]

It is possible to formulate straight-forward high-precision series estimators based upon the use of powers of the Gauss-Kummer Parameter h rather than the Eccentricity ε. But we must treat our source, whether paper publications or webpages, with our usual careful circumspection. Forgive the pleonasm. But trust *and verify*.

In this context, the MacLaurin term fragment function BCk(k) is a useful component, defined below:-

$$BCk(k) = \frac{(2k)!}{k!.(2k-k)!} = \frac{(2k)!}{k!.k!} = \frac{\prod_{l=k+1}^{2k} l}{k!}$$

Equation 1.46

Therefore, the development BChk() is defined by:-

$$BChk(k) = BCk(k) \times \frac{(-1)^{k+1}}{2^{2k}.(2k-1)}$$

Equation 1.47

whilst its square is:-

$$BChksqr(k) = \left(BCk(k) \times \frac{(-1)^{k+1}}{2^{2k}.(2k-1)}\right)^2$$

Equation 1.48

For the purposes of summation allow that n = 6.
Then:-

$$P_{wolf1hsqr} = \pi(a+b)\left(1 + \sum_{i=1}^{n} BChk(i)^2 . h^{2i}\right)$$

Equation 1.49

and:-

$$P_{wolf1h} = \pi(a+b)\left(1 + \sum_{i=1}^{n} BChk(i)^2 . h^{i}\right)$$

Equation 1.50

Equation 1.49 is the general form of the series in Equation 1.44 which we have already seen to be defective in admitting only even squares of h. Thus a discrepancy of SDF(P_{target},$P_{wolf1hsqr}$) = 0.497902026546732 (i.e. nigh on a half a percent error) is identical to that of Equation 1.44.

In contrast:-

$$P_{wolf1h} = \pi(a+b)\left(1 + \sum_{i=1}^{n} BChk(i)^2 . h^i\right) = \pi(a+b)\left(1 + \sum_{i=1}^{n} BChksqr(i) . h^i\right)$$

Equation 1.51

Both forms of Equation 1.51 give P_{wolf1h} as 22.1034921607095 so that SDF(P_{target},P_{wolf1h}) = 0.000000000000016, amongst the best perimeter formulae or processes we have seen and possibly the most economical if judicious cancellation of factorials is done.

By implicitly re-setting n to six we may examine the following three expansions of Equation 1.51:-

$$P_{wolf1h} = \pi(a+b)$$
$$\cdot\left(1 + \frac{h}{4} + \frac{h^2}{64} + \frac{h^3}{256} + \frac{h^4}{655.36} + \frac{h^{5.}}{1337.4693877551} + \frac{h^6}{2377.72335600907}\right)$$

Equation 1.52a

$$P_{wolf1h} = \pi(a+b) \cdot \left(1 + \frac{h}{4} + \frac{h^2}{64} + \frac{h^3}{256} + \frac{h^4}{655.36} + \frac{h^5}{1337.469} + \frac{h^6}{2377.723}\right)$$

Equation 1.52b

$$P_{wolf1h} = \pi(a+b) \cdot \left(1 + \frac{h}{4} + \frac{h^2}{64} + \frac{h^3}{256} + \frac{h^4}{655.36} + \frac{h^5}{1337.469} + \frac{h^6}{2378}\right)$$

Equation 1.52c

Equation 1.52a specifies the computed term denominators to full fifteen-figure precision, whilst 1.52b rounds them to three decimal places, and additionally Equation 1.52c rounds the last denominator to the integer 2378. Interestingly each equation, computed by Mathcad, yields the same perimeter of P_{wolf1h} = 22.1034921607095 so that SDF(P_{target},P_{wolf1h}) = 0.000000000000016. This is of course suggestive of further economies that may be investigated.

Approaches involving Ivory's Identity and its Ramanujan Approximation

The development of Equations 1.46 through 1.52 exemplify treatments of the series resulting from Ivory's Identity, which was first promulgated by James Ivory[10] in 1796AD. He offered his work explicitly as a contribution to the rectification of the ellipse, to borrow the concise and elegant phraseology of his age.

Ivory's Identity may be stated as:-

$$\frac{1}{\pi}\int_0^\pi \sqrt{1 + 2\sqrt{x}\cos(2\theta) + x}\,.d\theta = \sum_{j=0}^{n}\left\{\frac{1}{2j-1}\cdot\frac{1}{4^j}\cdot\binom{2j}{j}\right\}^2\cdot x^j$$

Equation 1.53

where:-

$$\binom{n}{k} = \frac{n!}{k!\,(n-k)!}$$

Equation 1.54

The RHS of Equation 1.53 is of course identical to the summations in Equations 1.50 and 1.51 in which the place of x is taken by the ellipse Gauss-Kummer parameter h.

Before we move forward it will be profitable, both for economy and numerical stability, to recollect that the binomial coefficient of Equation 1.54 can be simplified in these terms:-

$$\binom{2j}{k} = \frac{(2j)!}{j!\,(2j-j)!} = \frac{(2j)!}{(j!)^2} = \frac{\prod_{l=j+1}^{2j}l}{j!}$$

Equation 1.55

In terms of Ellipse Perimeter P_{Ivld} the identity is:-

$$\frac{1}{\pi}\int_0^\pi \sqrt{1 + 2\sqrt{h}\cos(2\theta) + h}\,.d\theta = \sum_{j=0}^{n}\left\{\frac{1}{2j-1}\cdot\frac{1}{4^j}\cdot\binom{2j}{j}\right\}^2\cdot h^j$$

Equation 1.56

from which it follows that:-

$$P_{Ivld} = \pi(a+b)\frac{1}{\pi}\int_0^\pi \sqrt{1 + 2\sqrt{h}\cos(2\theta) + h}\,.d\theta$$

Equation 1.57

or alternatively:-

$$P_{Ivld} = \pi(a+b)\cdot\sum_{j=0}^{n}\left\{\frac{1}{2j-1}\cdot\frac{1}{4^j}\cdot\binom{2j}{j}\right\}^2\cdot h^j$$

Equation 1.58

Many (or perhaps infinite) techniques of numerical integration may be employed to solve Equation 1.57 with more or less speed and precision, whilst appropriate summation algorithms may be employed to address numerically-stable versions of Equation 1.58.

Using Prime 3.1, the intrinsic definite integral form of Equation 1.57 (with π cancelled) gave SDF(P_{target},P_{Ivld}) = -0.000000000000016 for the a=4 b=3 test ellipse.

By the way, in an otherwise clear and well-written paper[11], Mark Vallarino of the University of Costa Rica offers the following as a formula for perimeter:-

$$P = 4 \int_0^{\pi/2} (a^2 . \sin^2 \theta + b^2 . \cos^2 \theta) . d\theta$$

Equation 1.59

This is incorrect.
But I was able empirically to establish that:-

$$a^2 + b^2 = \frac{4}{\pi} \int_0^{\pi/2} (a^2 . \sin^2 \theta + b^2 . \cos^2 \theta) . d\theta$$

Equation 1.60

The LHS is obviously the square of the chord between the quadrant axial extrema which is likely to be a systematic function of both ellipse area and perimeter, though I have not done the work definitively to state this. I have however managed to establish an analytical proof of Equation 1.60 in Appendix A.

Ramanujan's Second Approximation of Ellipse Perimeter

Technically, Ramanujan's Approximation is a truncation of the Ivory-Bessel Series given in Equation 1.58.
It is:-

$$P_{rama} = \pi \left\{ (a + b) + \frac{3 \cdot (a - b)^2}{10(a + b) + \sqrt{a^2 + 14 . a . b + b^2}} \right\}$$

Equation 1.61

It looks like there is scope for expressing this as a relation of Pythagorean Means, but again I have not done the work to move beyond speculation.

For the test ellipse of semi-axes a=4 b=3 the Ramanujan II outcomes were P_{rama} = 22.1034921607077 whilst SDF(P_{target},P_{rama}) = 0.000000000008326.

Therefore, Ramanujan II is not accurate enough for our purposes, given that we wish to drive towards the natural limit of our equipment, somewhere in the vicinity of h^8.

Sykora's 2005 Approximation

This is given in Stanislav Sykora's Internet paper as:-

$$P_{Syk} = 4 \cdot \frac{\pi ab + (a-b)^2}{a+b} - \frac{1}{2} \cdot \frac{ab}{a+b} \cdot \frac{(a-b)^2}{\pi ab + (a+b)^2}$$

Equation 1.62

Sykora states that the maximum error of this equation is 78.5 ppm.

My own EXCEL tests yielded an Average Perimeter %SDF Discrepancy $\mu(\delta)$ of 0.002349062862223 represented by Mean Clear Decimal Places of Approximation, dPlace, of 2.314960629921260: Something between four and five figure accuracy.

The Cantrell-Ramanujan Approximation

In his 2005 contribution, Stanislav Sykora's presents this as one of his final and best estimators of the ellipse perimeter (and implicitly the Complete Elliptic Integral of the Second Kind):-

$$P_{CR} = \pi(a+b) \left[1 + \frac{3h}{10 + \sqrt{4-3h}} + \left(\frac{4}{\pi} - \frac{14}{11} \right) \cdot h^{12} \right]$$

Equation 1.63

This is visibly a modification of Ramanujan II, itself said to be a truncation of MacLaurin's Theorem as configured for ellipse perimeter.

Sykora states that DW Cantrell contributed this modification in 2004, and alleges that this is the most accurate available approximator yielding a maximum error of 14.5 ppm.

My own EXCEL tests yielded an Average Perimeter %SDF Discrepancy, $\mu(\delta)$, of 0.000134932436860 represented by Mean Clear Decimal Places of Approximation, dPlace, of 7.409448818897640: This is in line with Sykora's claim for this estimator.

PART TWO

THE GUILLERA QUARTICALLY-CONVERGENT PROCESS
FOR THE RECTIFICATION OF THE ELLIPSE

In a brilliant paper[12] of 2017AD, Jesús Guillera of the University of Zaragoza offered two high-precision iterative processes for the length of ellipse perimeter, one quadratically-convergent and the other quartically.

Crudely speaking, a quartic process should quadruple its significant digits at each iteration. Accordingly, the sanguine researcher may hope for a good enough solution after a couple of passes.

We have reviewed several quadratic processes and will not detain ourselves with another. Instead we will examine the quartic in detail and move forward to define an adequate "closed-form" approximant by consolidating its terms.

Guillera's Quartic Process is defined in the following way:-

$$u_0 = \sqrt[4]{1 - \frac{1}{\omega}} \qquad\qquad v_0 = 2\omega \qquad\qquad w_0 = 1$$

Equation 2.1a $\qquad\qquad$ **Equation 2.1b** $\qquad\qquad$ **Equation 2.1c**

whilst each iterate is defined by:-

$$u_{i+1} = \frac{1 - \sqrt[4]{1 - u_i^4}}{1 + \sqrt[4]{1 - u_i^4}}$$

Equation 2.2

$$v_{i+1} = 4 \times \frac{v_i}{(1 + u_{i+1})^2}$$

Equation 2.3

$$w_{i+1} = w_i \cdot (1 + u_{i+1})^2 + \frac{1}{2} \cdot v_{i+1} \cdot \frac{u_{i+1}}{(1 + u_{i+1})} \cdot (1 - u_{i+1}^4)$$

Equation 2.4

Trials showed that no more than three iterations were sufficient to attain fifteen-figure accuracy. Table 2.1 shows the EXCEL® numerical outcomes using the 4-by-3 ellipse test data.

In this context, w_2 is wholly adequate and the Guillera Quartic Ellipse Perimeter, P_{G4}, may be specified as:-

$$P_{G4} = \frac{2\pi b^2}{a} \cdot w_2$$

Equation 2.5

a 4
b 3
ω = a2/b2 1.777777778
Limiting Iterate (w_3) 1.5635022238653280
Quartic Guillera Ellipse Perimeter 22.103492160709500

i	u_i	v_i	w_i	Δw
0	0.8132882808484893	3.5555555555555560	1.0000000000000000	0.5634096089885595
1	0.0717967697244491	12.3806250935782000	1.5634096089886500	0.0000926296464683
2	0.0000003321508765	49.5221713971390000	1.5635022386532800	0.0000000000000000
3	0.0000000000000000	198.0868585884560000	1.5635022386532800	0.0000000000000000
4	0.0000000000000000	792.3547423538220000	1.5635022386532800	0.0000000000000000

Table 2.1
Tabulation of the Guillera Quartic Iterates for the Test Ellipse

The Guillera Quartic Process Restated

As part of our condensation campaign we need to elaborate the various fragments and undertake systematic substitutions and simplifications, whilst keeping control of the numerical integrities of our program.

Such an iterative process should be terminated when the Difference Δw:-

$$\Delta w = w_i - w_{i-1}$$
Equation 2.6

becomes less than the (finite) tolerance acceptable: I terminated the process when the Difference became zero by inspection.

Phase 0

$$\varepsilon = \sqrt{1-\chi}$$

$$u_0 = \sqrt[4]{1-\chi} \qquad\qquad v_0 = 2\omega \qquad\qquad\qquad w_0 = 1$$

$$u_1 = \frac{1-\sqrt[4]{1-u_0^4}}{1+\sqrt[4]{1-u_0^4}} \qquad\qquad v_1 = 4 \times \frac{v_0}{(1+u_1)^2}$$

$$w_1 = w_0 \cdot (1+u_1)^2 + \frac{1}{2} \cdot v_1 \cdot \frac{u_1}{(1+u_1)} \cdot (1-u_1{}^4)$$

$$u_2 = \frac{1-\sqrt[4]{1-u_1^4}}{1+\sqrt[4]{1-u_1^4}} \qquad\qquad v_2 = 4 \times \frac{v_1}{(1+u_2)^2}$$

$$w_2 = w_1 \cdot (1+u_2)^2 + \frac{1}{2} \cdot v_2 \cdot \frac{u_2}{(1+u_2)} \cdot (1-u_2{}^4)$$

Phase 1

Using the previous iterative process as a fiducial source, set w_{true} to w_3 (we will use w_{true} to assess the accuracy of some approximate formulae):-

$$w_{true} = w_3$$
Equation 2.7

and define:-

$$q_1 = \frac{1-\psi^2}{1+\psi^2}$$
Equation 2.8

also:-

$$u_1 = q_1$$
Equation 2.9

$$v_1 = 8 \cdot \frac{\omega}{(1 + q_1)^2}$$
Equation 2.10

$$w_1 = 1 \cdot (1 + q_1)^2 + 4 \cdot \frac{\omega}{(1 + q_1)^2} \cdot \frac{q_1}{(1 + q_1)} \cdot (1 - q_1^4) = 1.5634096089886$$
Equation 2.11

With regard to this first "closed-form" approximation, Equation 2.11, SDF(w_{true},w_1) = 0.005924498372516, a discrepancy of some plus six thousands of one percent. This shows that one iteration of Guillera's Quartic Process places us within one ten-thousandth part of truth.

Phase 2

Noting that:-

$$q_2 = \frac{1 - \gamma}{1 + \gamma} = 0.000003321508765$$
Equation 2.12

and that accordingly:-

$$(1 + q_2)^2 \approx 1 + 2q_1$$
Equation 2.13

We may define:-

$$s = 1 + q_1$$
Equation 2.14

and:-

$$t = 1 + q_2$$
Equation 2.15

Set:-

$$u_2 = q_2$$
Equation 2.16

and:-

$$v_2 = 4 \cdot \frac{8 \cdot \frac{\omega}{(1+q_1)^2}}{(1+q_2)^2}$$

Equation 2.17

or alternatively:-

$$v_2 = 32 \cdot \omega \cdot \frac{(1+q_2)^2}{(1+q_1)^2}$$

Equation 2.18

Though Equations 2.17 and 2.18 are algebraically-equivalent, *using Equation 2.18 greatly increases numerical error when employing Mathcad Prime 3.1.*

We are now able to compose the second approximation of iterate w, w_2, in terms of the "closed form" structure:-

$$w_2 = (st)^2 \cdot \left\{ 1 + \left[\frac{2}{\eta s^2} \right]^2 \left[\frac{q_1(1-q_1^4)}{s} + \frac{4q_2}{t^5} \right] \right\}$$

Equation 2.19

$$\text{SDF}(w_{true}, w_2) = -0.000000000000028$$

Equation 2.19 may be rearranged as:-

$$w_2 = s^2 t^2 \cdot \left\{ 1 + \left[\frac{4}{s^4} \frac{a^2}{b^2} \right]^2 \left[\frac{q_1(1-q_1^4)}{s} + \frac{4q_2}{t^5} \right] \right\}$$

Equation 2.20

$$\text{SDF}(w_{true}, w_2) = -0.000000000000014$$

or alternatively:-

$$w_2 = s^2 t^2 \cdot \left\{ 1 + \left[\frac{2t}{\eta s} \right]^2 \left[\frac{(s-1)(1-q_1^4)}{s} + \frac{4(t-1)}{t^5} \right] \right\}$$

Equation 2.21

$$\text{SDF}(w_{true}, w_2) = -0.000000000000014$$

Though Equations 2.19, 2.20 and 2.21 are all algebraically-equivalent it can be seen that the preference of either 2.20 or 2.21 to 2.19 will halve numerical error.

We are now able to declare our fifteen-figure Composite Guillera Approximator (CGA) of the Ellipse Perimeter Length P_{G4} as:-

$$P_{G4} = \frac{2\pi b^2}{a} \cdot w_2 = \frac{2\pi b^2}{a} \cdot s^2 t^2 \cdot \left\{ 1 + \left[\frac{4}{s^4} \frac{a^2}{b^2} \right]^2 \left[\frac{q_1(1-q_1^4)}{s} + \frac{4q_2}{t^5} \right] \right\}$$

Equation 2.22

For the a=4, b=3 trial ellipse, the value of P_{G4} is 22.1034921607095, and SDF(P_{target},P_{G4}) = -0.000000000000064. The requisite accuracy is attained, but judicious simplifications of the algebraic structure may enable the further reduction of error.

With regard to Equation 2.21 other forms are of course possible and have differing effects upon numerical fidelity. For example the subtly different:-

$$w_{2approx} = s^2 t^2 \; | \; \left[\frac{2t}{\eta s}\right]^2 \left[\frac{(s-1)(1-q_1^4)}{s} + \frac{4(t-1)}{t^5}\right]$$

Equation 2.23
SDF(w_{true},$w_{2approx}$) = 0

Equation 2.23 is of course theoretically corrupt. Whether it would be good for other ellipse dimensions, or indeed for other computational systems, is a matter that can only be resolved empirically.

In my opinion, only the following further precision forms of the CGA are interesting: Equation 2.24 because of its versatile employment of substitutionary functions of a and b; Equation 2.25 because of its extreme numerical precision.

$$P_{G4} = 2\pi b \eta (st)^2 . \left\{1 + \left[\frac{2}{\eta s^2}\right]^2 \left[\gamma^4 \left(1 - \frac{1}{s}\right) + \frac{4(t-1)}{t^5}\right]\right\}$$

Equation 2.24
SDF(P_{target},P_{G4}) = -0.000000000000032
P_{G4} = 22.1034921607095
dPlace(SDF(P_{target},P_{G4})) = 13

$$P_{G4} = \frac{2\pi b^2}{a} \cdot \left\{[(1+q_2)(1+q_1)]^2 + \left[\frac{2}{\eta} \cdot \frac{1+q_2}{1+q_1}\right]^2 \cdot \left[\frac{q_1(1-q_1^4)}{1+q_1} + \frac{4q_2}{(1+q_2)^5}\right]\right\}$$

Equation 2.25
SDF(P_{target},P_{G4}) = -0.000000000000008
P_{G4} = 22.1034921607095
dPlace(SDF(P_{target},P_{G4})) = 13

A Debased Form of the CGA for Rough Estimation

The following rough form of the Composite Guillera Approximator may be employed for vicinitation or rough estimates:-

$$P_{G4ap} = 8\pi b \sqrt{1 - \varepsilon^2} \cdot \left[\frac{1}{\lambda^2} + \frac{\lambda^2}{4(1-\varepsilon^2)}\left(1 - \frac{\lambda}{2}\right)\right]$$

Equation 2.26
SDF(P_{target},P_{G4}) = 0.005219762870144
P_{G4} = 22.1023384108327
dPlace(SDF(P_{target},P_{G4})) = 2

This is nine orders of magnitude worse than Ramanujan II.

Curvature is defined as the second differential of a line (that is, the rate of change of the rate of change of its course). In Cartesian terms we can write:-

$$\kappa = \frac{d^2 y}{dx^2}$$
Equation 3.1

Technically, Equation 3.1 is an analytic infinitesimal. To define local curvature $\kappa step_i$ in grosser terms, we can use:-

$$\kappa step_i = \frac{\Delta y_i}{\Delta x_i}$$
Equation 3.2

The general conformation and the geometrical character of the variables are illustrated for three contiguous intervals in Figure 3.1:-

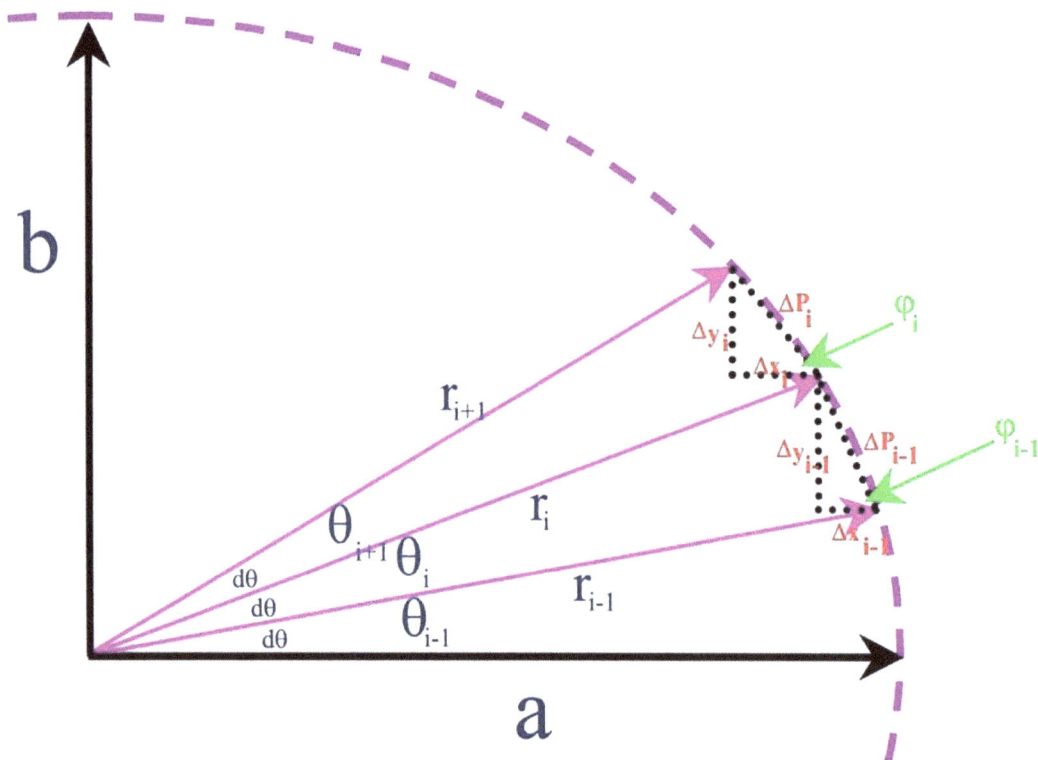

Figure 3.1
Geometry of Finite Curvature Variables

(The tilt of the semi-major axis a is an optical illusion).

As $\kappa step_i$ decreases in size we expect it to converge to the value of analytic κ (the calculus of limits).

Both κ and $\kappa step_i$ are capable of being positive or negative depending upon the direction of curvature.

We need to examine Curvature in at least two contexts:-

(a)   The bending or veering of the ellipse perimeter at a point in its course.

(b)   The bending distortion of ellipses of differing aspect ratios $\eta=b/a$ with regard to the ideal case, as resulting from different methods of perimeter computation. This is one possible metric of method quality.

Working in Cartesian terms we firstly need to define unpaced Change in Ellipse Perimeter Arc Length, ds_i, and corresponding Change in Angular Perimeter Course Azimuth, φ_i. It helps to think kinetically of a moving point scanning out the ellipse perimeter as it transits along its path, just like a planet orbiting the Sun.

(x_i, y_i) is the position of a specific (incremental) point along this course. The relevant equations in both analytical "exact" and finite difference forms are:-

$$ds_i = \sqrt{(x_{i+1} - x_i)^2 + (y_{i+1} - y_i)^2} \approx \sqrt{(\Delta x_{i+1} - \Delta x_i)^2 + (\Delta y_{i+1} - \Delta y_i)^2}$$

Equation 3.3

$$\varphi_i = -tan^{-1}\left(\frac{dy_i}{dx_i}\right) \approx -tan^{-1}\left(\frac{\Delta y_i}{\Delta x_i}\right)$$

Equation 3.4

$$\kappa = \frac{d\varphi}{ds} = \frac{1}{a^2 b^2} \cdot \left(\frac{x^2}{a^4} + \frac{y^2}{b^4}\right)^{-\frac{3}{2}} \approx \frac{\varphi_i - \varphi_{i-1}}{\frac{1}{2} \cdot ds_i + \frac{1}{2} \cdot ds_{i-1}} \approx \frac{2.\Delta\varphi}{\Delta s} = \kappa step_i$$

Equation 3.5

Equation 3.5 is the key curvature equation and the finite form of it should be understood as pertaining to the condition that all x and y are positive, as of course they shall be if we confine our computations to the north-east sector of the ellipse which naturally donates information to the other three quarters by symmetry.

When creating spreadsheet elaborations, for example with EXCEL®, you may need to "backstep" the $\kappa step_i$ entries to compare with κ formula results. The mean Percentage Specific Defect of κ against $\kappa backstep_i$ for a=1, b=0.75 was computed by EXCEL as -0.00676709). This fact may be surprising and unsought but is a natural corollary of the aforenoted convergence of approximation to the true local value. In my EXCEL® tabulations I tended to use the rather course interval size $\Delta x = (\pi/2)/64 = \pi/128$, but as Δx tends towards (but never achieves) zero, so the local approximations tend toward but never agree the local analytic value.

If required, Equation 3.5 can be expanded in this way:-

$$\kappa = \frac{\varphi_i - \varphi_{i-1}}{\frac{1}{2} \cdot ds_i + \frac{1}{2} \cdot ds_{i-1}}$$

$$= \frac{-tan^{-1}\left(\frac{dy_i}{dx_i}\right) - -tan^{-1}\left(\frac{dy_{i-1}}{dx_{i-1}}\right)}{\frac{1}{2} \cdot \sqrt{dx_i{}^2 + dy_i{}^2} + \frac{1}{2} \cdot \sqrt{dx_{i-1}{}^2 + dy_{i-1}{}^2}}$$

$$= \frac{tan^{-1}\left(\frac{dy_{i-1}}{dx_{i-1}}\right) - tan^{-1}\left(\frac{dy_i}{dx_i}\right)}{\frac{1}{2} \cdot \sqrt{dx_i{}^2 + dy_i{}^2} + \frac{1}{2} \cdot \sqrt{dx_{i-1}{}^2 + dy_{i-1}{}^2}}$$

$$= \frac{tan^{-1}\left(\frac{y_i - y_{i-1}}{x_i - x_{i-1}}\right) - tan^{-1}\left(\frac{y_{i+1} - y_i}{x_{i+1} - x_i}\right)}{\frac{1}{2} \cdot \sqrt{(x_{i+1} - x_i)^2 + (y_{i+1} - y_i)^2} + \frac{1}{2} \cdot \sqrt{(x_i - x_{i-1})^2 + (y_i - y_{i-1})^2}}$$

Equation 3.6

As aforementioned, Equation 3.6 may appropriately be re-arranged if all points are positive.

The idiom of Mathcad Express Prime 3.1 is notoriously eccentric. It is seldom workable to specify x_i as a function of x_{i+1} or x_{i-1}, or indeed vice versa. Rather we have to wrangle strange (incomplete) circumlocutions such as:-

$$i = 0 \dots n \qquad\qquad j = 0 \dots n - 1 \qquad\qquad k = 1 \dots n - 1$$
Equation 3.7a \qquad **Equation 3.7b** \qquad **Equation 3.7c**

$$x_i = r_i . cos(\theta_i) \qquad\qquad \varphi_j = tan^{-1}\left(\frac{dy_j}{dx_j}\right)$$
Equation 3.8a \qquad\qquad **Equation 3.8b**

and:-

$$\kappa step_k = \frac{\varphi_k - \varphi_{k-1}}{\frac{1}{2} . ds_k + \frac{1}{2} . ds_{k-1}}$$
Equation 3.9

Singularity Control

It is of course the case that normal (and economical) ellipse computations are defined only for the north-east quadrant in which no curvature singularities occur.

As part of my drafting of the complete ellipse, however, I defined 256 Cartesian points on the interval $0 \le \theta_i \le 2\pi$ (for graphical reasons one of the end-points is redundant). A complex singularity arises at the westernmost $i=128$; $\theta_i = \pi$ point. In EXCEL 2010® terms, this complex singularity manifests as a large number (5000-10000) rather than a VALUE? error or as an imaginary number.

We can analyse the situation in MATHCAD® Prime 3.1 (for example) in these following terms:-

$$\kappa_\pi = \frac{1}{a^2b^2}\cdot\left(\frac{x^2}{a^4}+\frac{y^2}{b^4}\right)^{-\frac{3}{2}} = \frac{1}{a^2b^2}\cdot\left(\frac{-a^2}{a^4}+\frac{0^2}{b^4}\right)^{-\frac{3}{2}} = u + vi$$

Equation 3.10

where in context i is the Square Root of Minus One, u is the Real Part of a Complex Number and v is the Imaginary Part of that Complex Number.

Interestingly, the powers enable us to re-state this singular curvature as the real variable:-

$$\kappa_\pi = \frac{1}{a^2b^2}\cdot\left(\frac{x^2}{a^4}+\frac{y^2}{b^4}\right)^{-\frac{3}{2}} = \frac{1}{a^2b^2}\cdot\left(\frac{a^2}{a^4}+\frac{0^2}{b^4}\right)^{-\frac{3}{2}} = v$$

Equation 3.11

From this we can move forward with:-

$$\kappa_\pi = \frac{1}{a^2b^2}\cdot\left(\frac{1}{a^2}\right)^{-\frac{3}{2}}$$

$$= \frac{1}{a^2b^2}\cdot\left(\sqrt{\frac{1}{a^2}}\right)^{-3}$$

$$= \frac{1}{a^2b^2}\cdot\left(\frac{1}{a}\right)^{-3}$$

$$= \frac{1}{a^2b^2}\cdot\left(\frac{1}{a^3}\right)^{-1}$$

$$= \frac{a}{b^2}$$

Equation 3.12

The expression a/b² can then be plugged in to the offending cell.

Perimeter Veering and Curvature Relations

A conceptual principal that should always be borne in mind, and which is reflected in the experimental outcomes, is that empirical Curvature is a function of a and b as Aspect Ratio *only* and that accordingly differences between methods reflects experimental aberrations rather than properties intrinsic to the approximators.

In terms of our Clause (b) objective above we shall briefly review the behaviors of our chosen "closed-form" approximants relative to the brute-force Simpson integrals which we adopt as fiducial. Such distortion is tabulated in terms of:-

$$\kappa(\kappa step_i) = f(\eta) = -log_{10}(|\kappa step_i|)$$
Equation 3.13

where $\eta = b/a$, a measure of Ellipse Aspect Ratio or, if you prefer, Prolation.

Appendix B tabulates the Empirical Curvature Metric, $\kappa step_i$, for:- CHECK SERIES 10- GaussKummer; Consolidated Guillera Process; Sykora Approximation; and the Cantrell-Ramanujan Approximation.

Figure 3.2 is a plot of that data that clearly demonstrates the near-identity of Curvature-Aspect Ratio relations for these four approximators. We should naturally hope that this were the case.

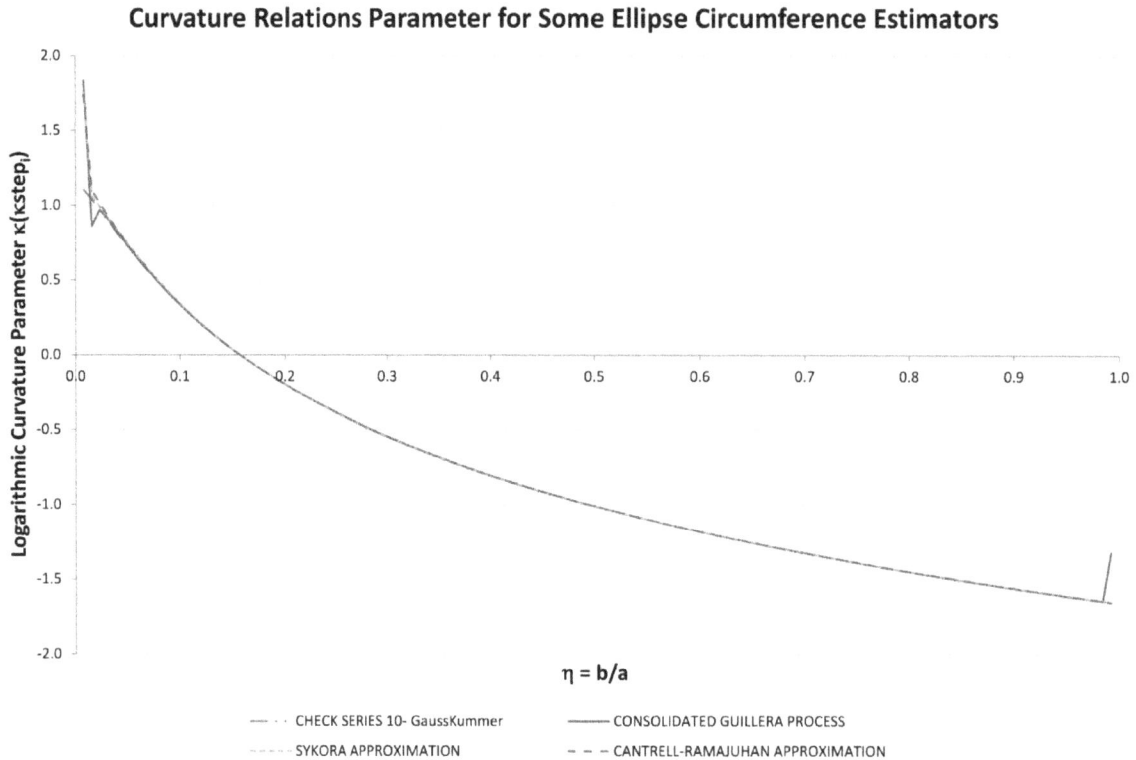

Curvature Relations Parameter for Some Ellipse Circumference Estimators

Figure 3.2
The Extreme Similarity of
Ellipse Perimeter Approximation Methods
as regards their Curvature-Prolation Relations

I am unable to explain why the Consolidated Guillera Process is notably deviant at the extremes of prolation.

Table 3.1 summarises certain averages arising from the approximator's behavior in the interval $0 \le \eta \le 1$. Notably, all three analytically-based precision methods, the formulaic Analytical Curvature Series, The Simpson Integration Fiducial list, and the Gauss-Kummer Series, all yield an Average Local Curvature $\mu(\kappa_j)$ between 0.8 and 0.9. The synthetic single-formula approximators are all very different to that, showing a variety of much higher values. Therefore we are entitled to think that $\mu(\kappa_j)$ is a sensitive discriminant of estimator type.

In this sense, Sykora and Cantrell-Ramanujan are very closely related as they have a $\mu(\kappa_j) \approx 1.175$.

Metric	Symbol	Value
Semi-Major Axis	a	1
Semi-Minor Axis	b	0.75
Gauss-Kummer Parameter, h	h	0.020408163

	Analytical Curvature Series	Fiducial Series (64- Simpson)	CHECK SERIES 10- GaussKummer	Consolidated Guillera Process	Sykora Approximation	Cantrell- Ramanujan Approximation
Average Local Curvature $\mu(\kappa_j)$	0.87751931	0.86154580	0.80729798	1.22111669	1.17980565	1.17138331
Average Perimeter %SDF Discrepancy $\mu(\delta)$			0.00063487	0.00019137	-0.00234906	0.00013493
Mean Clear Decimal Places of Approximation dPlace			9.75781250	10.40944882	2.31496063	7.40944882

η Series		
Lower Bound		0
Upper Bound		1
Number of Intervals		128
Interval Width		0.0078125

Table 3.1
Process Precision Averages

The Perimeter Fidelity Metric

The Perimeter Fidelity Metric, $\Phi(P(a,b))$, is a measure of how closely a given Approximation Process, defined in terms of its Perimeter Estimates $P(a,b)$, on the interval $0 \leq \eta \leq 1$, emulates the set of Fiducial Perimeters, P_{simp}.

The Perimeter Fidelity Metric is defined by:-

$$\Phi\big(P(a,b)\big) = -log_{10}(|SDF\%|) = -log_{10}\left\{\left|100\,\frac{P_{simp} - P(a,b)}{P_{simp}}\right|\right\}$$

Equation 4.1

Table 4.1 in Appendix C tabulates $\Phi(P(a,b))$ at $1/128$ intervals for the 129 η points for the four trial estimators CHECK SERIES 10- GaussKummer, Consolidated Guillera Process, Sykora Approximation and Cantrell-Ramanujan Approximation.

It is possible to read the entries in the table as nothing other than the precision of agreement in terms of the number of mantissa place-zeroes: So the bigger the number the better. For example, if the Φ entry is 1.72447514 there is about 0.02% difference in the fiducial and the process perimeter values; if the entry is 13.84798154 then the relevant difference is minus 1.419118×10^{-14} or in decimal terms -0.00000000000001419118.

This data is graphically represented by the plot Figure 4.1:-

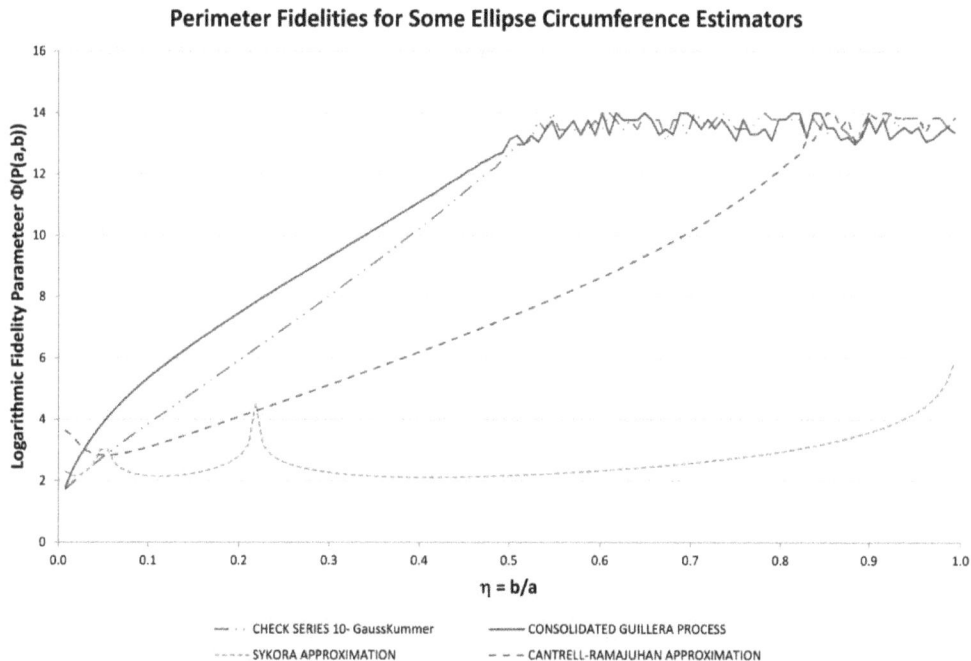

Perimeter Fidelities for Some Ellipse Circumference Estimators

Figure 4.1

The first thing that is clear from Figure 4.1 is that none of the approximators reach their full precision until a Prolation, η, of 0.5 is reached. Therefore, it is important to identify methods that can operate at large distortions of the circle.

The Second thing is that the two best estimators, the 10-term Gauss-Kummer Series and the Consolidated Guillera Process, reach a limit of $\Phi \approx 14$, and then return very erratic, noisy values around that limit. This is due to the limitations of the hardware and software I employ in these disquisitions.

Let L be the Limiting Number of (Denary) Digits expressible by a given machine system. I use a (binary) 64-bit Hewlett-Packard Pavilion 500400 microcomputer with appropriate 64-bit software. Let this Register Length, B, be 64 bits less (say) 2 bits for the signs of the mantissa and the exponent; and less a further six bits for the exponent value. Thus the modified Available Bits, B_{av}, is 56.

Then by the principle of inter-base equivalence:-

$$2^B = 10^L$$
Equation 4.2

Therefore, L is computable by:-

$$L = \frac{B_{av} \cdot log_{10}(2)}{log_{10}(10)} = B_{av} \cdot log_{10}(2) \approx 56 \times 0.30103 = 16.85768$$
Equation 4.3

Accordingly, the system can compute a number to something approximating sixteen denary digits (not allowing for the management of complex numbers, etc.).

This purely technical finitude accounts for the limit to which we may calculate our estimates.

On the other hand, I cannot explain the weird spikes in Sykora precision at η = 0.054688 and η = 0.21875

But Sykora's Formula is consistently dismissibly imprecise everywhere.

Whist Gauss-Kummer precision rises quasi-linearly from nothing to its 14-figure limit, the Consolidated Guillera Process gains precision more quickly, and is visibly superior to the other processes illustrated.

Timing Experiments

The (compiled) QB64 program RATEGAIN.exe, presented as source in Appendix D, was employed to time the five alternative ellipse perimeter processes[17].

The program depends upon the Composite and Carcase principle of algorithm isolation developed in TARIFF.bas and is illustrated for the fiducial 64-interval Simpson Integration by the segment HOLDERsimpson reproduced in Code Segment 4.1.

In the context of RATING it is important to separate Predicate Routines that are executed once or a limited number of times from the main solution algorithm which in our application is executed NETA=128 times.

```
      Sub HOLDERsimpson
      (KN,NETA,NS,ETAIW,DIW,LNC(),NA(),AA(),EPS(),EE2(),PAB(),TF())
' A Subroutine to Perform and Time a Process Elaboration
'     Arguments:
'         KN        The Process Serial Number
'         NETA      The Number of Ellipses
'         NS        The Number of Simpson Integration Terms
'         ETAIW     The Ellipse Aspect Ratio Interval width
'         DIW       The Simpson Integration Interval Width
'         LNC()     The Array of Required Cycles
'         NA()      The Array of Simpson Term Alternators
'         AA()      The Array of Simpson Integration Angles
'         EPS()     The Array of Ellipse Eccentricities
'         EE2()     The Array of Complete Elliptic Integrals of Second Kind
'         PAB()     The Array of Ellipse Perimeter Lengths
'         TF()      The Array of Process and Carcass Timings
'
' Do Process Predicate
      TS=TIMER(.001)
         Call ALTERNATOR( NS, NA())
         Call ANGLEFILL( NS, DIW, AA())
      TE=TIMER(.001)
      TF(0,KN)=TE-TS
' Do Process Elaboration Cycles
      TS=TIMER(.001)
      FOR LI=O TO LNC(KN)
         Call SIMPSON(NETA,ETAIW,DIW,NS,EPS(),AA(),NA(),EE2(),PAB())
      NEXT LI
      TE=TIMER(.001)
      TF(1,KN)=TE-TS
' Do Carcase Predicate
      TS=TIMER(.001)
      TE=TIMER(.001)
      TF(2,KN)=TE-TS
' Do Carcase Elaboration Cycles
      TS=TIMER(.001)
      FOR LI=O TO LNC(KN)
      NEXT LI
      TE=TIMER(.001)
      TF(3,KN)=TE-TS
' Calculate Summative Timings
      TF(4,KN)=TF(0,KN)+TF(1,KN)
      TF(5,KN)=TF(2,KN)+TF(3,KN)
      TF(6,KN)=TF(4,KN)-TF(5,KN)
      TF(7,KN)=TF(6,KN)/LNC(KN)
' Terminate
      End Sub
```

Code Segment 4.1
Subroutine HOLDERsimpson
Exemplary of the Composite and Carcase Timing Principle

The principal outputs of timing are TF(6,KN), the Total Time in Seconds for all NETA gainful elaborations; and TF(7,KN) the mean Mill Time for *each* Process Calculation.

By way of initial checking, the output PAB perimeter length values of RATINGs.exe were checked against the EXCEL®-indicated perimeters and, after excluding the very many exact zero differences, the mean discrepancy was found to be -9.9099×10^{-15} length units (where the Major Axis a is unity).

Table 4.1 presents a Raw 128 Rating Report as output as RATING.OUT by RATINGs.exe. I have used EXCEL to separate the data for clarity.

ELLIPSE PERIMETER LENGTH PROCESSES TIMING REPORT
All Times in Seconds

	S64	GAK	CGP	SYK	CAR	
Operation Number		1	2	3	4	5
Operation Code	S64	GAK	CGP	SYK	CAR	
Operation Description	64- Simpson Integration	10- Gauss Kummer Series	Consolidated Guillera Process	The Sykora Formula	The Cantrell-Ramanujan Formula	
Number of Cycles	128	128	128	128	128	128
Number of Process Cycles	128	128	128	128	128	128
Process Predicate Time	0	0	1.000000003841706D-03	0	0	0
Process Elaboration Time	0.398	0.028	1.699999999982538D-02	7.999999943538570D-03	3.0000000424916D-03	0
Carcase Predicate Time	0	0	0	0	0	0
Carcase Elaboration Time	0	0	0	0	0	0
Total Process Time	0.398	0.028	1.800000000366708D-02	7.999999943538570D-03	3.0000000424916D-03	0
Total Carcase Time	0	0	0	0	0	0
Nominal Mill Time	0.398	0.028	1.800000000366708D-02	7.999999943538570D-03	3.0000000424916D-03	0
Nominal Process Time	3.109375000008186D-03	2.18749999877218D-04	1.406250002864910D-04	6.249999955889951D-05	2.343750003313656D-05	
Overall Routine Duration =	0.455					

Table 4.1

MEAN NOMINAL PROCESS TIME
All Times in Seconds

	1	2	3	4	5
Operation Number					
Operation Code	S64	GAK	CGP	SYK	CAR
Operation Description	64- Simpson Integration	10- Gauss Kummer Series	Consolidated Guillera Proce	The Sykora Formula	The Cantrell-Ramanujan Formula
Number of Process Cycles	128.000000000000	128.000000000000	128.000000000000	128.000000000000	128.000000000000
Experiment 1	0.002367187499999	0.000156250000032	0.000078124999959	0.000039062500036	0.000023437499976
Experiment 2	0.002281250000010	0.000156249999918	0.000078125000073	0.000039062499923	0.000023437500090
Experiment 3	0.002210937499967	0.000164062499948	0.000078125000073	0.000039062499923	0.000023437500090
Experiment 4	0.002210937499967	0.000156250000032	0.000078124999959	0.000039062500036	0.000023437499976
Experiment 5	0.002210937499967	0.000156250000032	0.000085937499989	0.000039062499923	0.000023437500090
Experiment 6	0.002203125000051	0.000156250000032	0.000078125000073	0.000039062499923	0.000023437500090
Experiment 7	0.002187499999991	0.000156250000032	0.000085937499989	0.000039062500036	0.000023437499976
Experiment 8	0.002195312500021	0.000156249999918	0.000078125000073	0.000039062499923	0.000023437500090
Operation Code	S64	GAK	CGP	SYK	CAR
Mean Nominal Process Time $\mu(NPT)$	0.002233398437497	0.000157226562493	0.000080078125024	0.000039062499965	0.000023437500047
Population SD $\sigma(NPT)$	0.000057201893148	0.000002583741498	0.000003382911714	0.000000000000055	0.000000000000055
Coefficient of Variation, $C_v=\sigma(NPT)/\mu(NPT)$	0.025612041357213	0.016433237849477	0.042245141386885	0.000000001400471	0.000000002354774
Operation Code	S64	GAK	CGP	SYK	CAR
Log_{10} Mean Nominal Process Time $-Log_{10}(\mu(NPT))$	2.651033792033720	3.803474080627300	4.096486104128650	4.408239965696810	4.630088714052670
Experiment 1 ($-log_{10}MT$)	2.625767341	3.806179974	4.10720997	4.408239965	4.630088715
Experiment 2 ($-log_{10}MT$)	2.641827118	3.806179974	4.107209969	4.408239966	4.630088713
Experiment 3 ($-log_{10}MT$)	2.655423534	3.784990675	4.107209969	4.408239966	4.630088713
Experiment 4 ($-log_{10}MT$)	2.655423534	3.806179974	4.10720997	4.408239965	4.630088715
Experiment 5 ($-log_{10}MT$)	2.655423534	3.806179974	4.065817285	4.408239966	4.630088713
Experiment 6 ($-log_{10}MT$)	2.656960861	3.806179974	4.107209969	4.408239966	4.630088713
Experiment 7 ($-log_{10}MT$)	2.660051938	3.806179974	4.065817285	4.408239965	4.630088715
Experiment 8 ($-log_{10}MT$)	2.65850365	3.806179974	4.107209969	4.408239966	4.630088713
Maximum Experimental Mill Time $-Log_{10}(MT_{PRO})$	2.660051938	3.806179974	4.10720997	4.408239966	4.630088715
Minimum Experimental Mill Time $-Log_{10}(MT_{PRO})$	2.625767341	3.784990675	4.065817285	4.408239965	4.630088713
Mean Experimental Mill Time $-Log_{10}(MT_{PRO})$	2.651172689	3.803531312	4.096861798	4.408239966	4.630088714
Experiment 1 (MT/MT_{CAR})	101.0000001	6.666666675	3.333333335	1.66666667	1
Experiment 2 (MT/MT_{CAR})	97.33333296	6.666666638	3.333333324	1.666666657	1
Experiment 3 (MT/MT_{CAR})	94.33333297	6.999999971	3.333333324	1.666666657	1
Experiment 4 (MT/MT_{CAR})	94.33333343	6.666666675	3.333333335	1.66666667	1
Experiment 5 (MT/MT_{CAR})	94.33333297	6.666666642	3.666666652	1.666666657	1
Experiment 6 (MT/MT_{CAR})	93.99999964	6.666666642	3.333333324	1.666666657	1
Experiment 7 (MT/MT_{CAR})	93.33333343	6.666666675	3.66666667	1.66666667	1
Experiment 8 (MT/MT_{CAR})	93.66666631	6.666666638	3.333333324	1.666666657	1
Experiment 1 ($1+Log_{10}(MT/MT_{CAR})$) $=\rho_{PRO}$	3.004321374	1.823908741	1.522878745	1.22184875	1
Experiment 2 ($1+Log_{10}(MT/MT_{CAR})$) $=\rho_{PRO}$	2.988261595	1.823908739	1.522878744	1.221848747	1
Experiment 3 ($1+Log_{10}(MT/MT_{CAR})$) $=\rho_{PRO}$	2.974665179	1.845098038	1.522878744	1.221848747	1
Experiment 4 ($1+Log_{10}(MT/MT_{CAR})$) $=\rho_{PRO}$	2.974665181	1.823908741	1.522878745	1.22184875	1
Experiment 5 ($1+Log_{10}(MT/MT_{CAR})$) $=\rho_{PRO}$	2.974665179	1.823908739	1.564271429	1.221848747	1
Experiment 6 ($1+Log_{10}(MT/MT_{CAR})$) $=\rho_{PRO}$	2.973127852	1.823908739	1.522878744	1.221848747	1
Experiment 7 ($1+Log_{10}(MT/MT_{CAR})$) $=\rho_{PRO}$	2.970036777	1.823908741	1.564271431	1.22184875	1
Experiment 8 ($1+Log_{10}(MT/MT_{CAR})$) $=\rho_{PRO}$	2.971585064	1.823908739	1.522878744	1.221848747	1
Mean Relative Duration Metric, $\mu(\rho_{PRO})$	2.978916025	1.826557402	1.533226916	1.221848748	1

Table 4.2
Mean Nominal Process Time Table

It would of course be desirable to repeat several timing experiments both more reliably to identify a Mean Nominal Process Time $\mu(NPT)$ and to examine its dispersion.

Accordingly, I performed eight identical experiments using RATINGs and I present the raw and derivative results in Table 4.2.

As expected the expense of the fiducial Simpson routine is at $\mu(NPT) \approx 0.00223$ very great, but this drops to ≈ 0.000157 for Gauss-Kummer and ≈ 0.0003 for the three closed-form approximators. On the other hand, the Coefficients of Variation are relatively big for the Simpson and Gauss-Kummer routines and negligible for both Sykora and Cantrell-Ramanujan.

Obviously, the raw times tend to be very small, say 20 to 150 microseconds, and the management of the dater is facilitated by the taking of ratios and logarithms.

Perhaps the handiest of the metrics is the Ratio of Process Mill Time to Mill Time of the Swiftest Process (Cantrell-Ramanujan) defined by:-

$$R_{pro} = \frac{MT_{PRO}}{MT_{CAR}}$$
Equation 4.4

This is of course unity for the Cantrell-Ramanujan Approximator but we see instantly that the Simpson Routine is nearly a hundred times tardier than Cantrell-Ramanujan, whilst Gauss-Kummer is 6.6' more protracted and the CGP method is 3.33' times more protracted. If you are suspicious of the nice round figures I bet you are right to be: I think they are systemic artefacts, but are roughly in line with the facts.

Amongst the convenient logarithmatisations of Mill Time are:-

$$\lambda_{PRO} = -log_{10}(MT_{pro})$$
Equation 4.5

and:-

$$\rho_{PRO} = 1 + log_{10}\left(\frac{MT_{PRO}}{MT_{CAR}}\right)$$
Equation 4.6

ρ_{PRO} is another statistic with the obvious advantage that it has a unity value for the swiftest process, and is nearly three for the most protracted process, facilitating both illustrations and further computation.

Figure 4.2 is the Mill Time Averages Column Chart that shows the absolute Nominal Process Times for the five perimeter approximators. As you can see our new Consolidated Guillera Process equation holds an economical central place among the dramatically disparate methods.

Figure 4.3 controls the apparent disparities by plotting the Relative Duration Metric ρ_{PRO} of Equation 4.6.

Figure 4.4 shows the Coefficient of Variation defined by:-

$$C_V = \frac{\sigma(NPT)}{\mu(NPT)}$$
Equation 4.7

This notably interesting chart promotes the fact that for our new Consolidated Guillera Equation formula there is surprising instability in the results, more than twice that for the Gauss-Kummer Series and much more than for Simpson. Meanwhile, the Sykora and Cantrell-Ramanujan method are dead consistent.

Clearly, we are very much in a swings-and-roundabouts situation and as in life generally if we want precision we must sacrifice both economy and accuracy. Ours is a Fallen world.

Mill Time Averages for
Five Ellipse Perimeter Length Processes programmed in QB64

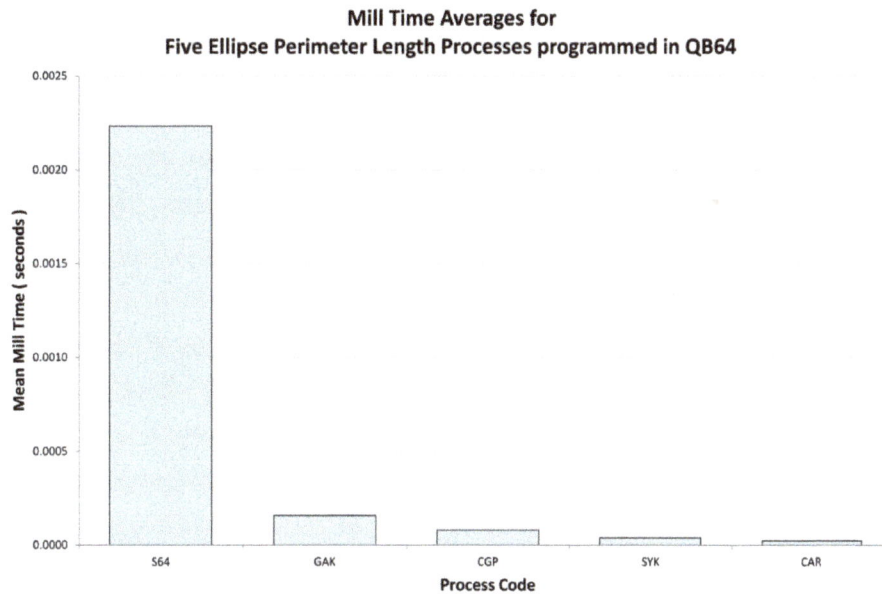

Figure 4.2
Mean Absolute Process Mill Time in Seconds

Mill Time Relative Duration Metric for
Five Ellipse Perimeter Length Processes programmed in QB64

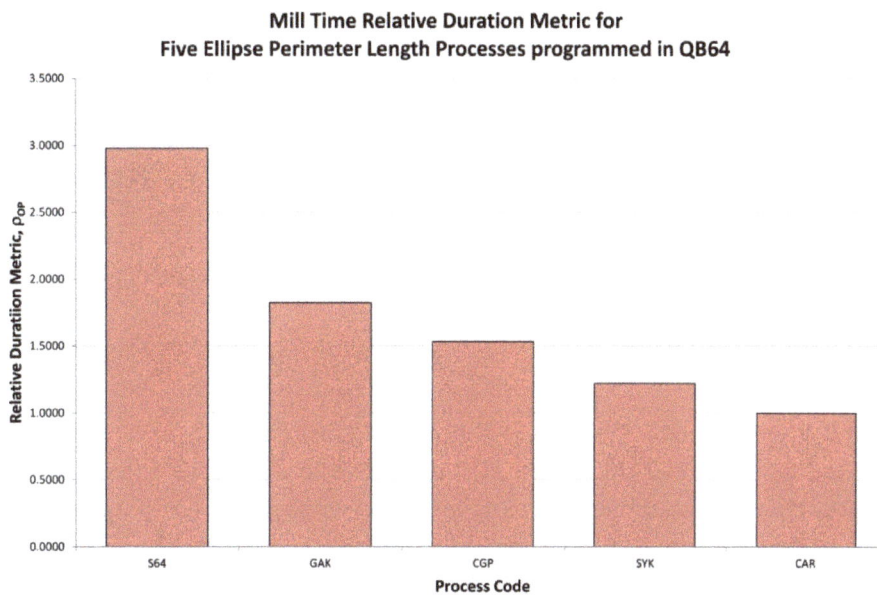

Figure 4.3
Relative Duration Metric ρ_{OP} for the Five Processes

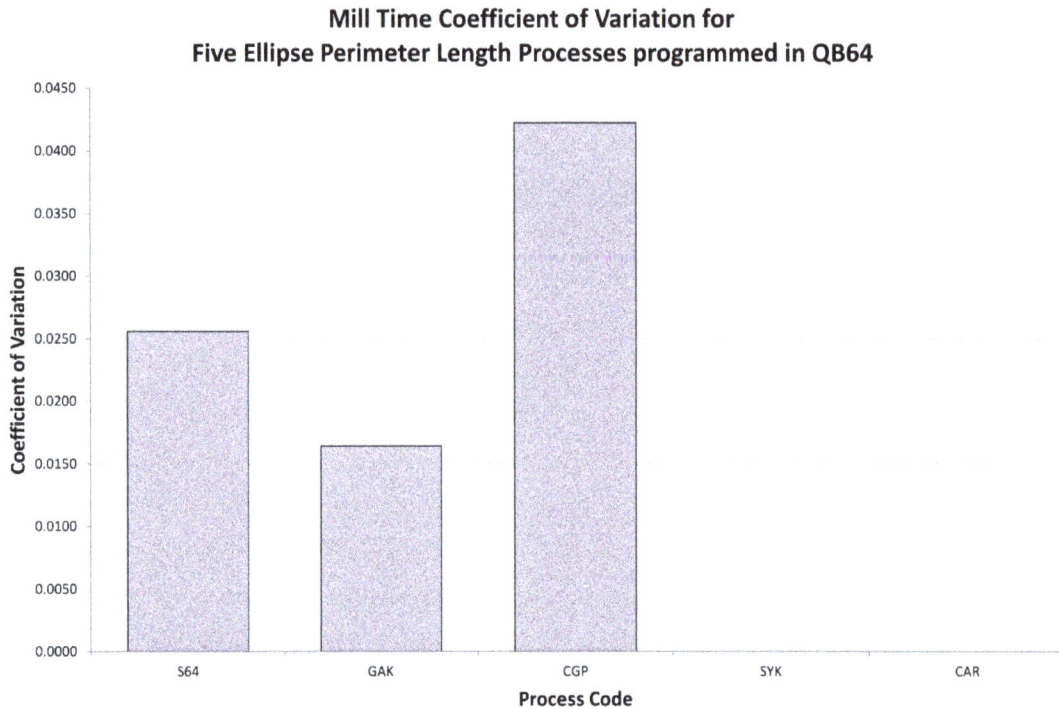

**Mill Time Coefficient of Variation for
Five Ellipse Perimeter Length Processes programmed in QB64**

**Figure 4.4
Coefficient of Variation in Mill Times
for the Five Processes**

The Variation Range of Process Lambda

Recall our definition of Process Lambda from Equation 4.5

Table 4.2 offers the Minimum, Maximum and Mean λ_{PRO} for the five processes and these are plotted, albeit obscurely, in Figure 4.5.

This is another way of looking at dispersion and this absolute treatment shows its true negligibility, even for very variable results such as those of the CGP method.

Relative Difference, Gain and Quality

To assess the quality of a given mathematical process we need to consider both its precision and its economy. In terms of the latter we are of course confining ourselves to questions of computational execution time, whatever the merits or otherwise of that approach.

Quality is a relative desideratum.

We measure the quality of a perimeter length estimation in comparison to a 64-interval Simpson Integration of the kernel of The Complete Elliptic Integral of the Second Kind. Whether this is a well-chosen fiducial is another matter that like Brother Jorge I shall leave to younger men.

As a first step, let us define the Discrepancy Difference, D_{PRO}, between the Ellipse Perimeter Lengths, PAB_{PRO}, in these terms:-

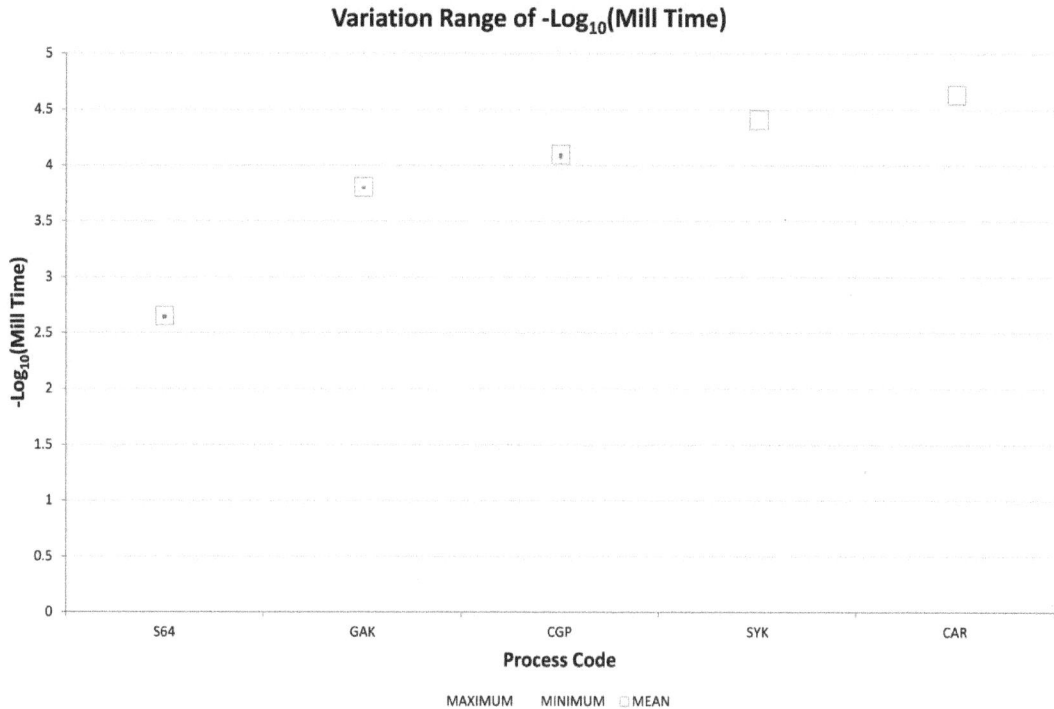

Variation Range of -Log₁₀(Mill Time)

MAXIMUM   MINIMUM   □MEAN

Figure 4.5
Absolute Variation Ranges of Mill Times
for the Five Processes

$$D_{PRO} = PAB_{S64} - PAB_{PRO}$$
Equation 4.8

where PAB_{S64} is the fiducial perimeter as yielded by the Simpsonian Fiducial Process and PAB_{PRO} is the relative Process under scrutiny. Clearly, D_{PRO} is zero for S64 and hopefully very small for anything else.

Gain, G, is the Number of Decimal Places gained upon completion of an estimate for a given Process at a given angular η-interval. It only really makes sense to define this parameter as an average for all the η-intervals in a finished execution, i.e. Mean Process Gain, $\mu(G_{PRO})$:-

$$\mu(G_{PRO}) = \frac{\sum_{i=0}^{n+1} G_{PRO}}{n+1}$$
Equation 4.9

where n is the Number of Intervals of η=b/a (called NETA in our computer programs).

Unless $\mu(G_{PRO})$ is $\approx 10^{-5}$, that is another extra five zeroes, then we feel we are wasting our time. Now this parameter can be negative so in order to establish the actual gain-of-zeroes it is useful to take the denary logarithm of the absolute value:-

$$\theta_{GPRO} = -log_{10}(|\mu(G_{PRO})|)$$
Equation 4.10

Correspondingly, for Time we may write:-

$$\theta_{TPRO} = -log_{10}(\mu(NPT_{PRO}))$$
Equation 4.11

Quality of course increases with the number of difference digits a Process contributes, and decreases with time expended. Therefore, the Quality of a Process may be defined as:-

$$Q_S(PRO) = \frac{\theta_{GPRO}}{\theta_{TPRO}} = \frac{-log_{10}(|\mu(G_{PRO})|)}{-log_{10}(\mu(NPT_{PRO}))} = \frac{log_{10}(|\mu(G_{PRO})|)}{log_{10}(\mu(NPT_{PRO}))}$$
Equation 4.12

Let me recapitulate, and maybe clarify our terms.

It is possible to assess the Quality, Q_{PRO}, of a given perimeter length estimator. Because quality is always a relative desideratum it is only meaningful to compute the metric by reference to the Fiducial Method, in our case the 64-interval Simpson Integration.

Firstly, we can establish the Raw Gain of precision, G, at each $\eta = b/a$ as:-

$$G_\eta = PAB_{FID} - PAB_{PRO}$$
Equation 4.13

where G_η is the Gain at the particular η value; PAB_{FID} is the relevant Ellipse Perimeter Length for the Fiducial Process; and PAB_{PRO} is the Perimeter Length according to the Process under examination.

It is quickly seen that since G_η for the Fiducial Process is inevitably zero, the Fiducial Gain is meaningless.

Since Quality increases with Gain and decreases with applied Mill Time it is natural to think of Quality as a velocity but, pedantically speaking, it has the dimensions of frequency (T^{-1}).

It is also intuitively appropriate to treat Gain as the Number of Further Significant Digits achieved by the Process under test, assessed in terms of (denary) mantissa zeroes in our experiments. So we may redefine Gain as:-

$$G_\eta = log_{10}(|(PAB_{FID} - PAB_{PRP})|)$$
Equation 4.14

It follows that a suitable metric of Process Quality, Q_{PRO}, is:-

$$Q_S(PRO) = \frac{-log_{10}(|\mu(G_{PRO})|)}{-log_{10}(\mu(NPT))} = \frac{log_{10}(|\mu(G_{PRO})|)}{log_{10}(\mu(NPT))}$$

Equation 4.15

If Q, $\mu(G_{PRO})$ and $\mu(NPT)$ are all positive.

μ denotes the Arithmetic Mean of the η-wise values; PRO denotes the Process Examined and NPT is the gross Nominal Process Time for the given Process and Experiment.

The subscript "S" denotes "Static" Quality defined for some particular Interval Size of η, which in turn is governed by the Number of Intervals, n.

It is reasonable to expect $0.5 < Q_S(PRO) < 1.5$, and the larger the $Q_S(PRO)$ the better.

I adapted the program RATINGs.bas as RATINGy.bas to include routines for computing and outputting gain and quality statistics.

Table 4.3 offers some highly provisional outcomes based upon the old RATINGn.bas PAB output as well as preliminary gain and quality results.

	Fiducial Series (64- Simpson)	CHECK SERIES 10- GaussKummer	Consolidated Guillera Process	Sykora Approximation	Cantrell -Ramanujan Approximation
Mean Gain (G_{PRO})	Not Applicable	2.52248E-05	7.59912E-06	-0.000116173	5.42546E-06
-Log$_{10}$(abs(μG_{PRO}))	Not Applicable	4.598171714	5.119236788	3.934893664	5.2655632
-Log$_{10}$(μ(NPT))	Not Applicable	3.803474081	4.096486104	4.408239966	4.630088714
Q$_S$(PRO)	Not Applicable	1.208939937	1.249665362	0.892622383	1.137248879

Table 4.3
Provisional Gain and Quality Indications

I should re-iterate that despite the nine places of decimals these figures are only approximate and those involving time especially unreliable. A suitable program of numerical experiments should be completed more accurately to establish the relevant metrics.

Notwithstanding that, it is clear that the Consolidated Guillera Process is qualitatively the best or rather optimal perimeter routine followed at some distance by the 10-term Gauss-Kummer Series. The Sykora Approximation is notably defective.

Figure 4.6 presents the Indicative Raw Mean Gain for the four processes whilst Figure 4.7 shows the Quality Metric $Q_S(PRO)$.

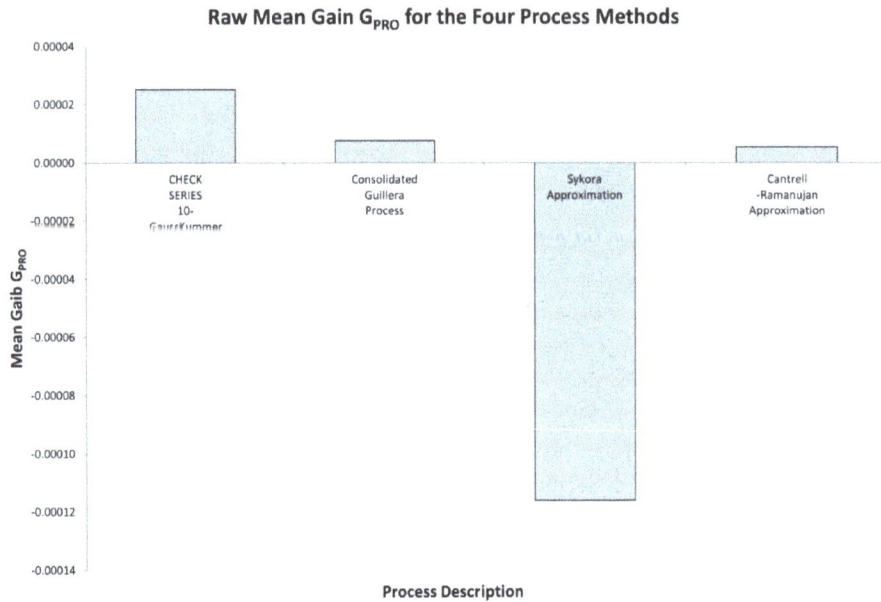

Figure 4.6
Indicative Raw Mean Gain
for the Four Comparative Processes

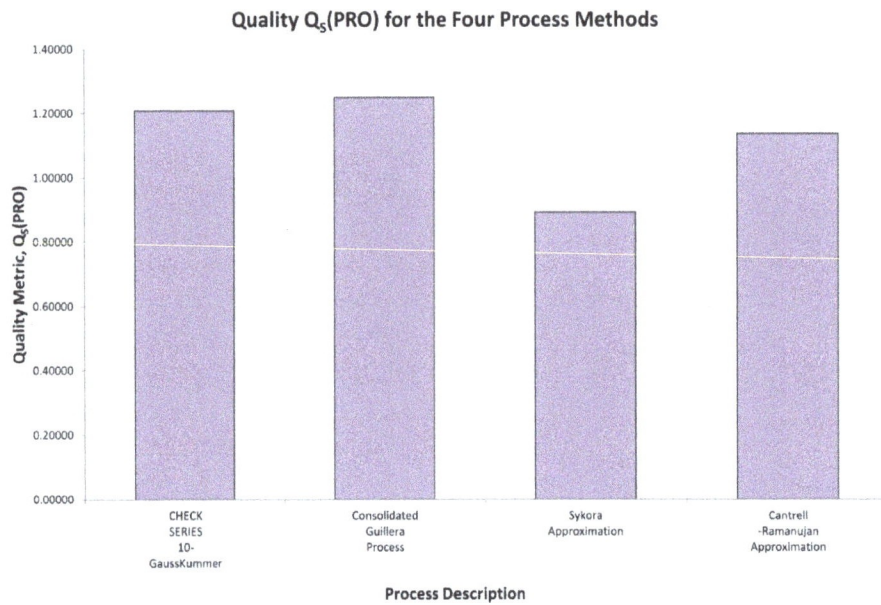

Figure 4.7
Indicative Quality Metric
for the Four Comparative Processes

Dynamic Quality

Dynamic Quality, $Q_D(PRO)$, varies with the Number of Ellipse Aspect Ratios, n_η.

The more the Number of Ellipse Aspect Ratios, which are equal to the number of straight-line segments with which the ellipse quadrant is defined, then the tighter the curvilinear definition of the quadrant's arc.

Static Quality, $Q_S(PRO)$, is merely the Dynamic Quality at a particular n_η, say $n_\eta = 128$.

It is important not to confuse n_η, the Number of Ellipse Aspect Ratios (NETA) treated; with the Rating Number of Process Cycles, LNC(KN): n_η (NETA) is intrinsic to ellipse definition whereas LNC(KN) is a rating repetition selected for a particular algorithm KN in order to mitigate computational celerity errors.

I ran my rating program RATEGAIN.exe at eight values of NETA in order to assess the quality of the four non-fiducial perimeter estimators as the ellipse circumference increased in resolution.

The chosen values of NETA were 8, 16, 32, 64, 128, 256, 512 and 1024.

Table 4.4 presents the behavior of $Q_D(PRO)$ with NETA.

Figure 4.8 shows the variation of Quality graphically.

The chart highlights some dramatic differences between the four processes. The 10-term Gauss-Kummer Series estimator (GAK) shown with a blue line and lozenge markers worsens exponentially until $n_\eta = 128$ and then collapses abruptly. I speculate that if more elaborative terms were superadded (perhaps 100 rather than 10) then precision of rectification would be maintained as resolution increased. The Sykora Formula, represented by a green line with triangular markers, starts badly at $n_\eta = 8$ but improves greatly by $n_\eta = 16$ and then settles to a steady increase as n_η rises to 1024. It is always about a third of an order of magnitude below the two best methods and this is in line with our expectations of it.

The Consolidated Guillera Process (CGP) and the Cantrell-Ramanujan Approximation (CAR) are by far the best of our "closed-form" perimeter approximators. CAR, represented by a purple line with circular markers, starts erratically and then rises consistently to 1.412370451 at $n_\eta = 1024$. The Consolidated Guillera Process, CGP, represented by a red line with square markers, declines exponentially to a low of 1.238020378 at $n_\eta = 64$ and then climes steadily to 1.440428278 at $n_\eta = 1024$.

Clearly the Consolidated Guillera Process is superior to the other three methods in the practical and precise approximation of ellipse perimeter length and by implication, the Complete Elliptic Integral.

ELLIPSE PERIMETER LENGTH PROCESSES GAIN REPORT
DYNAMIC QUALITY GAIN REPORT

	Committal Time	Committal Date	Number of Ellipse Aspect Ratios	Fiducial Series (64- Simpson)	CHECK SERIES 10- GaussKummer GAK	Consolidated Guillera Process CGP	Sykora Approximation SYK	Cantrell- Ramanujan Approximation CAR
Gain Experiment 1	11:26:15	07-29-2019	8	Not Applicable	1.762391300	1.611840236	0.265624385	1.094391678
Gain Experiment 2	11:26:30	07-29-2019	16	Not Applicable	1.487890143	1.351608972	0.776959261	0.352668272
Gain Experiment 3	11:26:38	07-29-2019	32	Not Applicable	1.358067191	1.294678109	0.772405562	1.031548685
Gain Experiment 4	11:26:48	07-29-2019	64	Not Applicable	1.288163314	1.238020378	0.850277629	1.031666587
Gain Experiment 5	11:26:58	07-29-2019	128	Not Applicable	1.259016934	1.246402503	0.892622383	1.137248879
Gain Experiment 6	11:27:06	07-29-2019	256	Not Applicable	-0.116335735	1.282416632	0.957927985	1.216043486
Gain Experiment 7	11:27:16	07-29-2019	512	Not Applicable	-0.154658536	1.356285976	1.027644315	1.306839839
Gain Experiment 8	11:27:28	07-29-2019	1024	Not Applicable	-0.169870839	1.440428278	1.122423150	1.412370451

Table 4.4
The Dynamic Change of Process Quality
with n_η Ellipse Perimeter Resolution

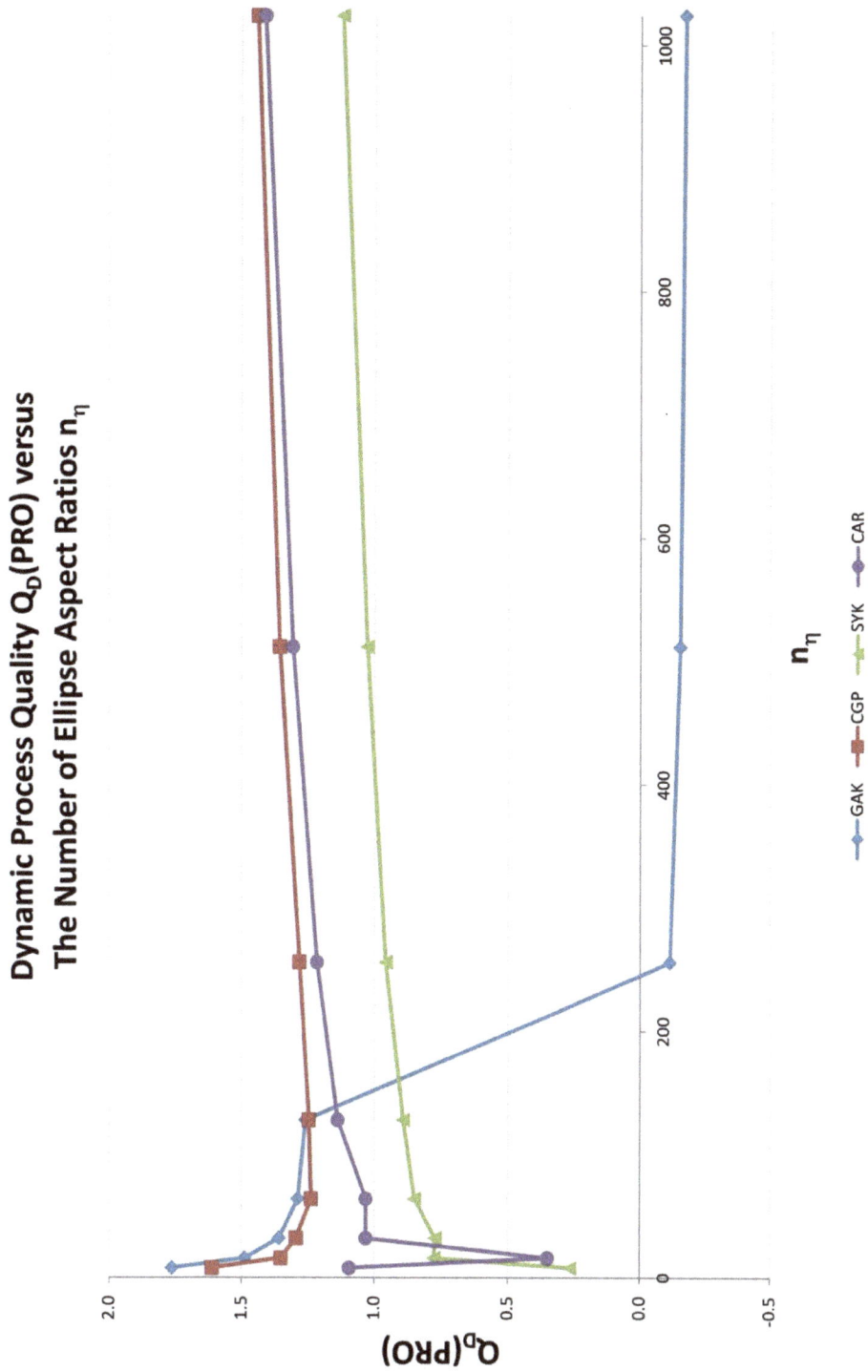

Figure 4.8
Dynamic Process Quality with Number of Ellipse Aspect Ratios

PART FIVE
INFINITESIMAL DISTORTIONS

This part concerns the cases when the Minor and Major Semi-Axes of an ellipse barely differ and the ellipse is not sensibly different to a circle. For example, the minor semi-axis may be 0.999 or 0.99999 when the major semi-axis is standardised to unity.

Under such conditions the First Eccentricity, c, becomes small and the Gauss-Kummer Parameter, h, tiny. Technically, when using fifteen-figure arithmetic, there is grave danger of many perimeter determination processes shoaling if η = b/a exceeds 0.999. Accordingly, we shall confine our experiments to grosser cases.

The Selection of Decimal Places of Semi-Axes Difference

To be particular, we will only examine the case of a=1 and b=0.999, knowing that elaborable oblations shall furnish very similar outcomes.

Therefore our chosen Number of Decimal Places of Difference, ID, is 3.

In general:-

$$a = 1$$
Equation 5.1

and:-

$$b = 1 - 10^{-ID}$$
Equation 5.2

So that when ID=3, b=0.999 and a-b=0.001

Eccentricity

$$\varepsilon = \sqrt{1 - \frac{b^2}{a^2}}$$
Equation 5.3

Hence by substitution:-

$$\varepsilon = \sqrt{1 - \frac{(1 - 10^{-ID})^2}{1}}$$

$$= \sqrt{1 - \frac{(1 - 2 \times 10^{-ID} + 10^{-2.ID})}{1}}$$

$$= \sqrt{2 \times 10^{-ID} - 10^{-2.ID}}$$

$$= \sqrt{10^{-ID}(2 - 10^{-ID})}$$
$$\varepsilon = 10^{\frac{-ID}{2}}\sqrt{(2 - 10^{-ID})}$$

Equation 5.4

and:-

$$\varepsilon^2 = 10^{-ID}(2 - 10^{-ID})$$

Equation 5.5

The Gauss-Kummer Parameter

$$h = \frac{(a - b)^2}{(a + b)^2}$$

Equation 5.6

Hence by substitution:-

$$h = \frac{\left(1 - (1 - 10^{-ID})\right)^2}{\left(1 + (1 - 10^{-ID})\right)^2}$$
$$= \frac{\left(1 - (1 - 10^{-ID})\right)^2}{4 - 4 \times 10^{-ID} + 10^{-2.ID}}$$

Equation 5.7

Further allow that c=10⁻ᴵᴰ, then the numerator of Equation 5.7 may be simplified as:-

$$\begin{aligned}
d &= (1 - (1 - 10^{-ID})^2) \\
&= (a - (a - c))^2 \\
&= (a - (a - c))^2 \\
&= a^2 - 2a(a - c) + (a - c)^2 \\
&= a^2 - 2a(a - c) + (a^2 - 2ac + c^2) \\
&= a^2 - 2a^2 + 2ac + a^2 - 2ac + c^2 \\
&= +2ac - 2ac + c^2 \\
d &= +c^2
\end{aligned}$$

Equation 5.8

Therefore:-

$$h = \frac{10^{-2.ID}}{4 - 4 \times 10^{-ID} + 10^{-2.ID}}$$

Equation 5.9

The Standard Circumference

None of our candidate Ellipse Perimeter Length Estimator Processes are trustworthy for barely-oblate ellipses and accordingly it is only meaningful to resort to brute-force intrinsic integral functions, given that our usual Newton-Cotes integrations are likely to suffer numerical instabilities if pushed to far (and they do).

So using the MATHCAD® Prime definite integral we may write for Ellipse Perimeter, P:-

$$P = 4a \int_0^{\frac{\pi}{2}} \sqrt{1 - \varepsilon^2 . \sin^2(\theta)} \, . \, d\theta = 4aE(\varepsilon)$$

Equation 5.10

We naturally expect that for unitary major semi-axis:-

$$P \approx 2\pi$$

Equation 5.11

Alternative forms of the Simpson's Rule Integral

Hithertofore we have relied upon a 64-interval Simpson's Rule integration for a fiducial basis of experiments upon circles grossly deformed to ellipses.

I had to exercise 512-interval Simpson to achieve anything like sensible results for subtly-elliptical structures and aware of likely numerical problems I essayed two alternative formulations

In both cases, allowing that the (even) Number of Intervals, n_s, is 512; the Interval Width, w, is given by:-

$$w = \frac{\pi}{2} \cdot \frac{1}{n_s}$$

Equation 5.12

(a)   The Alternator Form

This involves a single summation employing an alternator to yield the 4,2,4,2,4... alternation term multiplier series:-

$$P_{simp} = 4aE_{simp} = 4a \left\{ \frac{w}{3} \left[1 + \sum_{i=1}^{n_s-1} \left(3 - (-1)^i\right) \cdot \sqrt{1 - \varepsilon^2 . \sin^2(iw)} + 1 \right] \right\}$$

Equation 5.13

Though convenient, this is in theory one of the worst possible configurations for long series as it encourages errors of rounding and truncation, which may evolve into catastrophic numerical instabilities. Much depends of course upon register sizes and other low-level components of the software and hardware systems employed.

(b)   The Series Separation Form
        This technique involves the separation of the 4- terms and the 2- terms into two separate summations:-

$$P_{simp} = 4aE_{simp}$$

$$= 4a\left\{\frac{w}{3}\left[1 + 4\sum_{i=1}^{\frac{n_S}{2}}\sqrt{1 - \varepsilon^2.\sin^2\big((2i+1)w\big)} + 2\sum_{i=1}^{\frac{n_S}{2}-1}\sqrt{1 - \varepsilon^2.\sin^2(2iw)} + 1\right]\right\}$$

Equation 5.14

Because this dispenses with an alternator and splits the summation into two roughly equal sequences this is, in old-fashioned programming lore, the optimal choice.
        Reality is different.
        The Separated Series process was always markedly less precise than the Alternator Process for these small-ε experiments.
        I did not bother to pre-compute w/3 and ε² to constants.

Results for Candidate Processes

        In this section we will review the behaviors of all the potential Perimeter Length Estimator Processes, not just the key methods of interest discussed at length in Part Four.
        Table 5.1 presents the Fiducial Values for the MATHCAD experiment involving ID=3 selected Decimal Places of distortion, and Table 5.2 shows the Output Perimeter Lengths and Precisions for the fifteen methodologies discussed in Parts One and Two.
        The methods are sorted into descending order of exactitude.
        Table 5.1 confirms that whilst the Fiducial Perimeter is very near to twice the Ludolphine Constant it is not *at* that value.
        Table 5.2 notifies the experimenter that, for these barely-oblate objects, the Adlaj (quadratic) Iteration and Gauss-Kummer Series in virtually any state of truncation are "exact", as are any Guillera configurations (quartic) except the Lambda Form, a known rough-and-ready simplification. Cantrell-Ramanujan is also "exact".
        On the other hand 5-12 interval Simpson Integrations are now the worst methods, giving only four or five-figure accuracy, and the Series Separation is markedly worst of all.

Metric	Symbol	Value
Denary Order of Magnitude	ID	3
Major Semi Axis	a	1
Minor Semi Axis (1- 10^-ID)	b	0.999
Twice the Ludolphine Constant	2π	6.28318530717959
Fiducial Perimeter	P_{target}	6.28004410742153
Fiducial Eccentricity	ε_{target}	0.044710178
Fiducial Eccentricity Squared	ε_{target}	0.001999
Fiducial Gauss-Kummer Parameter	h_{target}	2.5025E-07

Table 5.1
Fiducial Values for the ID=3 Experiment

Method	Symbol	Value	SDF%	ABS(SDF%)	dPlace
Adlaj Process	ADL	6.28004410742153	0	0	exact
4-interval MacLaurin Series	MAC4	6.28004410742153	0	0	exact
32-interval Gauss-Kummer Series	GK32	6.28004410742153	0	0	exact
7-term Elaborated GK Series	GK7	6.28004410742153	0	0	exact
Michon's Rational GK Approximation	MIC	6.28004410742153	0	0	exact
Second Iteration of Guillera's Quartic Process	G4	6.28004410742153	0	0	exact
Consolidated Guillera Process (Eta Form)	CGPη	6.28004410742153	0	0	exact
Consolidated Guillera Process (q Form)	CGPq	6.28004410742153	0	0	exact
Cantrell-Ramanujan Formula	CAR	6.28004410742153	0	0	exact
From Target Elliptic Integral (32- Gauss-Kummer Series)	EGK	6.28004410742148	7.92001E-13	7.92001E-13	12
Consolidated Guillera Process (Lambda Form)	CGPλ	6.28004410742148	7.92001E-13	7.92001E-13	12
Sykora Formula	SYK	6.28004410892068	-2.3872E-08	2.38716E-08	7
Abbott's Formula	ABB	6.28004332163021	1.25125E-05	1.25125E-05	4
512-interval Simpson Integration (alternator form)	S512A	6.28004481980 3696	-6.5137E-05	6.51367E-05	4
512-interval Simpson Integration (series separation)	S512S	6.28003183588325	0.000195405	0.000195405	3

Table 5.2
Output Perimeter Lengths and Precisions for the ID=3 Experiment

PART SIX

THE OBLATION OF THE EARTH
AND THE ELASTIC DISTORTION OF SPINNING SPHEROIDS

In this context we do of course use "oblation" synonymously with "oblateness" to refer to the degree to which a sphere is deformed into an oval of revolution which may or may not be a "spheroid", which is an ellipse of revolution.

Many celestial bodies that approximate spheres or similar structures coalesced by force fields have the form of ovals of revolution very similar to, but seldom if ever identical with, ellipses of revolution.

Many such bodies which certainly include planets and probably include great stars such as neutron stars and pulsars are perturbed by immense tidal forces and their own rotations, and this physics causes them to depart from geometrically-ideal spheroidicity. Heterogeneities of mass concentration, both internal to the bodies and external further modify celestial dynamics and distort stars and planets.

Our Planet Earth is an oblate spheroid, sort of.

Table 6.1 lists some of the Earth's mathematical and physical properties relevant to our current disquisition[18].

The Oblateness Constant

The Oblateness Constant, q, is defined in this way:-

$$q = \frac{a^3 \omega^2}{GM}$$
Equation 6.1

where a is the Equatorial (Major) Semi-Axis of the Earth, ω is the Earth's Angular Frequency of Rotation ("angular velocity"), G is the Universal Gravitational Constant, and M is the Mass of the Earth.

Simple dimensional analysis confirms that q is dimensionless.

To six-figures, the value of q is 0.00346149 giving an inverse flattening 1/q of 288.893. This is discrepant to the measured flattening of the Earth which is 298.257, a matter of some 3.14% difference.

All Physical Measurements are in SI units

Metric	Symbol	Value	Dimensions M	L	T
Earth Equatorial Radius	a	6378137	0	1	0
Earth Polar Radius	b	6356752.314	0	1	0
Inverse Flattening	ff	298.2572236	0	0	0
Nominal Aspect Ratio of Meridional Ellipse	η	0.996647189	0	0	0
Scaled Equatorial Radius	1	1	0	1	0
Scaled Polar Radius	η	0.996647189	0	1	0
Ideal Meridional Ellipse Eccentricity	ε	0.081819191	0	0	0
Ideal Meridional Ellipse Eccentricity Squared	ε^2	0.00669438	0	0	0
Gauss-Kummer Parameter	h	0.00000281978111798139	0	0	0
Fiducial Perimeter	P_{target}	6.272656564	0	1	0
Fiducial Eccentricity	ε_{target}	0.081819191	0	0	0
Fiducial Gauss-Kummer Parameter	h_{target}	0.000002819781118	0	0	0
Computed Earth Meridional Perimeter		40007862.92	0	1	0
Measured Earth Meridional Perimeter		40008000	0	1	0
Angular Frequency of Rotation	ω	0.0000729211509	0	0	-1
Angular Velocity (WGS-84)	ω	0.0000729211509	0	0	-1
Universal Gravitational Constant	G	6.67408E-11	-1	3	-2
Mass of the Earth	M_{earth}	5.9722E+24	1	0	0
Volume of the Earth	V_{earth}	1.08321E+21	0	3	0
Density of the Earth (Wikipdia)	ρ_{earth}	5515	1	-3	0
Density of the Earth (computed)	ρ_{earth}	5513.427683	1	-3	0
Standard Gravitational Parameter	$\mu=GM$	3.986004418E+14	0	3	-2
Standard Gravitational Parameter (WGS-84)	$\mu=GM$	3.986004418E+14	0	3	-2
Geodetic Latitude of Bloxwich (degrees)	$\phi_{bloxwich}$	52.614	0	0	0
Geodetic Latitude of Bloxwich (radians)	$\phi_{bloxwich}$	0.918287532644296	0	0	0
Oblateness Constant	q	0.00346148782207746	0	0	0
Inverse Oblateness Constant	1/q	288.893115157584	0	0	0

Table 6.1
Selected Properties of the Planet Earth

Different Earth Radii at the Same Place

Knowing that the Earth is an imperfect sphere we are ready to believe that the Earth's Radius, R_{earth}, varies from location to location on the Earth's (ideal) Surface, even before we take into account hills and dales, or the depressions of the deep[19].

What may possibly surprise is that the Radius of the Earth differs *at the same place* depending upon which mathematical definition of Radius we find convenient.

Without considering the distortions of the small circles of latitude, also known to vary, we shall treat only of the great circles of longitude and the Latitude Values, φ, an obvious function of the Earth's quasi-elliptical Meridional Section. These latitude values plot along a given longitude.

Take for instance the Latitude of Bloxwich, a suburban town in Central England whose Latitude, $φ_{bloxwich}$, has a value 52.614°N, or an Angular Displacement of $φ_{bloxwich}$ = 0.918287532644296 radians measured from the Earth's Center. (We may like to think of the "Earth's Center" as a barycenter or some other unambiguous point in space, but even the concept of a planetary centre comes with its own baggage...).

We will review a few of the possible choices of Earth Radius available to us at Bloxwich.

The Ellipse Radius at Bloxwich

$$R_\varphi = \frac{ab}{\sqrt{(b\cos\varphi)^2 + (a\sin\varphi)^2}} = \frac{b}{\sqrt{1 - (\varepsilon\cos\varphi)^2}}$$
Equation 6.2

The value of $R_φ$ is 6364611.14269514 meters, say 6364.6 km, giving a great circle circumference P of some 40000 kilometers, a reasonably accurate figure.

But not a unique option, if there is such a thing.

The Geocentric Radius at Bloxwich

$$RG_\varphi = \sqrt{\frac{(a^2\cos\varphi)^2 + (b^2\sin\varphi)^2}{(a\cos\varphi)^2 + (b\sin\varphi)^2}}$$
Equation 6.3

The Geocentric Radius differs numerically, if subtly, from the ideal Ellipse Radius. Calculation or consultation of Table 6.2 will disclose that difference to be a mere 67 meters, so that SDF($R_φ$,$RG_φ$) defined in our usual percentage terms is -0.001051025.

Table 6.2 presents the Radius equations and data details as they pertain to Bloxwich for the Ellipse and Geocentric Radii and for five other kinds of Earth Radii. You can see that other than the Geocentric Radii the other species depart significantly from the elliptical ideal, in the case of the Prime Vertical by more than 27 Kilometers.

Species of Earth Radius		Value at Bloxwich	SDF (Rφ,R*)	Difference (meters)
Ellipse	$$R\varphi := \frac{a \cdot b}{\sqrt{(b \cdot \cos(\varphi))^2 + (a \cdot \sin(\varphi))^2}}$$	6364611.14269514	0.000000000	0.000000000
Geocentric	$$RG\varphi := \sqrt{\frac{(a^2 \cdot \cos(\varphi))^2 + (b^2 \cdot \sin(\varphi))^2}{(a \cdot \cos(\varphi))^2 + (b \cdot \sin(\varphi))^2}}$$	6364678.03632128	-0.001051025	-66.893626139
Meridional	$$M\varphi := \frac{(a \cdot b)^2}{\left((a \cdot \cos(\varphi))^2 + (b \cdot \sin(\varphi))^2\right)^{\frac{3}{2}}}$$	6375816.38466873	-0.176055406	-11205.241973590
Prime Vertical	$$N\varphi := \frac{(a)^2}{\left((a \cdot \cos(\varphi))^2 + (b \cdot \sin(\varphi))^2\right)^{\frac{1}{2}}}$$	6391658.04259503	-0.424957618	-27046.899899890
Azimuthal	$$Rz\varphi := \frac{1}{\frac{\cos(\varphi)^2}{M\varphi} + \frac{\sin(\varphi)^2}{N\varphi}}$$	6385808.55148883	-0.333051122	-21197.408793690
Gaussian	$$Rg\varphi := \sqrt{M\varphi \cdot N\varphi}$$	6383732.29961729	-0.300429303	-19121.156922149
Earth's Mean	$$Rm\varphi := \frac{2}{\frac{1}{M\varphi} + \frac{1}{N\varphi}}$$	6383727.38560648	-0.300352095	-19116.242911340

Table 6.2
Species of Earth Radii

It is now obvious that whatever this planet might or might not be, it is not an ellipsoid of revolution. The ellipsoid of revolution, or I should more correctly say, the ellipse of revolution, is only a model or an *approximation* of the shape of the Earth.

<u>Global Average Radii</u>

Evidently, rather than focus upon the Radii at Bloxwich or any other special radii we would often prefer to consider average radii, especially if we want to consider the Meridional Circumference of this or any planet.

But here again we have a variety of options where choice shall be colored by the application envisioned.

Mean Radius

This is the simplest, but not necessarily the most faithful, of the available options:-

$$R1 = \frac{2a + b}{3}$$
Equation 6.4

Equation 6.4 yields a Mean Radius of 6371008.7714 meters. Clearly, physical realities render this accurate only to the nearest kilometer, say 6371 kilometers.

Authalic Radius

Authalic Radius, R2, is the radius of a sphere that has the same *surface area* as the ellipsoid.

The Surface Area of an Ellipsoid, S_{oblate}, is defined by:-

$$S_{oblate} = 2\pi a^2 \left(1 + \frac{1 - \varepsilon^2}{\varepsilon} \cdot \tanh^{-1} \varepsilon\right) = 2\pi a^2 + \left[\pi \frac{b^2}{\varepsilon} \log_n \left\{\frac{1 + \varepsilon}{1 - \varepsilon}\right\}\right]$$
Equation 6.5

For the Earth, the value of S_{oblate} is $5.10065621721676 \times 10^{14}$, call it 5×10^8 square kilometers.

So the Authalic Radius is given by:-

$$R2 = \sqrt{\frac{a^2 + \frac{ab^2}{\sqrt{a^2 - b^2}} \cdot \log_n \left(\frac{a + \sqrt{a^2 - b^2}}{b}\right)}{2}}$$
Equation 6.6

The Authalic Radius may be expressed more succinctly as:-

$$R2 = \sqrt{\frac{a^2}{2} + \frac{b^2}{2} \frac{\tanh^{-1} \varepsilon}{\varepsilon}} = \sqrt{\frac{S_{oblate}}{4\pi}}$$
Equation 6.7

The value of the Authalic Radius is 6371007.18090341 meters.

The Volumetric Radius

The Volumetric Radius is the Volume of a sphere with the same volume as the ellipsoid. It is given by:-

$$R3 = \sqrt[3]{a^2b}$$
Equation 6.8

The value of the Volumetric Radius, R3, is 6371000.78999406 meters.

The Rectifying Radius

The Rectifying Radius, R4, is the radius of a sphere that has the same (meridional) radius as the ellipsoid. It is of course of particular interest to us in our discussion of ellipsoid circumference estimators.

The Rectifying Radius is identical to the Meridional Mean, M_r, which is yielded by:-

$$M_r = \frac{2}{\pi} \int_0^{\frac{\pi}{2}} \sqrt{a^2\cos^2(\varphi) + b^2\sin^2(\varphi)} \, . \, d\varphi$$
Equation 6.9

The Rectifying Radius (M_r) of the Earth is 6367449.14580084 meters. This of course implies that the Earth's Meridional Circumference is $2\pi M_r = 40007862.9171091$ meters.

Two ready approximations of M_r are available in the literature:-

$$M_{rap1} = \left(\frac{a^{\frac{3}{2}} + b^{\frac{3}{2}}}{2} \right)^{\frac{2}{3}} = 6367449.14580005$$
Equation 6.10

and:-

$$M_{rap2} = \frac{a + b}{2} = 6367444.6571$$
Equation 6.11

Because Equations 6.9 and 6.10 differ by less than one tenth of a millimeter it hardly makes sense to go to the expense of a definite integration when M_{rap1} will clearly do. Even the simple arithmetic average of Equation 6.11 gives us better than five-meter accuracy in the estimation. As noted above, it is rarely sensible to exceed a kilometer one way or the other when estimating the circumference of a geologically-active planet.

Rectification and the Meridional Mean

As we have seen there are a number of excellent formulae and processes for the estimation of ellipse perimeter lengths, and Newton-Cotes integration methods are by no means the best in all circumstances.

Notwithstanding that, I set up a 1024-interval Simpson's Rule in MATHCAD® Prime (in a worksheet I called REVIEWearthWSG84.mcdx) in order to compute a numerical fiducial Meridional Perimeter for the Earth.

This established the Earth Perimeter Fiducial P_{target} as 6.27266342121093 when the Major Semi-Axis is scaled to unity. Multiplication of P_{target} by Major Semi-Axis, a = 6378137 (meters) yields the Fiducial Circumference P_{earth} = 40007906.655372 meters that implies a Percentage Specific Defect SDF(P_{target},P_{earth}) = -1.09324167188193$\times 10^{-4}$.

A slightly different approach is to confront the definite integral:-

$$P_{earth1} = 4 \int_0^{\frac{\pi}{2}} \sqrt{a^2\cos^2(\varphi) + b^2\sin^2(\varphi)}\,.d\varphi = 40007862.9171091$$

Equation 6.12

which shortens the distance by some forty odd meters.

Whichever approach we take we can see that knowledge of the Meridional Mean, M_r, is a viable avenue to Perimeter Length so long as kilometer accuracies are not required.

These results are assembled on Table 6.3.

I computed M_r and Perimeter P using the MATHCAD® definite integral intrinsic function; and also I programmed 1024-interval Abramowitz and Stegun Rule 18 and also 1024-interval Simpson's Rule as EXCEL® spreadsheets. You can see that the results of all three avenues are essentially identical.

The integral and the indicial equation Rectifying Means are identical within limits of error and the simple average (a+b)/2 gives a result within 29 meters of the integral. As a practical man rather than an academic mathematician I consider these formulae tantamount:- Except that some are computationally a lot cheaper than others.

The (2a+b)/3 Mean Radius and the Authalic and Volumetric Radii are a different family of very nearly equivalent radius species. The (2a+b)/2 and the Authalic forms differ by three centimeters from each other and by 22.3 kilometers from the fiducial Rectifying Radius.

Earth Equatorial Radius	a	6378137
Earth Polar Radius	b	6356752.314
First Eccentricity	e	0.081819191
$2\pi a^2$		2.55604E+14
Ellipsoid Surface Area	S	5.10066E+14
Ellipsoid Surface Area	S	5.10066E+14
Newton-Cotes Interval Number		1024

MATHCAD INTRINSIC INTEGRAL FIDUCIALS

Meridional Mean	M_r	6367449.14580084
Indicated Earth Meridional Perimeter	P	40007862.91710910

RULE 18 FIDUCIALS

Meridional Mean	M_r	6367449.14580084
Indicated Earth Meridional Perimeter	P	40007862.91710900

SIMPSON'S RULE FIDUCIALS

Meridional Mean	M_r	6367449.14580084
Indicated Earth Meridional Perimeter	P	40007862.91710900

All Linear Values in Meters
All Areal Values in Square Meters

Global Radius	Equation Value	Computed Perimeter	SDF%	Difference in Meters
Mean	6371008.77140000	40030228.70437270	-0.05590348	-22365.78726365
Authalic	6371007.18090341	40030218.71098790	-0.05587850	-22355.79387882
Volumetric	6371000.78999406	40030178.55572020	-0.05577813	-22315.63861109
Rectifying Mean (by Rule 18)	6367449.14580084	40007862.91710900	0.00000000	0.00000007
Rectifying Mean (indicial approximation)	6367449.14580004	40007862.91710400	0.00000000	0.00000507
Mean (simple average)	6367444.65710000	40007834.71376990	0.00007049	28.20333923

Table 6.3
The Earth Radius Species Compared

The MacLaurin Spheroid as a Paradigm of the Oblate Earth

Sometime around 1742AD, when he was working at Marischal College, the Protestant University of Aberdeen, the Scottish genius Colin MacLaurin derived an equation for the Oblation of the Earth[20].

He was one of a remarkable brotherhood of eighteenth-century scientists to make startling innovations in the field of celestial mechanics, a fellowship that included the Cassini family, Maskelyne, Cavendish and of course Newton, amongst many others.

His equation is special to the case of a rotating biaxial ellipsoid (i.e. an oblate or prolate spheroid) and is this:-

$$\frac{\omega^2}{\pi G \rho} = \frac{2\sqrt{1 - \varepsilon_{mac}^2}}{\varepsilon_{mac}^3} \cdot (3 - 2\varepsilon_{mac}^2) \cdot \sin^{-1} \varepsilon_{mac} - \frac{6}{\varepsilon_{mac}^2} \cdot (1 - \varepsilon_{mac}^2)$$

Equation 6.13

Note the fact that if you move π to the RHS then the LHS is wholly physical and the RHS purely mathematical. Accordingly, postmodern thinkers might be forgiven for asserting that a universe even of Newtonian mechanics is really a purely mathematical phenomenon, and therefore a matter of textual analysis (no impiety intended).

Essentially, MacLaurin's treatment considers the Earth as a ball of fluid in which gravitational cohesion and centrifugal force stand in equipoise and the distortion of the sphere is a structural accommodation to these antagonistic powers. It also assumes that the planet is a mass of homogenous density.

Of course the Earth is a more or less rigid and elastic body, now known to vary widely in density from perhaps 2830 kg/m³ for the continental crustal rocks to maybe as high as 12000 kg/m³ for the deeper metallic core.

But I believe you will join me in wonder at MacLaurin's Spheroid, especially in remembrance of its place and time. If I may quote a later Georgian genius speaking of things very different "the wonder is not that it is done badly, but that it can be done at all". We shall see just how bad, or how good, MacLaurin's work was.

ε_{mac} is the First Eccentricity of the Earth that satisfies Equation 6.13. We may obviously compare it with the Eccentricity calculated from modern measurements of this planet, which is given by:-

$$\varepsilon = \sqrt{1 - \frac{b^2}{a^2}}$$

Equation 6.14

where ε is the Eccentricity of the Meridional Section of the Earth calculated from a, the measured Equatorial Radius and b the measured Polar Radius. (b is sometimes denoted c in the literature to clarify that a three-dimensional structure is under discussion).

My EXCEL® Equation Solver did not like Equation 6.13 so I resorted to trial-and-error to approximate ε_{mac}. I estimated the value to be near to 0.0928. Later, I used a diminishing-range heuristic to refine ε_{mac} to 0.092799102864000300.

At this value the Percentage Specific Defect discrepancy between the LHS and the RHS of Equation 6.12 is $7.26815402050799 \times 10^{-10}$. The precision of the %SDF is obviously spurious and is quoted only for checking purposes.

It is possible to compute a ratio parameter $\eta_{indicated}$ using ε_{mac} by reference to the following formula:-

$$\eta_{indicated} = \sqrt{1 - \varepsilon_{mac}^2}$$
Equation 6.15

and $\eta_{indicated}$ has the value 0.995684853007033000 whereas the surveyed $\eta = b/a$ aspect ratio is 0.996647189328169. Therefore, %SDF($\eta, \eta_{indicated}$) is 0.096557370697.

The agreement between MacLaurin's theoretical oblation and what we know to be the case is almost supernally neat. The discrepancy, such as it is, is in my opinion due to the non-Newtonian viscoelastic Earth competency and its geological heterogeneity.

Modern physicists and astronomers know the MacLaurin Spheroid to be an oversimplification but it is still worth knowing, not least because, like the orbit of Mercury, its anomalies light the way to subtler knowledge.

Table 6.4 presents a comparison of World Geodetic System and MacLaurin's terrestrial metrics. Note that the extremely close agreement of $\eta = b/a$ aspect ratios is vitiated by derived quantities that involve squares and square roots.

Metric	Symbol	Value
Earth Equatorial Radius	a	6378137
Actual Earth Polar Radius	b	6356752.314
Inverse Flattening	ff	298.2572236
Nominal Aspect Ratio of Meridional Ellipse	η	0.996647189
Scaled Equatorial Radius	1	1
Scaled Polar Radius	η	0.996647189
Eccentricity	ε	0.081819191
Eccentricity Squared	ε^2	0.00669438
Gauss-Kummer Parameter	h	2.81978E-06
Fiducial Perimeter	P_{target}	6.272656564
Fiducial Eccentricity	ε_{target}	0.081819191
Fiducial Gauss-Kummer Parameter	h_{target}	2.81978E-06
Computed Earth Meridional Perimeter		40007862.92
Measured Earth Meridional Perimeter		40008000

MacLaurin Spheroid

Metric	Symbol	Value	SDF% (WSG84,Mac)
Earth Equatorial Radius	a	6378137	0.00000000000
Implied Earth Polar Radius	b	6350614.401	0.09655737370697
Inverse Flattening	ff	230.7418159	22.636637886719
Nominal Aspect Ratio of Meridional Ellipse	η	0.995684353	0.09655737370697
Scaled Equatorial Radius	1	1	0.00000000000
Scaled Polar Radius	η	0.995684353	0.09655737370697
Eccentricity	ε	0.092795103	-13.419726851900
Eccentricity Squared	ε^2	0.008612673	-28.640344391597
Gauss-Kummer Parameter	h	4.67525E-06	-65.80280301449
	P_{target}		
	ε_{target}		
	h_{target}		

Table 6.4
The WSG84 and MacLaurin Spheroids Compared

References

1 Abramowitz and Stegun
   "Handbook of Mathematical Functions"
   "With Formulas, Graphs and Mathematical Tables"
   Milton Abramowitz and Irene A Stegun
   Dover Publications, Inc., New York   1965
   SBN 486-61272-4
   pp 1046

2 Wikipedia contributors. (2019, May 21). Elliptic integral.
   In *Wikipedia, The Free Encyclopedia*.
   Retrieved 13:33, June 20, 2019, from
 https://en.wikipedia.org/w/index.php?title=Elliptic_integral&oldid=898042689

3 Eqn.1 Page 1095
   "An Eloquent Formula for the Perimeter of an Ellipse"
   Semjon F Adlaj
   The Department of Mechanics
   Dorodnycin Computing Centre of the
     Institution of the Russian Academy of Sciences
   Vavilov St. 40, 119333, Moscow, Russia

4 Wikipedia contributors. (2019, June 13). Ellipse.
   In Wikipedia, The Free Encyclopedia.
   Retrieved 13:57, June 20, 2019, from
 https://en.wikipedia.org/w/index.php?title=Ellipse&oldid=901722991

5 "On the Perimeter of an Ellipse"
   Paul Abbott
   The Mathematical Journal V11 Issue 2
   Copyright: 2009, Wolfram Media, Inc
   http://www.mathematica-journal.com/issue/v11i2/contents/Abbott/Abbott.pdf

6 Wikipedia contributors. (2019, May 28). Arithmetic–geometric mean.
   In Wikipedia, The Free Encyclopedia.
   Retrieved 14:36, June 20, 2019, from
 https://en.wikipedia.org/w/index.php?title=Arithmetic%E2%80%93geometric_mean&oldid=899160154

7 Ellipse at Wolfram Mathworld
   http://mathworld.wolfram.com/Ellipse.html

8 Eqn.22 Page 15
   "The Arithmetic Geometric Mean of Gauss"

Tomack Gilmore
Queen Mary College, University of London
https://homepage.univie.ac.at/tomack.gilmore/papers/Agm.pdf

9   Michon's Approximation
        Ellipse Perimeter Discussion by Gerard P Michon in "Final Answers"
        Subtractive Series and MacLaurin Series for Ellipse Perimeter (1742AD)
        http://www.numericana-com/answer/ellipse.htm

10   "Ramanujan's Perimeter of an Ellipse"
        Mark B Vallarino
        Escuda de Matemática
        Universidad de Costa Rica
        2060 San José, Costa Rica
        1 February 2008
        arXiv: math/0506384v1 [math.CA]
        20 June 2005

        https://arxiv.org/abs/math/0506384

11   "Ramanujan's Perimeter of an Ellipse"
        Mark B Vallarino
        Escuda de Matemática
        Universidad de Costa Rica
        2060 San José, Costa Rica
        1 February 2008
        arXiv: math/0506384v1 [math.CA]
        20 June 2005

        https://arxiv.org/abs/math/0506384

12   "Self-Replication and Borwein-Like Algorithms"
        Jesús Guillera
        To the memory of Jonathan Borwein
        Department of Mathematics
        University of Zaragoza
        50009 Zaragoza, Spain
        arXiv: 1702.05378v2 [math.NT]
        21 Feb 2017

13   Table of Standard Integrals
        http://www.maths.usyd.edu.au/MATH1003/r/m1003/Table_of_Integrals.pdf

14   https://www.cymath.com/answer?q=int%28sin%28x%29%5E2%2Cx%29

15   "Approximations of Ellipse Perimeters"
Review by Stanislav Sykora
Stan's Library Volume One
Extrabyte
Via R.Sanzio 22C, Castano Primo, Italy 20022
December 27 2005
http://www.ebyte.it/library/docs/math05a/EllipsePerimeterApprox05.html

16   "Approximations of Ellipse Perimeters"
Review by Stanislav Sykora
Stan's Library Volume One
Extrabyte
Via R.Sanzio 22C, Castano Primo, Italy 20022
December 27 2005
http://www.ebyte.it/library/docs/math05a/EllipsePerimeterApprox05.html

17   QB64 website with compiler
QB64.org
https://www.portal.qb64.org/

18   "Figure of the Earth" wiki
Wikipedia contributors. (2019, August 5). Figure of the Earth. In *Wikipedia, The Free Encyclopedia*. Retrieved 09:08, August 6, 2019, from
https://en.wikipedia.org/w/index.php?title=Figure_of_the_Earth&oldid=909476844

19   "Earth Radius" wiki
Wikipedia contributors. (2019, August 1). Earth radius. In *Wikipedia, The Free Encyclopedia*. Retrieved 09:09, August 6, 2019, from
https://en.wikipedia.org/w/index.php?title=Earth_radius&oldid=908826468

20   "MacLaurin Spheroid" wiki
Wikipedia contributors. (2019, May 13). Maclaurin spheroid. In *Wikipedia, The Free Encyclopedia*. Retrieved 09:10, August 6, 2019, from
https://en.wikipedia.org/w/index.php?title=Maclaurin_spheroid&oldid=896946047

21   "A New Algorithm for Computing a Single Root of
 a Real Continuous Function"
CJF Ridders
IEEE Transactions on Circuits and Systems
Vol. CAS-26, No.11, November 1979
pp 979-980

22   Wikipedia contributors. (2017, June 1). Ridders' method. In *Wikipedia, The Free Encyclopedia*. Retrieved 09:05, August 6, 2019, from
https://en.wikipedia.org/w/index.php?title=Ridders%27_method&oldid=783258694

APPENDIX A

AN ANALYTICAL PROOF OF EQUATION 1.60

We shall prove the relation:-

$$a^2 + b^2 = \frac{4}{\pi} \int_0^{\pi/2} (a^2.\sin^2\theta + b^2.\cos^2\theta).d\theta$$

Equation 1.60

that we were content to accept on empirical grounds in Part One of this disquisition. Our prove will rely upon the theorem of the Linearity of Integrals[13]:-

$$\int_0^{\theta_{max}} [\lambda.f(\theta) + \mu.g(\vartheta)].d\theta = \lambda \int_0^{\theta_{max}} f(\theta).d\theta + \mu \int_0^{\theta_{max}} g(\theta).d\theta$$

Equation A1

In our context, treating of the ellipse quarter-sector, θ_{max} is $\pi/2$ whilst:-

$$\lambda = a^2 \qquad\qquad \textbf{Equation A2a}$$
$$\mu = b^2 \qquad\qquad \textbf{Equation A2b}$$

and:-

$$f(\theta) = \int_0^{\pi/2} \sin^2\theta.d\theta$$

Equation A3a

$$g(\theta) = \int_0^{\pi/2} \cos^2\theta.d\theta$$

Equation A3b

Also, the General Multiplier, z, is:-

$$z = \frac{4}{\pi}$$

Equation A4

<u>Standard Integrals</u>[14]

The required standard integrals are:-

$$\int \sin^2 x = \frac{1}{2}x - \frac{1}{4}\sin 2x + c$$

Equation A5

and:-

$$\int \cos^2 x = \frac{1}{2}x + \frac{1}{4}\sin 2x + c$$

Equation A6

where c is a constant of integration.

<u>The Elaboration of Equation 1.20 with Definite Integrals</u>

In terms of Equation A1 it is now possible to write the two definite integrals as:-

$$
\begin{aligned}
If\theta &= {}_{0}^{\pi/2}\left[\frac{1}{2}x - \frac{1}{4}\sin 2x + c\right] \\
&= \left[\frac{1}{2}\frac{\pi}{2} - \frac{1}{4}\sin \pi + c\right] - \left[\left[\frac{1}{2}0 - \frac{1}{4}\sin 0 + c\right]\right] \\
&= \left[\frac{\pi}{4} + c\right] - [0 + c] \\
&= \frac{\pi}{4}
\end{aligned}
$$

Equation A7

and-

$$
\begin{aligned}
Ig\theta &= {}_{0}^{\pi/2}\left[\frac{1}{2}x + \frac{1}{4}\sin 2x + c\right] \\
&= \left[\frac{1}{2}\frac{\pi}{2} + \frac{1}{4}\sin \pi + c\right] - \left[\left[\frac{1}{2}0 + \frac{1}{4}\sin 0 + c\right]\right] \\
&= \left[\frac{\pi}{4} + c\right] - [0 + c] \\
&= \frac{\pi}{4}
\end{aligned}
$$

Equation A8

Accordingly:-

$$Soln = z\left[\int_0^{\theta_{max}} [\lambda.f(\theta) + \mu.g(\vartheta)].d\theta\right]$$

$$= z\left[\lambda \int_0^{\theta_{max}} f(\theta).d\theta + \mu \int_0^{\theta_{max}} g(\theta).d\theta\right]$$

$$= z[\lambda.If\theta + \mu.Ig\theta]$$

$$= \frac{4}{\pi}\left[a^2.\frac{\pi}{4} + b^2.\frac{\pi}{4}\right]$$

$$= \frac{4}{\pi}.\frac{\pi}{4}[a^2 + b^2]$$

$$= a^2 + b^2$$

Equation A9

Therefore:-

$$a^2 + b^2 = \frac{4}{\pi}\int_0^{\pi/2} (a^2.\sin^2\theta + b^2.\cos^2\theta).d\theta$$

Equation 1.60

which was to be proven.

APPENDIX B

κSTEP CURVATURE RELATIONS

Curvature Metric

$\kappa(\kappa step_i) = -\log10(ABS(\kappa step_i))$

Serial	$\eta = b/a$	CHECK SERIES 10-GaussKummer	Consolidated Guillera Process	Sykora Approximation	Cantrell-Ramanujan Approximation
0	0.00000000				
1	0.00781250	1.10358046	1.83281715	1.75469274	1.73715710
2	0.01562500	1.03804819	0.86093152	1.05981389	1.10281543
3	0.02343750	0.96800238	0.96924913	0.99185108	1.01613313
4	0.03125000	0.89612135	0.91092171	0.92062443	0.93084145
5	0.03906250	0.82424746	0.83760999	0.84836430	0.84867089
6	0.04687500	0.75357788	0.76374618	0.77659713	0.77035584
7	0.05468750	0.68484478	0.69221259	0.70631241	0.69608555
8	0.06250000	0.61846017	0.62370255	0.63811321	0.62577095
9	0.07031250	0.55462279	0.55832576	0.57233529	0.55919437
10	0.07812500	0.49339346	0.49600189	0.50913491	0.49608979
11	0.08593750	0.43474642	0.43658261	0.44855099	0.43618402
12	0.09375000	0.37860366	0.37989665	0.39054793	0.37921616
13	0.10156250	0.32485755	0.32576874	0.33504454	0.32494583
14	0.10937500	0.27338551	0.27402825	0.28193342	0.27315536
15	0.11718750	0.22405934	0.22451318	0.23109365	0.22364928
16	0.12500000	0.17675108	0.17707183	0.18239916	0.17625254
17	0.13281250	0.13133653	0.13156342	0.13572389	0.13080834
18	0.14062500	0.08769727	0.08785790	0.09094517	0.08717595
19	0.14843750	0.04572173	0.04583551	0.04794557	0.04522871
20	0.15625000	0.00530553	0.00538617	0.00661401	0.00485227
21	0.16406250	-0.03364840	-0.03359122	-0.03315380	-0.03405696
22	0.17187500	-0.07123008	-0.07118953	-0.07145524	-0.07159311
23	0.17968750	-0.10752283	-0.10749407	-0.10838091	-0.10784197
24	0.18750000	-0.14260365	-0.14258325	-0.14401490	-0.14288192
25	0.19531250	-0.17654370	-0.17652923	-0.17843515	-0.17678482
26	0.20312500	-0.20940872	-0.20939846	-0.21171383	-0.20961664
27	0.21093750	-0.24125949	-0.24125222	-0.24391773	-0.24143810
28	0.21875000	-0.27215223	-0.27214708	-0.27510869	-0.27230520
29	0.22656250	-0.30213899	-0.30213534	-0.30534398	-0.30226969
30	0.23437500	-0.33126798	-0.33126540	-0.33467664	-0.33137946
31	0.24218750	-0.35958396	-0.35958213	-0.36315584	-0.35967890
32	0.25000000	-0.38712846	-0.38712717	-0.39082721	-0.38720924
33	0.25781250	-0.41394013	-0.41393922	-0.41773312	-0.41400880

34	0.26562500	-0.44005493	-0.44005429	-0.44391294	-0.44011327
35	0.27343750	-0.46550639	-0.46550593	-0.46940332	-0.46555592
36	0.28125000	-0.49032578	-0.49032546	-0.49423836	-0.49036782
37	0.28906250	-0.51454232	-0.51454210	-0.51844988	-0.51457800
38	0.29687500	-0.53818337	-0.53818321	-0.54206756	-0.53821363
39	0.30468750	-0.56127451	-0.56127440	-0.56511914	-0.56130018
40	0.31250000	-0.58383975	-0.58383968	-0.58763057	-0.58386153
41	0.32031250	-0.60590163	-0.60590157	-0.60962612	-0.60592009
42	0.32812500	-0.62748130	-0.62748126	-0.63112855	-0.62749695
43	0.33593750	-0.64859868	-0.64859865	-0.65215922	-0.64861195
44	0.34375000	-0.66927251	-0.66927249	-0.67273819	-0.66928376
45	0.35156250	-0.68952048	-0.68952047	-0.69288430	-0.68953002
46	0.35937500	-0.70935925	-0.70935924	-0.71261529	-0.70936733
47	0.36718750	-0.72880457	-0.72880456	-0.73194787	-0.72881142
48	0.37500000	-0.74787132	-0.74787132	-0.75089779	-0.74787712
49	0.38281250	-0.76657360	-0.76657359	-0.76947993	-0.76657850
50	0.39062500	-0.78492472	-0.78492472	-0.78770831	-0.78492887
51	0.39843750	-0.80293734	-0.80293734	-0.80559621	-0.80294085
52	0.40625000	-0.82062344	-0.82062344	-0.82315620	-0.82062641
53	0.41406250	-0.83799441	-0.83799441	-0.84040016	-0.83799692
54	0.42187500	-0.85506106	-0.85506106	-0.85733938	-0.85506318
55	0.42968750	-0.87183368	-0.87183368	-0.87398453	-0.87183547
56	0.43750000	-0.88832205	-0.88832205	-0.89034578	-0.88832356
57	0.44531250	-0.90453549	-0.90453549	-0.90643276	-0.90453676
58	0.45312500	-0.92048288	-0.92048288	-0.92225464	-0.92048395
59	0.46093750	-0.93617269	-0.93617269	-0.93782016	-0.93617359
60	0.46875000	-0.95161299	-0.95161299	-0.95313761	-0.95161375
61	0.47656250	-0.96681150	-0.96681150	-0.96821492	-0.96681214
62	0.48437500	-0.98177560	-0.98177560	-0.98305962	-0.98177614
63	0.49218750	-0.99651233	-0.99651233	-0.99767892	-0.99651278
64	0.50000000	-1.01102844	-1.01102844	-1.01207969	-1.01102881
65	0.50781250	-1.02533037	-1.02533037	-1.02626851	-1.02533068
66	0.51562500	-1.03942432	-1.03942432	-1.04025165	-1.03942458
67	0.52343750	-1.05331619	-1.05331619	-1.05403511	-1.05331641
68	0.53125000	-1.06701168	-1.06701168	-1.06762465	-1.06701186
69	0.53906250	-1.08051623	-1.08051623	-1.08102578	-1.08051638
70	0.54687500	-1.09383507	-1.09383507	-1.09424376	-1.09383519
71	0.55468750	-1.10697322	-1.10697322	-1.10728366	-1.10697332
72	0.56250000	-1.11993551	-1.11993551	-1.12015033	-1.11993560
73	0.57031250	-1.13272658	-1.13272658	-1.13284843	-1.13272665
74	0.57812500	-1.14535089	-1.14535089	-1.14538243	-1.14535095
75	0.58593750	-1.15781273	-1.15781273	-1.15775662	-1.15781278
76	0.59375000	-1.17011624	-1.17011624	-1.16997515	-1.17011628

77	0.60156250	-1.18226539	-1.18226539	-1.18204198	-1.18226542
78	0.60937500	-1.19426402	-1.19426402	-1.19396093	-1.19426404
79	0.61718750	-1.20611582	-1.20611582	-1.20573569	-1.20611584
80	0.62500000	-1.21782436	-1.21782436	-1.21736980	-1.21782437
81	0.63281250	-1.22939307	-1.22939307	-1.22886667	-1.22939308
82	0.64062500	-1.24082528	-1.24082528	-1.24022961	-1.24082529
83	0.64843750	-1.25212418	-1.25212418	-1.25146177	-1.25212419
84	0.65625000	-1.26329287	-1.26329287	-1.26256624	-1.26329288
85	0.66406250	-1.27433434	-1.27433434	-1.27354595	-1.27433435
86	0.67187500	-1.28525148	-1.28525148	-1.28440378	-1.28525149
87	0.67968750	-1.29604708	-1.29604708	-1.29514247	-1.29604708
88	0.68750000	-1.30672384	-1.30672384	-1.30576469	-1.30672384
89	0.69531250	-1.31728437	-1.31728437	-1.31627301	-1.31728438
90	0.70312500	-1.32773121	-1.32773121	-1.32666993	-1.32773122
91	0.71093750	-1.33806681	-1.33806681	-1.33695786	-1.33806681
92	0.71875000	-1.34829353	-1.34829353	-1.34713912	-1.34829353
93	0.72656250	-1.35841367	-1.35841367	-1.35721598	-1.35841368
94	0.73437500	-1.36842947	-1.36842947	-1.36719062	-1.36842947
95	0.74218750	-1.37834308	-1.37834308	-1.37706516	-1.37834308
96	0.75000000	-1.38815659	-1.38815659	-1.38684165	-1.38815659
97	0.75781250	-1.39787204	-1.39787204	-1.39652209	-1.39787204
98	0.76562500	-1.40749139	-1.40749139	-1.40610838	-1.40749139
99	0.77343750	-1.41701656	-1.41701656	-1.41560242	-1.41701656
100	0.78125000	-1.42644940	-1.42644940	-1.42500602	-1.42644940
101	0.78906250	-1.43579172	-1.43579172	-1.43432093	-1.43579172
102	0.79687500	-1.44504527	-1.44504527	-1.44354887	-1.44504527
103	0.80468750	-1.45421175	-1.45421175	-1.45269149	-1.45421175
104	0.81250000	-1.46329281	-1.46329281	-1.46175042	-1.46329281
105	0.82031250	-1.47229006	-1.47229006	-1.47072722	-1.47229006
106	0.82812500	-1.48120506	-1.48120506	-1.47962340	-1.48120506
107	0.83593750	-1.49003934	-1.49003934	-1.48844046	-1.49003934
108	0.84375000	-1.49879436	-1.49879436	-1.49717984	-1.49879436
109	0.85156250	-1.50747158	-1.50747158	-1.50584293	-1.50747158
110	0.85937500	-1.51607239	-1.51607239	-1.51443110	-1.51607239
111	0.86718750	-1.52459816	-1.52459816	-1.52294567	-1.52459816
112	0.87500000	-1.53305020	-1.53305020	-1.53138794	-1.53305020
113	0.88281250	-1.54142982	-1.54142982	-1.53975916	-1.54142982
114	0.89062500	-1.54973828	-1.54973828	-1.54806056	-1.54973828
115	0.89843750	-1.55797680	-1.55797680	-1.55629332	-1.55797680
116	0.90625000	-1.56614657	-1.56614657	-1.56445862	-1.56614657
117	0.91406250	-1.57424876	-1.57424876	-1.57255758	-1.57424876
118	0.92187500	-1.58228452	-1.58228452	-1.58059131	-1.58228452
119	0.92968750	-1.59025494	-1.59025494	-1.58856089	-1.59025494

120	0.93750000	-1.59816112	-1.59816112	-1.59646735	-1.59816112
121	0.94531250	-1.60600409	-1.60600409	-1.60431173	-1.60600409
122	0.95312500	-1.61378490	-1.61378490	-1.61209503	-1.61378490
123	0.96093750	-1.62150455	-1.62150455	-1.61981821	-1.62150455
124	0.96875000	-1.62916402	-1.62916402	-1.62748224	-1.62916402
125	0.97656250	-1.63676426	-1.63676426	-1.63508803	-1.63676426
126	0.98437500	-1.64430620	1.64430620	-1.64263649	-1.64430620
127	0.99218750	-1.32131510	-1.32131510	-1.65012852	-1.65179078
128	1.00000000				

Table 3.1
Curvature Relations

APPENDIX C

THE PERIMETER FIDELITY METRIC $\Phi(P(a,b))$

Perimeter Fidelity Metric

$$\Phi(P(a,b))=-Log10(Abs(SDF\%))$$

Serial	η = b/a	CHECK SERIES -10 Gauss-Kummer	Consolidated Guillera Process	Sykora Approximation	Cantrell-Ramanujan Approximation
0	0	1.52730252			
1	0.007813	1.72447514	1.7825272	2.30188996	3.62822762
2	0.015625	1.93207198	2.33957854	2.16494169	3.47971161
3	0.023438	2.13641863	2.75724606	2.1882761	3.17979771
4	0.03125	2.3328136	3.12377656	2.30353381	2.98747078
5	0.039063	2.52130624	3.45869	2.52218268	2.88687247
6	0.046875	2.70337487	3.7672993	3.01255603	2.84343141
7	0.054688	2.8805592	4.05194668	3.03279959	2.83611715
8	0.0625	3.05409156	4.3148363	2.58599723	2.85268643
9	0.070313	3.22487739	4.55861345	2.39908626	2.88572075
10	0.078125	3.3935645	4.78618808	2.29199693	2.93050456
11	0.085938	3.56061712	5.00038626	2.22542247	2.98390658
12	0.09375	3.72637402	5.2036915	2.1840258	3.04376742
13	0.101563	3.89108933	5.3981394	2.16023782	3.10854922
14	0.109375	4.05495998	5.58532814	2.14982508	3.17712662
15	0.117188	4.21814368	5.76648637	2.15025859	3.24865714
16	0.125	4.38077064	5.94255529	2.16000242	3.32249823
17	0.132813	4.54295132	6.1142626	2.17817316	3.39815231
18	0.140625	4.70478152	6.2821811	2.20437615	3.47522953
19	0.148438	4.86634591	6.44677183	2.23864345	3.55342174
20	0.15625	5.02772045	6.60841449	2.28144954	3.63248394
21	0.164063	5.18897422	6.76742849	2.33381231	3.71222071
22	0.171875	5.35017066	6.92408793	2.39752532	3.79247611
23	0.179688	5.5113686	7.07863081	2.4756445	3.87312601
24	0.1875	5.67262308	7.23126872	2.57356188	3.95407213
25	0.195313	5.83398594	7.38218837	2.70170209	4.03523742
26	0.203125	5.99550634	7.53155688	2.88400362	4.11656225
27	0.210938	6.15723121	7.67952974	3.19911598	4.19800153
28	0.21875	6.31920562	7.82624548	4.5413797	4.27952213
29	0.226563	6.48147309	7.9718268	3.17250291	4.36110096
30	0.234375	6.64407566	8.11640022	2.88796247	4.44272325
31	0.242188	6.80705467	8.26006777	2.72307715	4.52438117
32	0.25	6.97045023	8.40294615	2.60855953	4.60607266

Perimeter Fidelity Metric

$$\Phi(P(a,b))=-Log10(Abs(SDF\%))$$

Serial	η = b/a	CHECK SERIES -10 Gauss-Kummer	Consolidated Guillera Process	Sykora Approximation	Cantrell-Ramanujan Approximation
33	0.257813	7.13430202	8.54509796	2.52215496	4.68780047
34	0.265625	7.29864922	8.68663968	2.45382259	4.76957132
35	0.273438	7.46353024	8.82763605	2.3981557	4.85139525
36	0.28125	7.62898365	8.96822287	2.35190069	4.933285
37	0.289063	7.79504664	9.10838057	2.31294293	5.01525559
38	0.296875	7.96176056	9.24828215	2.27982795	5.09732387
39	0.304688	8.12915921	9.38783839	2.25151077	5.17950825
40	0.3125	8.29728771	9.52722546	2.22721436	5.26182837
41	0.320313	8.46617395	9.6665409	2.2063445	5.34430496
42	0.328125	8.63588312	9.80600674	2.18843618	5.42695959
43	0.335938	8.80641785	9.94485624	2.17311837	5.50981458
44	0.34375	8.97782383	10.08446112	2.16009016	5.5928929
45	0.351563	9.15020214	10.22370767	2.14910403	5.67621804
46	0.359375	9.32345144	10.36258114	2.13995388	5.75981399
47	0.367188	9.49780046	10.50360103	2.13246631	5.84370526
48	0.375	9.67304593	10.64278971	2.12649407	5.92791668
49	0.382813	9.84961486	10.78290729	2.12191109	6.01247364
50	0.390625	10.02735979	10.92643305	2.11860869	6.09740188
51	0.398438	10.20590558	11.06580587	2.11649261	6.18272749
52	0.40625	10.38687432	11.20936014	2.11548063	6.26847732
53	0.414063	10.56650849	11.34465974	2.11550071	6.35467824
54	0.421875	10.74940356	11.48883628	2.11648947	6.4413583
55	0.429688	10.93567536	11.64912318	2.11839096	6.52854554
56	0.4375	11.12067136	11.7882329	2.12115564	6.61626881
57	0.445313	11.30285941	11.91154998	2.12473959	6.70455768
58	0.453125	11.49575191	12.08274815	2.12910374	6.79344264
59	0.460938	11.68655081	12.19646458	2.13421334	6.88295448
60	0.46875	11.88459299	12.3679632	2.14003746	6.97312529
61	0.476563	12.06868531	12.47617063	2.1465486	7.06398776
62	0.484375	12.21468558	12.61925617	2.15372227	7.15557584
63	0.492188	12.4796874	12.69356722	2.16153679	7.24792467
64	0.5	12.69533132	13.13466402	2.16997295	7.34107058

Perimeter Fidelity Metric
Φ(P(a,b))=-Log10(Abs(SDF%))

Serial	η = b/a	CHECK SERIES -10 Gauss-Kummer	Consolidated Guillera Process	Sykora Approximation	Cantrell-Ramanujan Approximation
65	0.507813	12.96034033	13.26137033	2.17901383	7.43505036
66	0.515625	12.96211111	12.96211111	2.18864464	7.52990264
67	0.523438	13.13997612	13.26491485	2.19885251	7.62566816
68	0.53125	13.74381258	13.04484258	2.20962638	7.72238965
69	0.539063	13.44456155	13.44456155	2.22095691	7.8201098
70	0.546875	14	13.74737279	2.23283634	7.91887466
71	0.554688	13.4481261	13.1470961	2.24525843	8.01873076
72	0.5625	13.44991125	13.44991125	2.25821841	8.11972787
73	0.570313	13.45169804	13.75272804	2.27171291	8.22192024
74	0.578125	13.4534863	13.27739504	2.2857399	8.32535693
75	0.585938	13.45527581	13.75630581	2.3002987	8.43009697
76	0.59375	13.7580964	13.28097515	2.31538997	8.5362103
77	0.601563	13.45885791	14	2.33101563	8.64374525
78	0.609375	14	13.28455888	2.34717895	8.75277351
79	0.617188	13.76347294	14	2.36388449	8.86338997
80	0.625	13.46423615	13.76526615	2.38113819	8.97560789
81	0.632813	13.76705961	13.76705961	2.39894734	9.08958953
82	0.640625	13.46782318	13.76885317	2.41732067	9.20534836
83	0.648438	14	14	2.43626836	9.3230236
84	0.65625	13.77244004	13.77244004	2.45580218	9.44265584
85	0.664063	13.77423306	13.2971118	2.47593547	9.56439646
86	0.671875	13.07705563	13.47499563	2.49668335	9.68820873
87	0.679688	13.77781763	13.30069637	2.51806274	9.81454937
88	0.6875	13.77960892	14	2.54009254	9.94328481
89	0.695313	13.30427815	14	2.56279375	10.07468162
90	0.703125	14	13.78318895	2.58618968	10.20846337
91	0.710938	13.78497745	13.48394746	2.61030614	10.34548686
92	0.71875	13.48573481	13.78676481	2.63517164	10.48508386
93	0.726563	13.7885509	13.48752091	2.66081773	10.62838261
94	0.734375	14	13.79033565	2.68727922	10.77455689
95	0.742188	13.79211894	13.49108894	2.71459462	10.92524112
96	0.75	13.49287069	13.09493068	2.74280651	11.07957093

Perimeter Fidelity Metric
$\Phi(P(a,b))=-\text{Log10(Abs(SDF\%))}$

Serial	$\eta = b/a$	CHECK SERIES -10 Gauss-Kummer	Consolidated Guillera Process	Sykora Approximation	Cantrell-Ramanujan Approximation
97	0.757813	13.4946508	13.7956808	2.77196206	11.23817359
98	0.765625	13.49642919	13.32033793	2.80211357	11.39951918
99	0.773438	13.79923577	13.32211451	2.83331914	11.57134906
100	0.78125	14	13.49998046	2.86564348	11.74031262
101	0.789063	13.80278318	13.10381318	2.89915883	11.92772192
102	0.796875	13.80455386	13.80455386	2.93394607	12.09698369
103	0.804688	13.80632243	13.80632243	2.97009604	12.30117245
104	0.8125	13.8080888	14	3.00771116	12.46566612
105	0.820313	13.80985292	14	3.04690736	12.66372489
106	0.828125	13.20955473	13.20955473	3.08781648	13.11264472
107	0.835938	13.81337413	14	3.1305892	13.11440413
108	0.84375	13.5141011	14	3.17539869	13.5141011
109	0.851563	13.3397643	13.51585556	3.22244519	14
110	0.859375	13.51760746	13.51760746	3.27196189	14
111	0.867188	13.51935674	13.12141673	3.3242224	13.51935674
112	0.875	13.34501209	13.22007335	3.3795506	13.82213334
113	0.882813	13.04572598	12.97877919	3.43833353	13.04572598
114	0.890625	13.52458835	13.22355836	3.50103888	13.34849709
115	0.898438	14	13.82735665	3.5682389	14
116	0.90625	13.82909208	13.35197083	3.64064391	13.82909208
117	0.914063	13.52979461	13.83082461	3.71915042	14
118	0.921875	14	13.13358418	3.80491226	14
119	0.929688	13.53325078	13.35715952	3.89944937	13.83428078
120	0.9375	13.83600434	13.53497434	4.0048208	13.83600434
121	0.945313	13.83772484	13.36060358	4.123914	13.83772484
122	0.953125	13.83944223	13.53841224	4.26095851	13.83944223
123	0.960938	13.5401265	13.06300524	4.4225123	13.36403524
124	0.96875	13.84286758	13.14389758	4.6195628	13.84286758
125	0.976563	13.84457547	13.36745422	4.87270875	13.84457547
126	0.984375	13.54525014	13.54525014	5.22819662	13.54525014
127	0.992188	13.84798154	13.37086028	5.83359777	13.84798154
128	1				

APPENDIX D

THE QB64 SOURCE PROGRAM
RATEGAIN.bas

```
'       Program RATEGAIN
' A Program
' To Compute a Set of Empirical Ellipse Perimeter Estimation Mill Times
' ( with check matrix of lengths referent to EXCEL worksheet PERIMETER
LENGTHS )
' Based upon Data Located from data text file RATING.INP
'       Written by:-
'
'               James R Warren BSc MSc PhD PGCE
'               "Southgate"
'               31 Victoria Avenue
'               Bloxwich
'               Walsall
'               WS3 3HS
'
'               29 July 2019
'
'       This Program is Written in QBasic for elaboration as a compiled
'       QB64 RATEGAIN*.exe object
'
'       Compiled QB64 on a 64-bit Hewlett-Packard Pavilion 500400:
'       Total Execution Time = Not Known
'
'       REM $DYNAMIC
' Initialise the Random Number Generator
'       Randomize (Timer)
' VARIABLE TYPE DEFAULTS
'       DEFDBL A-H, O-R, T-Z
'       DEFSTR S
'       DEFINT I-K, M-N
'       DEFLNG L
' Declare Dynamic Array Dimensions
'       COMMON SHARED NUnits
'       COMMON SHARED NRX, NRY, NAR, NP
' Assign the Dynamic Arrays' Maxima and Dimension the Dynamic Arrays
'       ( None )
' Pseudo-Static Array Declarations
'       COMMON SHARED A(), B(), ETA(), H()
'       COMMON SHARED NA(),EPS(),AA()
'       COMMON SHARED EE2(), PAB()
'       COMMON SHARED D()
'       COMMON SHARED TF(), DL()
'       COMMON SHARED DGKT()
' Assign the Pseudo-Static Array Dimensions Maxima and Dimension the Pseudo-
Static Arrays
'       NETAMAX=1024
'       NS=64:NGK=10:NPRO=5
'       ReDim A(NETAMAX),B(NETAMAX),H(NETAMAX),ETA(NETAMAX),EPS(NETAMAX)
'       ReDim NA(NS),AA(NS)
'       ReDim EE2(5,NETAMAX), PAB(5,NETAMAX)
'       ReDim D(NPRO,NETAMAX)
'       ReDim TF(7,NPRO), DL(NPRO,5)
'       ReDim DGKT(7,NGK)
' Object Definitions
'       ( None )
' Public Object Instantiations
'       ( None )
' Static Array Declarations
'       ( None )
' Dynamic Array Declarations
'       ( None )
' Constant Definitions
'       Const PI = 3.14159265358979
```

```
        Const PI2 = 6.28318530717959
        Const HP = 1.5707963267949
        Const RC = 2.506628274631
        Const EMC = 0.577215664901533
        Const ENAP = 2.71828182845905
        Const DL10 = 2.30258509299405
' Declare Logical Unit Number Holders
        COMMON SHARED ISW, IW, IX
' Declare String Constant Data ( for Common Shared emulation )
        COMMON SHARED SC, SM, SCR
' Declare Formats ( for Common Shared emulation )
        COMMON SHARED SI4, SF10P8, SF11P6, SF15P12, SF13P8, SF20P15
' Define Random Number Range Parameters
'       ( none )
' String Constant Definitions
        SC = ":": SM = ",": SCR = Chr$(13) + Chr$(10)
' Format Definitions
        SI4 = "####": SF10P8 = "#.########"
        SF11P6 = "####.######":SF15P12="##.############":SF13P8=
"####.########"
        SF20P15="####.##############"
' Perform Actions
' Input Data
        PRINT
        PRINT "INPUT NUMBER OF ELLIPSE ASPECT RATIOS:";
        INPUT NETA
        PRINT
        Call PROIN(NP, KO(), SOC(), SOP(), LNC())
' Predication of Ellipse Aspect Ratios
        Call ETAFILL( NETA, ETAIW, A(), B(), H(), ETA(), EPS())
' Do Timings
        PRINT "Entering Sequence:-" :PRINT
        DStart = Timer(.001)
        Call HOLDERs64( 1, NETA, NS, ETAIW, DIW, LNC(), NA(), AA(), EPS(),
EE2(), PAB(), TF())
        Call HOLDERgak( 2, NETA, NGK, ETAIW, DIW, LNC(), NA(), AA(), EPS(),
EE2(), PAB(), TF())
        Call HOLDERcgp(3,NETA,ETAIW,DIW,LNC(),EPS(),EE2(),PAB(),TF())
        Call HOLDERsyk(4,NETA,A(),B(),ETAIW,DIW,LNC(),EPS(),EE2(),PAB(),TF())
        Call
HOLDERcar(5,NETA,A(),B(),H(),ETAIW,DIW,LNC(),EPS(),EE2(),PAB(),TF())
        Call GAIN(NETA,ETA(),PAB(),D(),DL(),TF())
' Output the Results
        DFinish = Timer(.001)
        PMT = DFinish - DStart
        PRINT "Total Execution Time = ";PMT
' Output and Terminate
        Call PABOUT( NETA, ETA(), PAB())
        Call TOUT1( PMT, NETA, NOP, KO(), SOC(), SOP(), LNC(), DLTM(), TF())
        Call GAINOUT( NETA, DL() )
        End

        Function DLOG10(X)
' A Function to Return the Double-Precision Base Ten Logarithm of Argument X
'       Arguments:-


'       X       The Argument whose Base Ten Logarithm is to be Determined
'
        IF X=0 THEN
            DLOG10=-15
        ELSE
            DLOG10 = Log(X) / DL10
```

```
        END IF
        End Function

        Function FACT(JX)
' A Function to Return A Double-Precision Factorial of the Short Integer JX
'       Arguments:-
'           JX      The Argument whose Factorial is to be Determined
'
        FACT=1
        IF JX>0 THEN
            FOR I=1 TO JX
                FACT=FACT*I
            NEXT I
        END IF
        End Function

        Function ELLIPTIC(I,J)
' A Function to Compute the Complete Elliptic Integral of the Second Kind
Kernal
'       Arguments:
'           I       The Serial Subscript of the Ellipse Aspect Ratio
'           J       The Serial Subscript of the Simpson Integral Term
'           EPS()   The Array of Ellipse Eccentricitys
'           AA()    The Array of Sweep Angles
'
        ELLIPTIC=SQR(1-EPS(I)^2*SIN(AA(J))^2)
        End Function

        Sub FILEOPEN(ISW,NETA,SPath,SFile,SExt)
'A Subroutine to Open a Sequential File for Transput
'       Arguments:-
'           ISW     The Logical Unit Number
'           NETA    The Number of Ellipse Aspect Ratios
'           SPath   The Directories' Path
'           SFile   The Filename
'           SExt    The Filename Extension
'
        IF ISW=1 THEN
            SConcat=SPath+SFile+SExt
            OPEN SConcat FOR INPUT AS ISW
        ELSE
            SNETA=MID$(STR$(NETA),2)
            STime=TIME$
            SHour=LEFT$(Time$,2)
            SMinute=MID$(Time$,4,2)
            SSecond=RIGHT$(Time$,2)
            Snocolon=SHour+SMinute+SSecond
            SConcat=SPath+SFile+"x"+SNETA+"x"+Snocolon+SExt
            OPEN SConcat FOR OUTPUT AS ISW
        END IF
        End Sub

        Sub PROIN(NP, KO(), SOC(), SOP(), LNC())
' A Subroutine to Input Operations' Descriptors Data
'       Arguments:-
'           NP      The Number of Processes
'           KO()    The Array  of Operations Serial Numbers
'           SOC()   The Array  of Operation Codes
'           SOP()   The Array  of Operation Descriptions
'           LNC()   The Array  of Required Cycles
'
' Open the File
        ISW=1
```

```
        Call FILEOPEN(ISW,NETA,"C:\RGFILES\INPUT\","RATING",".INP")
' Input the Data
        Input #ISW, NP
        For I = 1 To NP
            Input #ISW, KO(I), SOC(I), SOP(I), LNC(I)
        Next I
' Terminate
        Close ISW
        End Sub


        Sub PABOUT( NETA, ETA(), PAB())
' A Subroutine to Output Operations' Descriptors Data with
' Operation or Function Timings
'       Arguments:-
'           NETA    The Number of Ellipse Aspect Ratios
'           ETA()   The Array of Ellipse Aspect Ratios
'           PAB()   The Array of Ellipse Perimeter Lengths
'
'
' Open the File
        ISW=2
        Call FILEOPEN(ISW,NETA,"C:\RGFILES\OUTPUT\","RATEGAIN",".PER")
' Output the Data
        Write #ISW,SM,SM,"Perimeter Length Metric"
        Write #ISW,SM,SM,"P(a,b)"
        Write #ISW,
        Write #ISW,
        Write #ISW,SM,SM,"Fiducial Series (64- Simpson)","CHECK SERIES 10-
GaussKummer","Consolidated Guillera Process","Sykora
Approximation","Cantrell-Ramanujan Approximation"
        Write #ISW,"Serial","Eta=b/a"
        FOR I=0 TO NETA
            Write #ISW,I,ETA(I),PAB(1,I),PAB(2,I),PAB(3,I),PAB(4,I),PAB(5,I)
        NEXT I
' Terminate
        Close ISW
        End Sub


        Sub TOUT( PMT, NETA, NP, KO(), SOC(), SOP(), LNC(), DLTM(), TF())
' A Subroutine to Output Processes' Descriptors Data with
' Process Timings
'       Arguments:-
'           PMT     The Total  Execution Time
'           NETA    The Number of Ellipse Aspect Ratios
'           NP      The Number of Processes
'           KO()    The Array  of Operations Serial Numbers
'           SOC()   The Array  of Operation Codes
'           SOP()   The Array  of Operation Descriptions
'           LNC()   The Array  of Required Cycles
'           DLTM()  The Array  of Operation or Function Timings
'           TF()    The Array  of Process and Carcass Timings
'
' Open the File
        ISW=2
        Call FILEOPEN(ISW,NETA,"C:\RGFILES\OUTPUT\","RATEGAIN",".TIM")
' Output the Data
        Write #ISW,"ELLIPSE PERIMETER LENGTH PROCESSES TIMING REPORT"
        Write #ISW,"All Times in Seconds"
        Write #ISW,
        Write #ISW,"Operation Number",KO(1),KO(2),KO(3),KO(4),KO(5)
        Write #ISW,"Operation Code",SOC(1),SOC(2),SOC(3),SOC(4),SOC(5)
        Write #ISW,"Operation Description",SOP(1),SOP(2),SOP(3),SOP(4),SOP(5)
        Write #ISW,"Number of Cycles",LNC(1),LNC(2),LNC(3),LNC(4),LNC(5)
```

```
        Write #ISW,
        Write #ISW,"Number of Process
Cycles",LNC(1),LNC(2),LNC(3),LNC(4),LNC(5)
        Write #ISW,
        Write #ISW,"Process Predicate
Time",TF(0,1),TF(0,2),TF(0,3),TF(0,4),TF(0,5)
        Write #ISW,"Process Elaboration
Time",TF(1,1),TF(1,2),TF(1,3),TF(1,4),TF(1,5)
        Write #ISW,
        Write #ISW,"Carcase Predicate
Time",TF(2,1),TF(2,2),TF(2,3),TF(2,4),TF(2,5)
        Write #ISW,"Carcase Elaboration
Time",TF(3,1),TF(3,2),TF(3,3),TF(3,4),TF(3,5)
        Write #ISW,
        Write #ISW,"Total    Process
Time",TF(4,1),TF(4,2),TF(4,3),TF(4,4),TF(4,5)
        Write #ISW,"Total    Carcase
Time",TF(5,1),TF(5,2),TF(5,3),TF(5,4),TF(5,5)
        Write #ISW,
        Write #ISW,"Nominal Mill
Time",TF(6,1),TF(6,2),TF(6,3),TF(6,4),TF(6,5)
        Write #ISW,"Nominal Process
Time",TF(7,1),TF(7,2),TF(7,3),TF(7,4),TF(7,5)
        Write #ISW,
        Write #ISW,"Overall Routine Duration = ",PMT
' Terminate
        Close ISW
        End Sub


        Sub TOUT1( PMT, NETA, NP, KO(), SOC(), SOP(), LNC(), DLTM(), TF())
' A Subroutine to Output Processes' Descriptors Data with
' Process Timings
'       Arguments:-
'           PMT      The Total   Execution Time
'           NETA     The Number of Ellipse Aspect Ratios
'           NP       The Number of Processes
'           KO()     The Array  of Operations Serial Numbers
'           SOC()    The Array  of Operation Codes
'           SOP()    The Array  of Operation Descriptions
'           LNC()    The Array  of Required Cycles
'           DLTM()   The Array  of Operation or Function Timings
'           TF()     The Array  of Process and Carcass Timings
'
' Dimension Caption Array
        Dim SCAP(12)
' Fill Caption Array
        SCAP(0)="Process Predicate Time"
        SCAP(1)="Process Elaboration Time"
        SCAP(2)="Carcase Predicate Time"
        SCAP(3)="Carcase Elaboration Time"
        SCAP(4)="Total Process Time"
        SCAP(5)="Total Carcase Time"
        SCAP(6)="Nominal Mill Time"
        SCAP(7)="Nominal Process Time"
' Open the File
        ISW=2
        Call FILEOPEN(ISW,NETA,"C:\RGFILES\OUTPUT\","RATEGAIN",".TIM")
' Output the Data
        Write #ISW,"ELLIPSE PERIMETER LENGTH PROCESSES TIMING REPORT"
        Write #ISW,"All Times in Seconds"
        Write #ISW,"Committal Time:",TIME$
        Write #ISW,"Committal Date:",DATE$
        Write #ISW,
```

```
        Write #ISW,"Operation Number",KO(1),KO(2),KO(3),KO(4),KO(5)
        Write #ISW,"Operation Code",SOC(1),SOC(2),SOC(3),SOC(4),SOC(5)
        Write #ISW,"Operation Description",SOP(1),SOP(2),SOP(3),SOP(4),SOP(5)
        Write #ISW,
        Write #ISW,"Number of Process
Cycles",LNC(1),LNC(2),LNC(3),LNC(4),LNC(5)
        Write #ISW,
        FOR I = 0 TO 7
            PRTNT #ISW,SCAP(I);
            FOR J= 1 TO 5
                PRINT #ISW,SM;
                PRINT #ISW,USING SF20P15; TF(I,J);
            NEXT J
            PRINT #ISW,
        NEXT I
        Write #ISW,
        DSMT=0.0
        FOR J=1 TO 5
            DSMT=DSMT+TF(6,J)
        NEXT J
        Write #ISW,"Overall Routine Duration",PMT
        Write #ISW,"Sum of Mill Times",DSMT
' Terminate
        Close ISW
        End Sub


        Sub GAIN(NETA,ETA(),PAB(),D(),DL(),TF())
' A Subroutine to Compute Process Denary Digit Gain and Process Quality
'       Argumenta:
'           NETA      The Number of Ellipse Aspect Ratios
'           ETA()     The Array of Ellipse Aspect Ratios
'           PAB()     The Array of Ellipse Perimeter Lengths
'           D()       The Array of Perimeter Length Differences
'           DL()      The Array of Process Gain Statistics
'           TF()      The Array of Process Timings
'
' Compute Perimeter Differences
        FOR I=1 TO NETA
            FOR J=2 TO 5
                D(J,I)=PAB(1,I)-PAB(J,I)
            NEXT J
        NEXT I
' Compute Mean Gains and Process Qualities
        FOR J=2 TO 5
            TOT=0.0
            FOR I=1 TO NETA
                TOT=TOT+D(J,I)
            NEXT I
            DL(J,1)=TOT/NETA
            DL(J,2)=TF(7,J)
            DL(J,3)=-DLOG10(ABS(DL(J,1)))
            DL(J,4)=-DLOG10(TF(7,J))
            DL(J,5)=DL(J,3)/DL(J,4)
        NEXT J
' Terminate
        End Sub


        Sub GAINOUT(NETA,DL())
' A Subroutine to Output Process Gain Results
' A Subroutine to Compute Process Denary Digit Gain and Process Quality
'       Argumenta:
'           NETA      The Number of Ellipse Aspect Ratios
'           DL()      The Array of Process Gain Statistics
```

```
'
' Dimension Caption Array
      Dim SCAP(12)
' Fill Caption Array
      SCAP(0)="Mean NPT"
      SCAP(1)="Mean Gain"
      SCAP(2)="NPT"
      SCAP(3)="Log10 Absolute Mean Gain"
      SCAP(4)="Log10 NPT"
      SCAP(5)="Process Quality"
      SCAP(6)="Not Applicable"
' Open the File
      ISW=2
      Call FILEOPEN(ISW,NETA,"C:\RGFILES\OUTPUT\","RATEGAIN",".GAN")
' Output the Data
      PRINT #ISW,"ELLIPSE PERIMETER LENGTH PROCESSES GAIN REPORT"
      PRINT #ISW,"All Times in Seconds"
      PRINT #ISW,"Committal Time:",SM,TIME$
      PRINT #ISW,"Committal Date:",SM,DATE$
      PRINT #ISW,"Number of Ellipse Aspect Ratios:",SM,NETA
      PRINT #ISW,
      PRINT #ISW,SM;"Fiducial Series (64- Simpson)";SM;"CHECK SERIES 10-
GaussKummer";SM;"Consolidated Guillera Process";
      PRINT #ISW,SM;"Sykora Approximation";SM;"Cantrell-Ramanujan
Approximation"
      PRINT #ISW,
      FOR I=1 TO 5
          PRINT #ISW,SCAP(I);SM;SCAP(6);
          FOR J=2 TO 5
              PRINT #ISW,SM;
              PRINT #ISW,USING SF15P12;DL(J,I);
          NEXT J
          PRINT #ISW,
      NEXT I
' Terminate
      Close ISW
      End Sub

      Sub METAtrailingPABEE2(I1,I2,ETA(),A(),B(),EPS(),H(),EE2(),PAB())
' A Subroutine to Display Selected Output Values
'      Arguments:-
'          I1        The Start  Subscript of an Output Row
'          I2        The Finish Subscript of an Output Row
'          ETA()     The Array of Ellipse Aspect Ratios
'          A()       The Array of Ellipse Major Semi-Axes
'          B()       The Array of Ellipse Minor Semi-Axes
'          EPS()     The Array of Ellipse Eccentricities
'          H()       The Array of Ellipse Gauss-Kummer Parameters
'          EE2()     The Array of Complete Elliptic Integrals of the Second Kind
'          PAB()     The Array of Ellipse Perimeter Lengths
'
      PRINT
      FOR I=I1 TO I2
          PRINT
          PRINT "Serial ";I
          PRINT" ETA = ";ETA(I);" A = ";A(I);" B = ";B(I);" EPS = ";EPS(I)
          PRINT " H = "; H(I)
          PRINT
          PRINT " EE2 smsn = "; EE2(1,I); " PAB smsn = "; PAB(1,I)
          PRINT " EE2 gk   = "; EE2(2,I); " PAB gk   = "; PAB(2,I)
          PRINT " EE2 cgp  = "; EE2(3,I); " PAB cgp  = "; PAB(3,I)
          PRINT " EE2 syk  = "; EE2(4,I); " PAB syk  = "; PAB(4,I)
          PRINT " EE2 cr   = "; EE2(5,I); " PAB cr   = "; PAB(5,I)
```

```
        NEXT I
        PRINT
' Terminate
        End Sub


        Sub ETAFILL( NETA, ETAIW, A(), B(),H(), ETA(), EPS())
' A Subroutine to Fill the Ellipse Prolation Arrays
'       Arguments:
'           NETA       The Number of Ellipse Aspect Ratios
'           ETAIW      The Ellipse Aspect Ratio Interval Width
'           A()        The Array of Ellipse Major Semi-Axes
'           B()        The Array of Ellipse Minor Semi-Axes
'           H()        The Array of Ellipse Gauss-Kummer Parameters
'           ETA()      The Array of Ellipse Aspect Ratios
'           EPS()      The Array of Ellipse Eccentricities
'
        ETAIW=1/NETA
        FOR I=0 TO NETA
            A(I)=1:ETA(I)=ETAIW*I:B(I)=ETA(I)
            EPS(I)=SQR(1-(B(I)/A(I))^2)
            H(I)=(A(I)-B(I))^2/(A(I)+B(I))^2
        NEXT I
        End Sub


        Sub ALTERNATOR( NS, NA())
' A Subroutine to Fill the Simpson Integration
' Term Coefficient Alternator Array NA(NS)
'       Arguments:
'           NS         The Number of Simpson Integration Terms
'           NA()       The Array  of Simpson Term Alternators
'
        NA(0)=1
        FOR J=1 TO NS-1
            NA(J)=3-(-1)^J
        NEXT J
        NA(NS)=1
        End Sub


        Sub ANGLEFILL(NS,DIW,AA())
' A Subroutine to Fill the Angle Array AA(NA)
'       Arguments:
'           NS         The Number of Simpson Integration Terms
'           DIW        The Simpson Integration Interval Width
'           AA()       The Array of Simpson Integration Angles
'
        DIW=HP/NS
        FOR J=0 TO NS
            AA(J)=J*DIW
        NEXT J
        End Sub


        Sub SIMPSON( NETA, ETAIW, DIW, NS, EPS(), AA(), NA(), EE2(), PAB())
' A Subroutine to Compute a Simpson Integral Ellipse Perimeter
'       Arguments:
'           NETA       The Number of Ellipse Aspect Ratios
'           ETAIW      The Ellipse Aspect Ratio Interval Width
'           DIW        The Simpson Integration Interval Width
'           NS         The Number of Simpson Integration Terms
'           EPS()      The Array of Ellipse Eccentricities
'           AA()       The Array of Simpson Integration Angles
'           NA()       The Array of Simpson Term Alternators
'           EE2()      The Array of Complete Elliptic Integrals of the Second Kind
'           PAB()      The Array of Ellipse Perimeter Lengths
```

```
        FOR I=0 TO NETA
            TOT=1.0
                FOR J=1 TO NS-1
                    DNA=NA(J)
                    FRAG=DNA * SQR(1-(EPS(I)^2)*(SIN(AA(J))^2))
                    TOT=TOT + FRAG
                NEXT J
            TOT=TOT+SQR(1-EPS(I)^2*SIN(AA(NS))^2)
            EE2(1,I)=TOT*DIW/3
            PAB(1,I)=4*A(I)*EE2(1,I)
        NEXT I
        End Sub

        Sub GKFACTS(NGK,DGKT())
' A Subroutine to Compute the Factorial Components of
' Gauss-Kummer Series Terms
'    Arguments:
'        NGK        The Number of Gauss-Kummer Series Terms
'        DGKT()    The Array of Functional Components
'
        FOR J=0 TO NGK
            DGKT(0,J)=J
            DGKT(1,J)=FACT(J)
            DGKT(2,J)=FACT(2*DGKT(0,J))
            DGKT(3,J)=DGKT(1,J)^2
            DGKT(4,J)=DGKT(2,J)/DGKT(3,J)
            DGKT(5,J)=((-1)^(DGKT(0,J)+1))/(2^(2*DGKT(0,J))*(2*DGKT(0,J)-1))
            DGKT(6,J)=DGKT(4,J)*DGKT(5,J)
            DGKT(7,J)=DGKT(6,J)^2
        NEXT J
        End Sub

        Sub GKTERMS(NETA,NGK,DGKT(),A(),B(),H(),PAB(),EE2())
' A Subroutine to Compute the Gauss-Kummer Series Terms
' and hence the Ellipse Gauss-Kummer Parameter,
' the Ellipse Perimeter Length and the
' Ellipse Complete Elliptic Integral of the Second Kind
'    Arguments:
'        NETA        The Number of Ellipse Aspect Ratios
'        NGK         The Number of Gauss-Kummer Series Terms
'        DGKT()      The Array of Factorial Components
'        A()         The Array of Ellipse Major Semi-Axes
'        B()         The Array of Ellipse Minor Semi-Axes
'        H()         The Array of Gauss-Kummer Parameters
'        PAB()       The Array of Ellipse Perimeter Lengths
'        EE2()       The Array of Ellipse Complete Elliptic Integrals of the
Second Kind
'
        FOR I=0 TO NETA
            TOT=1
            FOR J=1 TO NGK
                TOT=TOT+DGKT(7,J)*H(I)^DGKT(0,J)
            NEXT J
            PAB(2,I)=PI*(A(I)+B(I))*TOT
            EE2(2,I)=PAB(2,I)/(4*A(I))
        NEXT I
        End Sub

        Sub CGP(NETA,A(),B(),PAB(),EE2())
' A Subroutine to Compute the Ellipse Perimeter Lengths
' and the Ellipse Complete Elliptic Integrals of the Second Kind
' according to the Consolidated Guillera Process
```

```
'        Arguments:
'          NETA      The Number of Ellipse Aspect Ratios
'          A()       The Array  of Ellipse Major Semi-Axes
'          B()       The Array  of Ellipse Minor Semi-Axes
'          PAB()     The Array  of Ellipse Perimeter Lengths
'          EE2()     The Array  of Ellipse Complete Elliptic Integrals of the
' Second Kind
'
        FOR I=1 TO NETA
          ETAL=B(I)/A(I)
          PSI=SQR(ETAL)
          Q1=(1-PSI)/(1+PSI)
          Q4M1=1-Q1^4
          GAM=Q4M1^0.25
          Q2=(1-GAM)/(1+GAM)
          PQ1=1+Q1
          PQ2=1+Q2
          PF1=2*PI*B(I)^2/A(I)
          PF2= (PQ1*PQ2)^2
          PF3=((2*PQ2)/(ETAL*PQ1))^2
          PF4=(Q1*Q4M1/PQ1)
          PF5=4*Q2/PQ2^5
          PAB(3,I)=PF1*(PF2+PF3*(PF4+PF5))
          EE2(3,I)=PAB(3,I)/(4*A(I))
        NEXT I
        End Sub

        Sub SYKORA(NETA,A(),B(),PAB(),EE2())
' A Subroutine to Compute the Ellipse Perimeter Lengths
' and the Ellipse Complete Elliptic Integrals of the Second Kind
' according to the Sykora Formula
'        Arguments:
'          NETA      The Number of Ellipse Aspect Ratios
'          A()       The Array  of Ellipse Major Semi-Axes
'          B()       The Array  of Ellipse Minor Semi-Axes
'          PAB()     The Array  of Ellipse Perimeter Lengths
'          EE2()     The Array  of Ellipse Complete Elliptic Integrals of the
' Second Kind
'
        FOR I=1 TO NETA
          AB=A(I)*B(I)
          AMB=A(I)-B(I)
          APB=A(I)+B(I)
          PF1=4*(PI*AB+AMB^2)/APB
          PF2=0.5*(AB/APB)*(AMB^2/(PI*AB+APB^2))
          PAB(4,I)=PF1-PF2
          EE2(4,I)=PAB(4,I)/(4*A(I))
        NEXT I
        End Sub

        Sub CR(NETA,A(),B(),H(),PAB(),EE2())
' A Subroutine to Compute the Ellipse Perimeter Lengths
' and the Ellipse Complete Elliptic Integrals of the Second Kind
' according to the Cantrell-Ramanujan Formula
'        Arguments:
'          NETA      The Number of Ellipse Aspect Ratios
'          A()       The Array  of Ellipse Major Semi-Axes
'          B()       The Array  of Ellipse Minor Semi-Axes
'          H()       The Array  of Gauss-Kummer Parameters
'          PAB()     The Array  of Ellipse Perimeter Lengths
'          EE2()     The Array  of Ellipse Complete Elliptic Integrals of the
' Second Kind
'
```

```
        FOR I=1 TO NETA
           APB=A(I)+B(I)
           PAB(5,I)=PI*APB*(1+(3*H(I)/(10+SQR(4-3*H(I))))+(4/PI-
14/11)*H(I)^12)
           EE2(5,I)=PAB(5,I)/(4*A(I))
        NEXT I
        End Sub


' ****** START OF HOLDER SECTION

        Sub
HOLDERs64(KN,NETA,NS,ETAIW,DIW,LNC(),NA(),AA(),EPS(),EE2(),PAB(),TF())
' A Subroutine to Perform and Time a Process Elaboration
'     Arguments:
'        KN        The Process Serial Number
'        NETA      The Number of Ellipse Aspect Ratios
'        NS        The Number of Simpson Integration Terms
'        ETAIW     The Ellipse Aspect Ratio Interval Width
'        DIW       The Simpson Integration Interval Width
'        LNC()     The Array of Required Cycles
'        NA()      The Array of Simpson Term Alternators
'        AA()      The Array of Simpson Integration Angles
'        EPS()     The Array of Ellipse Eccentricities
'        EE2()     The Array of Complete Elliptic Integrals of the Second Kind
'        PAB()     The Array of Ellipse Perimeter Lengths
'        TF()      The Array of Process and Carcass Timings
'
' Do Process Predicate
        TS=TIMER(.001)
           Call ALTERNATOR( NS, NA())
           Call ANGLEFILL( NS, DIW, AA())
        TE=TIMER(.001)
        TF(0,KN)=TE-TS
' Do Process Elaboration Cycles
        TS=TIMER(.001)
        FOR LI=O TO LNC(KN)
           Call SIMPSON( NETA, ETAIW, DIW, NS, EPS(), AA(), NA(),EE2(), PAB())
        NEXT LI
        TE=TIMER(.001)
        TF(1,KN)=TE-TS
' Do Carcase Predicate
        TS=TIMER(.001)
        TE=TIMER(.001)
        TF(2,KN)=TE-TS
' Do Carcase Elaboration Cycles
        TS=TIMER(.001)
        FOR LI=O TO LNC(KN)
        NEXT LI
        TE=TIMER(.001)
        TF(3,KN)=TE-TS
' Calculate Summative Timings
        TF(4,KN)=TF(0,KN)+TF(1,KN)
        TF(5,KN)=TF(2,KN)+TF(3,KN)
        TF(6,KN)=TF(4,KN)-TF(5,KN)
        TF(7,KN)=TF(6,KN)/LNC(KN)
' Terminate
        End Sub


        Sub
HOLDERgak(KN,NETA,NGK,ETAIW,DIW,LNC(),NA(),AA(),EPS(),EE2(),PAB(),TF())
' A Subroutine to Perform and Time a Process Elaboration
'     Arguments:
'        KN        The Process Serial Number
```

```
'          NETA      The Number of Ellipse Aspect Ratios
'          NGK       The Number of Gauss-Kummer Series Terms
'          ETAIW     The Ellipse Aspect Ratio Interval Width
'          DIW       The Simpson Integration Interval Width
'          LNC()     The Array of Required Cycles
'          NA()      The Array of Simpson Term Alternators
'          AA()      The Array of Simpson Integration Angles
'          EPS()     The Array of Ellipse Eccentricities
'          EE2()     The Array of Complete Elliptic Integrals of the Second Kind
'          PAB()     The Array of Ellipse Perimeter Lengths
'          TF()      The Array of Process and Carcass Timings
'
' Do Process Predicate
      TS=TIMER(.001)
          Call GKFACTS(NGK,DGKT())
      TE=TIMER(.001)
      TF(0,KN)=TE-TS
' Do Process Elaboration Cycles
      TS=TIMER(.001)
      FOR LI=O TO LNC(KN)
          Call GKTERMS(LNC(2),NGK,DGKT(),A(),B(),H(),PAB(),EE2())
      NEXT LI
      TE=TIMER(.001)
      TF(1,KN)=TE-TS
' Do Carcase Predicate
      TS=TIMER(.001)
      TE=TIMER(.001)
      TF(2,KN)=TE-TS
' Do Carcase Elaboration Cycles
      TS=TIMER(.001)
      FOR LI=O TO LNC(KN)
      NEXT LI
      TE=TIMER(.001)
      TF(3,KN)=TE-TS
' Calculate Summative Timings
      TF(4,KN)=TF(0,KN)+TF(1,KN)
      TF(5,KN)=TF(2,KN)+TF(3,KN)
      TF(6,KN)=TF(4,KN)-TF(5,KN)
      TF(7,KN)=TF(6,KN)/LNC(KN)
' Terminate
      End Sub

      Sub HOLDERcgp(KN,NETA,ETAIW,DIW,LNC(),EPS(),EE2(),PAB(),TF())
' A Subroutine to Perform and Time a Process Elaboration
'    Arguments:
'          KN        The Process Serial Number
'          NETA      The Number of Ellipse Aspect Ratios
'          ETAIW     The Ellipse Aspect Ratio Interval Width
'          DIW       The Simpson Integration Interval Width
'          LNC()     The Array of Required Cycles
'          EPS()     The Array of Ellipse Eccentricities
'          EE2()     The Array of Complete Elliptic Integrals of the Second Kind
'          PAB()     The Array of Ellipse Perimeter Lengths
'          TF()      The Array of Process and Carcass Timings
'
' Do Process Predicate
      TS=TIMER(.001)
      TE=TIMER(.001)
      TF(0,KN)=TE-TS
' Do Process Elaboration Cycles
      TS=TIMER(.001)
      FOR LI=O TO LNC(KN)
          Call CGP(NETA,A(),B(),PAB(),EE2())
```

```
        NEXT LI
        TE=TIMER(.001)
        TF(1,KN)=TE-TS
' Do Carcase Predicate
        TS=TIMER(.001)
        TE=TIMER(.001)
        TF(2,KN)=TE-TS
' Do Carcase Elaboration Cycles
        TS=TIMER(.001)
        FOR LI=O TO LNC(KN)
        NEXT LI
        TE=TIMER(.001)
        TF(3,KN)=TE-TS
' Calculate Summative Timings
        TF(4,KN)=TF(O,KN)+TF(1,KN)
        TF(5,KN)=TF(2,KN)+TF(3,KN)
        TF(6,KN)=TF(4,KN)-TF(5,KN)
        TF(7,KN)=TF(6,KN)/LNC(KN)
' Terminate
        End Sub


        Sub HOLDERsyk(KN,NETA,A(),B(),ETAIW,DIW,LNC(),EPS(),EE2(),PAB(),TF())
' A Subroutine to Perform and Time a Process Elaboration
'       Arguments:
'           KN      The Process Serial Number
'           NETA    The Number of Ellipse Aspect Ratios
'           A()     The Array  of Ellipse Major Semi-Axes
'           B()     The Array  of Ellipse Minor Semi-Axes
'           ETAIW   The Ellipse Aspect Ratio Interval Width
'           DIW     The Simpson Integration Interval Width
'           LNC()   The Array of Required Cycles
'           EPS()   The Array of Ellipse Eccentricities
'           EE2()   The Array of Complete Elliptic Integrals of the Second Kind
'           PAB()   The Array of Ellipse Perimeter Lengths
'           TF()    The Array of Process and Carcass Timings
'
' Do Process Predicate
        TS=TIMER(.001)
        TE=TIMER(.001)
        TF(O,KN)=TE-TS
' Do Process Elaboration Cycles
        TS=TIMER(.001)
        FOR LI=O TO LNC(KN)
            Call SYKORA(NETA,A(),B(),PAB(),EE2())
        NEXT LI
        TE=TIMER(.001)
        TF(1,KN)=TE-TS
' Do Carcase Predicate
        TS=TIMER(.001)
        TE=TIMER(.001)
        TF(2,KN)=TE-TS
' Do Carcase Elaboration Cycles
        TS=TIMER(.001)
        FOR LI=O TO LNC(KN)
        NEXT LI
        TE=TIMER(.001)
        TF(3,KN)=TE-TS
' Calculate Summative Timings
        TF(4,KN)=TF(O,KN)+TF(1,KN)
        TF(5,KN)=TF(2,KN)+TF(3,KN)
        TF(6,KN)=TF(4,KN)-TF(5,KN)
        TF(7,KN)=TF(6,KN)/LNC(KN)
' Terminate
```

```
        End Sub

        Sub
HOLDERcar(KN,NETA,A(),B(),H(),ETAIW,DIW,LNC(),EPS(),EE2(),PAB(),TF())
' A Subroutine to Perform and Time a Process Elaboration
'       Arguments:
'           KN          The Process Serial Number
'           NETA        The Number of Ellipse Aspect Ratios
'           A()         The Array  of Ellipse Major Semi-Axes
'           B()         The Array  of Ellipse Minor Semi Axes
'           H()         The Array  of Gauss-Kummer Parameters
'           ETAIW       The Ellipse Aspect Ratio Interval Width
'           DIW         The Simpson Integration Interval Width
'           LNC()       The Array of Required Cycles
'           EPS()       The Array of Ellipse Eccentricities
'           EE2()       The Array of Complete Elliptic Integrals of the Second Kind
'           PAB()       The Array of Ellipse Perimeter Lengths
'           TF()        The Array of Process and Carcass Timings
'
' Do Process Predicate
        TS=TIMER(.001)
        TE=TIMER(.001)
        TF(0,KN)=TE-TS
' Do Process Elaboration Cycles
        TS=TIMER(.001)
        FOR LI=O TO LNC(KN)
            Call CR(NETA,A(),B(),H(),PAB(),EE2())
        NEXT LI
        TE=TIMER(.001)
        TF(1,KN)=TE-TS
' Do Carcase Predicate
        TS=TIMER(.001)
        TE=TIMER(.001)
        TF(2,KN)=TE-TS
' Do Carcase Elaboration Cycles
        TS=TIMER(.001)
        FOR LI=O TO LNC(KN)
        NEXT LI
        TE=TIMER(.001)
        TF(3,KN)=TE-TS
' Calculate Summative Timings
        TF(4,KN)=TF(0,KN)+TF(1,KN)
        TF(5,KN)=TF(2,KN)+TF(3,KN)
        TF(6,KN)=TF(4,KN)-TF(5,KN)
        TF(7,KN)=TF(6,KN)/LNC(KN)
' Terminate
        End Sub
```

On the Morphology and Dynamics of
Spinning Ellipsoids

by
James R Warren BSc MSc PhD PGCE

PART ONE
ASPECTS OF ELLIPSOIDAL GEOMETRY

Some Definitions

The ellipsoid is a modified sphere whose orthogonal sections are ellipses. Therefore, in general, its Principal Radii a, b and c differ along the respective x, y and z axes. The general equation of an ellipsoid is:-

$$\frac{a^2}{x^2} + \frac{b^2}{y^2} + \frac{c^2}{z^2} = 1$$
Equation 1.1

where it is understood that Equation 1.1 defines any point upon the *surface* of the ellipsoid.

For the TRIAXIAL ellipsoid all of a, b and c differ. On the other hand a BIAXIAL ellipsoid has two common values amongst its principal axes.

Because spinning bodies tend to form flattened spheres (i.e. approximations of biaxial ellipsoids) our attention will focus upon the geometry and mensuration of biaxial ellipsoids, though we shall review treatments of the general, triaxial, case.

A BIAXIAL ellipsoid shorted along z is called an OBLATE ellipsoid and a biaxial ellipsoid extended along z is called a PROLATE ellipsoid.

Biaxial Eccentricity

A biaxial ellipsoid, which our spinning planet approximates, has a shortened height c whilst its equatorial dimensions a and b are the same.

Accordingly we may define Flattening f in this way:-

$$f = \frac{a - c}{a} = 1 - \frac{c}{a} = 1 - \lambda = 1 - s$$
Equation 1.2

and Eccentricity e is:-

$$e = \sqrt{1 - \lambda^2}$$
Equation 1.3

Clearly, the concept of Flattening is only appropriate to oblate ellipsoids.

The Argument φ

φ is a function of λ (sometimes noted as s) and is defined by:-

$$\varphi = \cos^{-1} \lambda$$
Equation 1.4

It is an argument of the elliptic integrals defined below.

The Argument k

The second argument of elliptic integrals of interest to us is k, defined as:-

$$k = \sqrt{\frac{a^2(b^2 - c^2)}{b^2(a^2 - c^2)}}$$
Equation 1.5

From this relation it is of course the case that:-

$$k^2 = \frac{a^2(b^2 - c^2)}{b^2(a^2 - c^2)}$$
Equation 1.6

k must always be unity or greater than unity for real solutions.

When we discuss barely-oblate objects, for example the planets Earth and Mars, we compute that k only differs from one in the thirteenth or fourteenth decimal position and this can of course engender numerical problems. For the Martian case k = 1.00000000000022 and k^2 = 1.00000000000044. With these tiny differences it is of course the case that:-

$$(1 + \Delta x)^2 \approx 1 + 2.\Delta x$$
Equation 1.7

I have not investigated the effects of this approximation.

Some Convenient Functions

Specific Defect

As is our usual convention we will assess the accuracy of any present trial method against a fiducial equation or algorithm using (Percentage) Specific Defect, PSD(x_{fido}, x_{test}), where x_{fido} is the Fiducial Value and x_{test} the trial method in hand.

The relevant function is:-

$$PSD(x_{fido}, x_{test}) = 100\left(\frac{x_{fido} - x_{test}}{x_{fido}}\right)$$

Equation 1.8

Specific Defect is essentially the same as Standard Deviation and Root Mean Square when averaged for multiple data, but unlike those it preserves the *sign* of discrepancy, and is convenient for pairs of results.

Trigonometrical Inverse Sine

We may safely use intrinsic ASIN functions with any respectable programming language or software utility, but where convenient we may revert (for example) to the MacLaurin Series definition for ArcSine:-

$$AS(x, n) = \sum_{i=0}^{n}\left(\frac{(2i)!}{2^{2i}.(i!)^2} \cdot \frac{x^{2i+1}}{2i+1}\right)$$

Equation 1.9

Writing in 2020, available 64-bit software packages usually yield fifteen-figure accuracy, though ellipsoidal mensuration is very sensitive to the characteristics of the algorithm applied.

The Incomplete Elliptic Integral of the First Kind

The Incomplete Elliptic Integral of the First Kind $F(\varphi, k)$ is defined by the function:-

$$FT(\varphi, k) = \int_{0}^{\varphi}\frac{1}{\sqrt{1 - k^2.(\sin\theta)^2}}.d\theta$$

Equation 1.10

The Incomplete Elliptic Integral of the Second Kind

The Incomplete Elliptic Integral of the Second Kind $E(\varphi, k)$ is defined by the function:-

$$ET(\varphi, k) = \int_{0}^{\varphi}\sqrt{1 - k^2.(\sin\theta)^2}.d\theta$$

Equation 1.11

For Mars, and fifteen decimal places:-

$$ET(\varphi,k) = ET(\varphi,k^2) = 0.108365999426112$$
$$FT(\varphi,k) = FT(\varphi,k^2) = 0.10879320101671$$

Though these integrals differ in the fourth decimal place it is important that we recognise (and apply) their distinctness.

The Volume of the Ellipsoid

The Volume of an Ellipsoid, V, is given by:-

$$V = \frac{4}{3}\pi abc$$
Equation 1.12

which readily simplifies to:-

$$V = \frac{4}{3}\pi a^2 c$$
Equation 1.13

for the case of a biaxial ellipsoid.
For the sphere r = a = b = c, and the volume is clearly:-

$$V = \frac{4}{3}\pi r^3$$
Equation 1.14

where r is the Sphere Radius.

The Volumetric Radius, R_3

The Volumetric Radius, R_3, is that length that is the *radius of a sphere which has the same volume* as the ellipsoid that it characterises. Accordingly, the Volumetric Radius is defined as:-

$$R_3 = \sqrt[3]{abc}$$
Equation 1.15

When we later address the role of R_3 in celestial mechanics it is important to remember that body volume is never truly conserved because of gravitational compression and elastic distension amongst other vagaries.
Also we may here note that:-

$$a = \frac{R_3}{bc}$$
Equation 1.16a

$$b = \frac{R_3}{ac}$$
Equation 1.16b

$$c = \frac{R_3}{ab}$$
Equation 1.16c

Surface Area of the Triaxial Ellipsoid

The Wikipedia[0] Expression, S_1

The Surface Area of an Ellipsoid is S_j where S is the Ellipsoid Surface Area and j is an arbitrary Serial Number. For the Wikipedia Version, S_1 is yielded by:-

$$S_1 = 2\pi c^2 + \frac{2\pi ab}{\sin\varphi}[ET(\varphi,k).(\sin\varphi)^2 + FT(\varphi,k).(\cos\varphi)^2]$$
$$= 2\pi\left\{c^2 + \frac{ab}{\sin\varphi}[ET(\varphi,k).(\sin\varphi)^2 + FT(\varphi,k).(\cos\varphi)^2]\right\}$$
Equation 1.17

which for a biaxial ellipsoid may be re-expressed as:-

$$S_1 = 2\pi\left\{c^2 + \frac{ab}{\sin\varphi}[ET(\varphi,k).(1-\lambda^2) + FT(\varphi,k).\lambda^2]\right\}$$
Equation 1.18

The Wolfram[7] Expression, S_2

In the Wolfram idiom, triaxiality is treated in the following way:-

$$e_1 = \sqrt{\frac{a^2 - c^2}{a^2}}$$
Equation 1.19a

$$e_2 = \sqrt{\frac{b^2 - c^2}{b^2}}$$
Equation 1.19b

$$k_w = \frac{e_2}{e_1}$$
Equation 1.20

$$\varphi = \sin^{-1}\sqrt{1 - \frac{c^2}{a^2}}$$
Equation 1.21

whilst the Wolfram Surface Area, S_2, is given by:-

$$S_2 = 2\pi\left[c^2 + b\sqrt{a^2 - c^2}.ET(\varphi, k_w) + \frac{bc^2}{\sqrt{a^2 - c^2}}FT(\varphi, k_w)\right]$$
Equation 1.22

The Wolfram form is self-evidently a "programmers'" algorithmic form and there is clear scope for extensive simplification.

The Rektorys[4] Expression, S_3

The Rektorys expression for Ellipsoid Surface Area, S_3, is prepared in the following manner:-

$$k_{Rek} = \sqrt{\frac{a^2(b^2 - c^2)}{b^2(a^2 - c^2)}}$$
Equation 1.23

k_{Rek} is the same as ordinary k, the argument of the elliptic integral. Thus:-

$$S_3 = 2\pi\left\{c^2 + \frac{b}{\sqrt{a^2 - c^2}} \cdot [c^2.FT(\varphi, k_{Rek}) + (a^2 - c^2).ET(\varphi, k_{Rek})]\right\}$$
Equation 1.24

Table 1.1 demonstrates that $S_1 = S_2 = S_3$ for a diversity of trial ellipsoids, some triaxial and some biaxial.

The merits of the three identities hinge upon their computational economy and after obvious algebraic simplifications they may be compared for a given object system.

The Elliptic Integrals were computed by means of A&S Rule 18 extended to 128 intervals in EXCEL 2010®. This yielded something superior to fifteen-figure accuracy. Several astrophysical inputs are currently only known to three or four figures.

a	b	c	Biax/Triax	V	φ	k	F(φ,k)	E(φ,k)	S_1	S_2	S_3	S_4
6	3	2	TRIAXIAL	1.50796447E+02	1.23095942	0.625	1.34081979	1.13536125	1.64066708E+02	1.64066708E+02	1.64066708E+02	2.73184747E+02
1	0.94	0.89	TRIAXIAL	3.50434189E+00	0.47345116	0.498093527	0.47774873	0.46922192	1.11423718E+01	1.11423718E+01	1.11423718E+01	1.16565546E+01
1000	2	1	TRIAXIAL	8.37758041E+03	1.56979633	0.75000075	1.90947902	1.31781008	1.65663887E+04	1.65663887E+04	1.65663887E+04	6.28233307E+06
6.001	6	5.999	TRIAXIAL	9.04778659E+02	0.02581845	0.500125	0.02581917	0.02581774	4.52383064E+02	4.52383064E+02	4.52383064E+02	4.52439607E+02
3.141593	2.718282	2.718282	TRIAXIAL	9.72360808E+01	0.52513558	0	0.52513558	0.52513558	1.02631387E+02	1.02631387E+02	1.02631387E+02	1.13046679E+02
3.141593	2.718282	1.618034	TRIAXIAL	5.78789447E+01	1.02974643	0.878800191	1.19339322	0.90259998	7.77787977E+01	7.77787977E+01	7.77787977E+01	8.66258856E+01
3.141593	2.718282	0.618034	TRIAXIAL	2.21077896E+01	1.37277845	0.986484712	2.18439095	0.99751084	5.95036858E+01	5.95036858E+01	5.95036858E+01	6.76651777E+01
2.718282	0.618034	0.618034	TRIAXIAL	4.34918429E+00	1.34142845	0	1.34142845	1.34142845	1.69404805E+01	1.69404805E+01	1.69404805E+01	5.17530482E+01
1	1	1	SPHERICAL	4.18879020E+00	0.00000000	0	0.00000000	0.00000000	1.25663706E+01	1.25663706E+01	1.25663706E+01	1.25663706E+01
1	0.5	0.333333	TRIAXIAL	6.98131701E-01	1.23095942	0.625	1.34081979	1.13536125	4.55740856E+00	4.55740856E+00	4.55740856E+00	7.58846520E+00
1	0.94	0.89	TRIAXIAL	3.50434189E+00	0.47345116	0.498093527	0.47774873	0.46922192	1.11423718E+01	1.11423718E+01	1.11423718E+01	1.16565546E+01
1	0.002	0.001	TRIAXIAL	8.37758041E-06	1.56979633	0.75000075	1.90947902	1.31781008	1.65663887E-02	1.65663887E-02	1.65663887E-02	6.28233307E+00
1	0.999833	0.999667	TRIAXIAL	4.18669639E+00	0.02581845	0.500125	0.02581917	0.02581774	1.25620085E+01	1.25620085E+01	1.25620085E+01	1.25635786E+01
1	0.865256	0.865256	TRIAXIAL	3.13601281E+00	0.52513558	0	0.52513558	0.52513558	1.03987336E+01	1.03987336E+01	1.03987336E+01	1.14540233E+01
1	0.865256	0.515036	TRIAXIAL	1.86668478E+00	1.02974643	0.878800191	1.19339322	0.90259998	7.88063984E+00	7.88063984E+00	7.88063984E+00	8.77703727E+00
1	0.865256	0.196726	TRIAXIAL	7.13010139E-01	1.37277845	0.986484712	2.18439095	0.99751084	6.02898387E+00	6.02898387E+00	6.02898387E+00	6.85591590E+00
1	0.227362	0.227362	TRIAXIAL	2.16533136E-01	1.34142845	0	1.34142845	1.34142845	2.29264472E+00	2.29264472E+00	2.29264472E+00	7.00401344E+00
1	0.999833	0.999833	TRIAXIAL	4.18739429E+00	0.01825615	0.01825615	0.01825615	0.01825615	1.25635787E+01	1.25635787E+01	1.25635787E+01	1.25649746E+01
6378.137	6356.752	6356.752	TRIAXIAL	1.07957553E+12	0.08191076	0.08191076	0.08191078	0.08191076	5.08924880E+08	5.08924880E+08	5.08924880E+08	5.10065622E+08
1	0.996647	0.996647	TRIAXIAL	4.16074883E+00	0.08191078	0	0.08191078	0.08191078	1.25102502E+01	1.25102502E+01	1.25102502E+01	1.25382916E+01
6378.137	6378.137	6356.752	BIAXIAL	1.08320732E+12	0.08191076	1	1.08200250	0.08181919	5.10065622E+08	5.10065622E+08	5.08924880E+08	5.10065622E+08
6378137	6378137	6356752	BIAXIAL	1.08320732E+21	0.08191076	1	1.08200250	0.08181919	5.10065622E+14	5.10065622E+14	5.10065622E+14	5.10065622E+14
1	1	0.996647	BIAXIAL	4.17474598E+00	0.08191076	1	1.08200250	0.08181919	1.25382916E+01	1.25382916E+01	1.25382916E+01	1.25382916E+01

Table 1.1
Selected Measures of Arbitrary Ellipsoids

Cote's Formula, S_4

Rektorys gives a Surface Area formula special to the case of an oblate (biaxial) ellipsoid. This formula is identical to that known as Cote's Formula, discovered in 1714 by Roger Cotes and as good as any recipe for the oblate case:-

$$S_4 = 2\pi \left\{ a^2 + \frac{c^2}{2e_1} \log_n \left(\frac{1+e_1}{1-e_1} \right) \right\}$$

Equation 1.25

Cote's Formula applies to the sphere and to the biaxial ellipsoid but not to the triaxial case.

Demonstration of the Independence of $E(\varphi,k)$ and $F(\varphi,k)$ from R_3

Recollect the definitions of Equation 1.15 and Equations 1.16a,b,c.
It is accordingly the case that:-

$$\varphi = \cos^{-1} \lambda = \frac{c}{a} = \frac{\frac{R_3}{ab}}{\frac{R_3}{bc}} = \frac{R_3}{ab} \cdot \frac{bc}{R_3} = \frac{c}{a} = \lambda$$

Equation 1.4

Also, for Equation 1.5:-

$$k = \sqrt{\frac{a^2(b^2 - c^2)}{b^2(a^2 - c^2)}}$$

Equation 1.5

And by substitution:-

$$k = \sqrt{\frac{\left(\frac{R_3^{\,3}}{bc}\right)^2 \left(\left(\frac{R_3^{\,3}}{ac}\right)^2 - \left(\frac{R_3^{\,3}}{ab}\right)^2\right)}{\left(\frac{R_3^{\,3}}{ac}\right)^2 \left(\left(\frac{R_3^{\,3}}{bc}\right)^2 - \left(\frac{R_3^{\,3}}{ab}\right)^2\right)}}$$

Equation 1.26

Now note that:-

$$\left(\frac{R_3^{\,3}}{ac}\right)^2 - \left(\frac{R_3^{\,3}}{ab}\right)^2 = \frac{R_3^6}{a^2}\left(\frac{1}{c^2} - \frac{1}{b^2}\right)$$

Equation 1.27a

whilst:-

$$\left(\frac{R_3^{\,3}}{bc}\right)^2 - \left(\frac{R_3^{\,3}}{ab}\right)^2 = \frac{R_3^6}{b^2}\left(\frac{1}{c^2} - \frac{1}{a^2}\right)$$

Equation 1.27b

Therefore by substitution in Equation 1.26:-

$$k = \sqrt{\frac{\left(\frac{R_3^{\,3}}{bc}\right)^2\left(\frac{R_3^6}{a^2}\left(\frac{1}{c^2} - \frac{1}{b^2}\right)\right)}{\left(\frac{R_3^{\,3}}{ac}\right)^2\left(\frac{R_3^6}{b^2}\left(\frac{1}{c^2} - \frac{1}{a^2}\right)\right)}}$$

Equation 1.28

$$k = \sqrt{\frac{\dfrac{R_3^{12}}{a^2 \cdot b^2 \cdot c^2}\left(\frac{1}{c^2} - \frac{1}{b^2}\right)}{\dfrac{R_3^{12}}{a^2 \cdot b^2 \cdot c^2}\left(\frac{1}{c^2} - \frac{1}{a^2}\right)}}$$

Equation 1.29

$$k = \sqrt{\frac{\dfrac{1}{c^2} - \dfrac{1}{b^2}}{\dfrac{1}{c^2} - \dfrac{1}{a^2}}}$$

Equation 1.30

Therefore φ is a function of a and c only and k is a function of a, b and c.

And it is demonstrated that no Triaxial Surface Area equation requires knowledge of radius, in particular Volumetric Radius R₃.

Where a>b=c the ellipsoid is of a (rotated) prolate biaxial form and k is zero. Accordingly, E(φ,0) = φ and F(φ,0) = φ because:-

$$ET(\varphi, 0) = \int_0^{\varphi} \sqrt{1 - k^2.(\sin\theta)^2}\,.d\theta$$

$$= \int_0^{\varphi} \sqrt{1 - 0^2.(\sin\theta)^2}\,.d\theta$$

$$= \int_0^{\varphi} \sqrt{1} \, . \, d\theta$$

$$= \int_0^{\varphi} 1 \, . \, d\theta$$

$$= [\varphi]_0^1$$

$$= \varphi$$

Equation 1.31a

and:-

$$FT(\varphi, 0) = \int_0^{\varphi} \frac{1}{\sqrt{1 - k^2 . (\sin \theta)^2}} \, . \, d\theta$$

$$= \int_0^{\varphi} \frac{1}{\sqrt{1 - 0^2 . (\sin \theta)^2}} \, . \, d\theta$$

$$= \int_0^{\varphi} \frac{1}{\sqrt{1}} \, . \, d\theta$$

$$= \int_0^{\varphi} 1 \, . \, d\theta$$

$$= [\varphi]_0^1$$

$$= \varphi$$

Equation 1.32b

These simplifications can of course lead to significant savings, but caution should be exercised because a rotated prolate ellipsoid is not the same thing as an oblate ellipsoid.

<u>The Elimination of Elliptic Integrals</u>

It is possible to simplify the equations for surface area by the important economy of eliminating the elliptic integrals *when the figure is an oblate biaxial spheroid.*

Celestial bodies that display inertial oblation are usually oblate biaxial spheroids that spin about their shorter z-axis, 2c.

Recall that:-

$$e = \sqrt{1 - \lambda^2} = \sqrt{1 - s^2}$$

Equation 1.3

$$\varphi = \cos^{-1} \lambda$$

Equation 1.4

$$k = \sqrt{\frac{a^2(b^2 - c^2)}{b^2(a^2 - c^2)}}$$

Equation 1.5

It is obvious that if a = b then k = 1 whatever the value of c. Therefore, the two incomplete elliptic integrals may be written:-

$$ET(\varphi, 1) = \int_0^\varphi \sqrt{1 - 1^2.(\sin\theta)^2}\,.d\theta = \int_0^\varphi \sqrt{1 - (\sin\theta)^2}\,.d\theta$$

Equation 1.33

and:-

$$FT(\varphi, 1) = \int_0^\varphi \frac{1}{\sqrt{1 - 1^2.(\sin\theta)^2}}\,.d\theta = \int_0^\varphi \frac{1}{\sqrt{1 - (\sin\theta)^2}}\,.d\theta$$

Equation 1.34

In both cases the analytic integrals are straight forward and may be stated in these terms:-

$$ET(\varphi, 1) = \int_0^\varphi \sqrt{1 - (\sin\theta)^2}\,.d\theta = \sin(\cos^{-1}\varphi) = \sin\varphi$$

Equation 1.35

and:-

$$FT(\varphi, 1) = \int_0^\varphi \frac{1}{\sqrt{1 - (\sin\theta)^2}}\,.d\theta = \tan(\cos^{-1}\varphi) = Log_n(\tan\varphi + \sec\varphi)$$

Equation 1.36

In the course of my MATHCAD® experiment utilising NASA data for the planet Earth, k was computed to be 1.00000000000078. In my opinion, this tiny deviation from unity is a numerical artefact and was treated as such in analysis, though as aforenoted all due circumspection must be exercised, especially when dealing with astrophysical outcomes that are either vanishingly small or almost unthinkably immense. The algebraic value of k is of course unity, as also aforementioned.

By the same token:-

PSD(E(φ,1),sin(φ)) = 0.000000000000017

PSD(F(φ,1),log_n(tan(φ)+sec(φ))) = 0.000000000000068

And I consider the respective 17 and 68 parts in a myriad milliard to be adventitious errors.

Given that (for example) the formula for the Surface Area of a general ellipsoid is:-

$$S = 2\pi \left\{ c^2 + ab \left[ET(\varphi, k) . \sqrt{1 - s^2} + FT(\varphi, k) . \frac{s^2}{\sqrt{1 - s^2}} \right] \right\}$$

Equation 1.37

For the oblate spheroidal case we may make appropriate substitutions to yield:-

$$S = 2\pi \left\{ c^2 + ab \left[\sin\varphi . \sqrt{1 - s^2} + \ln(\tan(\varphi) + \sec(\varphi)) . \frac{s^2}{\sqrt{1 - s^2}} \right] \right\}$$

Equation 1.38

$$S = 2\pi \left\{ c^2 + ab \left[\sin\varphi . \sqrt{1 - s^2} + \ln\left[\tan\varphi + \frac{1}{\cos\varphi} \right] . \frac{s^2}{\sqrt{1 - s^2}} \right] \right\}$$

Equation 1.39

or:-

$$S_5 = 2\pi \left\{ c^2 + ab \left[e . \sin\varphi + \ln\left[\tan\varphi + \frac{1}{\cos\varphi} \right] . \frac{s^2}{e} \right] \right\}$$

Equation 1.40

A complex alternative is:-

$$S_{5i} = 2\pi \left\{ c^2 + ab \left[e . \sin\varphi + \ln(\tan(\varphi) + \sec(\varphi)) . \frac{s^2}{e} \right] \right\}$$

Equation 1.41

An Alternative Non-Trigonometric Statement of Oblate Biaxial Ellipsoid Surface Area

The Cotes-Rektorys Formula provides a straight-forward and accurate expression of Surface Area, S, without multiple potentially-expensive trigonometric calls.

$$S_4 = 2\pi \left\{ a^2 + \frac{c^2}{2e_1} \cdot \log_n \left(\frac{1 + e_1}{1 - e_1} \right) \right\}$$

Equation 1.25

where:-

$$e_1 = \sqrt{\frac{a^2 - c^2}{a^2}} = \sqrt{1 - \left(\frac{c}{a}\right)^2} = e$$

Equation 1.19a

These days, it is often misleading to apply tariff-based costing exercises to mathematical algorithms applied with modern utility packages. Rather, one is better advised to trial the actual composite function on the particular object system that you intend to employ. Accordingly, one should prepare functional applications segments elaborating (for example) S_1, S_2, S_3 and S_4 and assess their timings and accuracies holistically. By the same token, Equation 1.40 S_5 can be compared with tolerable approximants, such as that of S_6 which we shall next examine.

Let us begin by defining S_6 in terms of the earlier Equation 1.34 S_5 in this way:-

$$S_6 \approx 2\pi\{c^2 + ab[A + B]\}$$
Equation 1.42

where the local variables A and B are defined as:-

$$A = \sin\varphi \cdot \sqrt{1 - s^2}$$
Equation 1.43a

$$B = \ln\left[\tan\varphi + \frac{1}{\cos\varphi}\right] \cdot \frac{s^2}{\sqrt{1 - s^2}}$$
Equation 1.43b

Now we may note that:-

$$\cos\varphi = \cos(\cos^{-1} s) = s$$
Equation 1.44a

$$\sin\varphi = \sqrt{1 - s^2}$$
Equation 1.44b

$$\tan\varphi = \frac{\sqrt{1 - s^2}}{s}$$
Equation 1.44c

and simplify A and B in these terms:-

$$A = (1 - s^2)(1 - s^2) = e^2$$
Equation 1.45

$$A^2 = e^4$$
Equation 1.46

$$B = Log_n\left[\frac{\sqrt{1 - s^2}}{s} + \frac{1}{s}\right] \cdot \frac{s^2}{\sqrt{1 - s^2}}$$

$$= Log_n\left[\frac{1}{s}\left(1 + \sqrt{1 - s^2}\right)\right] \cdot \frac{s^2}{\sqrt{1 - s^2}}$$

$$= Log_n\left[\frac{1}{s}(1 + e)\right] \cdot \frac{s^2}{e}$$

$$= Log_n\left[\frac{1 + e}{s}\right] \cdot \frac{s^2}{e}$$

Equation 1.47

$$B^2 = \frac{s^4}{e^2}\left(Log_n\left[\frac{1 + e}{s}\right]\right)^2$$

Equation 1.48

Further note that:-

$$A + B \approx 1 + \frac{1 - \sqrt{A^2 + B^2}}{2} = \frac{3 - \sqrt{A^2 + B^2}}{2}$$

Equation 1.49

And therefore:-

$$S_6 \approx 2\pi\left\{c^2 + ab\left[\frac{3 - \sqrt{A^2 + B^2}}{2}\right]\right\}$$

Equation 1.50

At this juncture we may for terrestrial data note that:-

$PSD(S_{triwiki}, S_4) = 0.000000000000515$
$PSD(S_{triwiki}, S_5) = 0.000000000000343$
$PSD(S_{triwiki}, S_6) = 0.000113508183096$

On the face of it, S_6 is nine orders of magnitude worse than S_5. The discrepancy of S_5 can arguably be condoned as numerical error, but that of S_6 is clearly due to some analytic omission. Notwithstanding that, in a world where the most precise Equatorial Radius c is known to nine figures and the Universal Gravitational Constant G is known to six an error of $\sim 10^6$ may be tolerable for rough estimates.

We shall continue. By substitution:-

$$S_6 \approx 2\pi\left\{c^2 + ab.\frac{1}{2}\left[3 - \sqrt{A^2 + B^2}\right]\right\}$$

$$\approx 2\pi\left\{c^2 + ab.\frac{1}{2}\left[3 - \sqrt{e^4 + \frac{s^4}{e^2}\left(Log_n\left[\frac{1 + e}{s}\right]\right)^2}\right]\right\}$$

$$\approx 2\pi\left\{c^2 + ab.\frac{1}{2}\left[3 - \sqrt{e^4 + \frac{1 - 2e^2 + e^4}{e^2}\left(Log_n\left[\frac{1+e}{\sqrt{1-e^2}}\right]\right)^2}\right]\right\}$$

Equation 1.51

From which:-

$$S_6 \approx 2\pi\left\{c^2 + \frac{ab}{2}\left[3 - \sqrt{e^4 + \left(e^2 + \frac{1}{e^2} - 2\right)\left(Log_n\left[\frac{1+e}{\sqrt{1-e^2}}\right]\right)^2}\right]\right\}$$

Equation 1.52

PART TWO
ASPECTS OF BIAXIAL SPHEROID STRAIN

We recollect of Part One the facts that:-

$$V = \frac{4}{3}\pi abc = \frac{4}{3}\pi a R_3^3$$
Equation 1.12

where:-

$$R_3 = \sqrt[3]{abc}$$
Equation 1.15

Let us now separate the Volumes and the Surface Areas.
The Implied Unstrained Sphere Surface Area, S_0, is:-

$$S_0 = 4\pi R_3^2$$
Equation 2.1

whilst the Unstrained Sphere Volume, V_0, is:-

$$V_0 = \frac{4}{3}\pi R_3^3$$
Equation 2.2

The Implied Strained Sphere Surface Area, S_4, is:-

$$S_4 = 2\pi \left\{ a^2 + \frac{c^2}{2e_1} \cdot \log_n \left(\frac{1 + e_1}{1 - e_1} \right) \right\}$$
Equation 2.3(1.25)

And the Strained Sphere Volume, S_e, is:-

$$V_e = \frac{4}{3}\pi abc$$
Equation 2.4(1.12)

It is of course not strictly the case that planetary volume is conserved under strain. Gravitating matter compresses, even if crystalline, and thermal effects cause volumetric changes, whilst centrifugal effects engage differential stresses. Also, planets and stars are internally differentiated.

Notwithstanding that, we shall assume Conservation of Volume to assist approximations of the planetary distortions we wish to assess.

Accordingly the Volumetric Ratio, VR, of a rotating spheroid under stress may be approximated as:-

$$VR = \frac{V_e}{V_0} = \left(\frac{4}{3}\pi abc\right) \times \left(\frac{3}{4\pi R_3^3}\right) = \frac{abc}{R_3^3} = 1$$

Equation 2.5

whilst the non-conserved Surface Area Ratio, SR, is:-

$$SR = \frac{S_4}{S_0} = \left(2\pi\left\{a^2 + \frac{c^2}{2e_1}\cdot\log_n\left(\frac{1+e_1}{1-e_1}\right)\right\}\right) \times \left(\frac{1}{4\pi R_3^2}\right) = \frac{1}{2R_3^2}\left\{a^2 + \frac{c^2}{2e_1}\cdot\log_n\left(\frac{1+e_1}{1-e_1}\right)\right\}$$

Equation 2.6

For trials and comparisons we may list the Principal Radii of the Planet Earth as:-

a	$6.37813699999997\times10^6$	meters
b	6.378137×10^6	meters
c	6.3567523142×10^6	meters

Computational artefacts aside, radii a and b are clearly the same for our oblately biaxial Earth (6378137 meters).

For the case of the Earth, these MATHCAD®-computed values were found to apply:-

R_3	$6.37100078999405\times10^6$	meters
S_0	$5.10064598404192\times10^{14}$	square meters
S_4	$5.1006562172167\times10^{14}$	square meters
V_0	$1.0832073197937\times10^{21}$	cubic meters
V_e	$1.0832073197937\times10^{21}$	cubic meters
VR	1	
SR	1.00000200625075	

The Oblation Number, Ob(V,S)

The dimensionless ratio of Volume and Surface Area, Ob(V,S), may conveniently be stated as:-

$$Ob(V,S) = \frac{V^2}{S^3}$$

Equation 2.7

The Limiting Value of Ob(V,S) for the perfect sphere may by substitution be elaborated as:-

$$Ob(V,S) = \frac{V^2}{S^3}$$

$$Ob_{sphere} = \frac{\left(\frac{4}{3}\pi R_3^3\right)^2}{(4\pi R_3^2)^3}$$

$$= \frac{\left(\frac{4}{3}\pi\right)^2}{(4\pi)^3}$$

$$= \frac{\frac{16}{9}\pi^2}{64\pi^3}$$

$$= \frac{\frac{16}{9}}{64\pi}$$

$$= \frac{1}{9 \times 4\pi}$$

$$= \frac{1}{36\pi}$$

$$Ob_{sphere} = 0.008841941282883$$

Equation 2.8

This is the maximum value of the Oblation Number. The value for Planet Earth is marginally less at:-

$$Ob_{earth} = 0.008841888065643$$

Equation 2.9

Infinitesimal Strain Under Volume Conservation

Allow that *relative to R_3* the Infinitesimal Equatorial Strain Δa (or Δb) is unity: Then the corresponding Infinitesimal Polar Strain is Δc. It can be shown that for an oblate biaxial spheroid Δc must be two.

Accordingly, we may write the Infinitesimally-Strained Spheroidal Volume in these terms:-

$$V_e = \frac{4}{3}\pi abc$$

Equation 2.10

and:-

$$V_{\Delta e} = \frac{4}{3}\pi(R_3 + \Delta a)(R_3 + \Delta a)(R_3 - \Delta c)$$

Equation 2.11

The Volumetric Infinitesimal Strain Ratio, T, is defined as:-

$$T = \frac{V_e}{V_{\Delta e}}$$

$$T = \frac{\frac{4}{3}\pi abc}{\frac{4}{3}\pi(R_3 + \Delta a)(R_3 + \Delta a)(R_3 - \Delta c)}$$

$$= \frac{abc}{(R_3 + \Delta a)(R_3 + \Delta a)(R_3 - \Delta c)}$$

$$= \frac{abc}{(R_3 + \Delta a)(R_3 + \Delta a)(R_3 - 2 * \Delta a)}$$

Equation 2.12

Within computational error, T is unity. Hence the Equatorial Strain, ES, may be calculated as:-

$$ES = R_3 - 2.\Delta a = \frac{ab}{(R_3 + \Delta a)(R_3 + \Delta a)} = \frac{ab}{(R_3 + \Delta a)^2}$$

Equation 2.13

ES is of course a dimensionless ratio of areas.
For Planet Earth, ES is 1.00224115634688

Aspects of Linear Spheroidal Strain (a = b > c: The Oblate Case)

The Principal Infinitesimals may be defined as:-

$$\Delta c = R_3 - c$$

Equation 2.14a

and:-

$$\Delta a = a - R_3$$

Equation 2.14b

Whilst the Strain in the Polar Radius, ε_c, is:-

$$\varepsilon_c = R_3 - \Delta c$$

Equation 2.15a

and the Strain in the Equatorial Radius, ε_a, is:-

$$\varepsilon_a = \Delta a + R_3$$

Equation 2.15a

whilst the Linear Strain Ratio, ε_R, is:-

$$\varepsilon_R = \frac{\Delta c}{\Delta a}$$

Equation 2.16

For Planet Earth, these results apply:-

Δc	$1.4248475794049 \times 10^4$	meters
Δa	$7.13621000591759 \times 10^3$	meters
ε_c	6.3567523142×10^6	meters
ε_a	$6.37813699999997 \times 10^6$	meters
ε_R	1.99664468705849	

The Linear Strain Treatment of the Eccentricity, e

Recall that Eccentricity, e, may be defined using:-

$$e = \sqrt{1 - \lambda^2} = \sqrt{1 - \left(\frac{c}{a}\right)^2}$$

Equation 1.3

Eccentricity may be expressed in a strained form as:-

$$e = \sqrt{1 - \lambda^2} = \sqrt{1 - \left(\frac{\varepsilon_c}{\varepsilon_a}\right)^2}$$

Equation 2.17

So:-

$$e = \sqrt{1 - \frac{(R_3 - \Delta c)^2}{(\Delta a + R_3)^2}}$$

$$= \sqrt{1 - \frac{(R_3^2 - 2.\Delta c.R_3 + \Delta c^2)}{(\Delta a^2 + 2.\Delta a.R_3 + R_3^2)}}$$

Equation 2.18

When the Eccentricity is small it is feasible to approximate Equation 2.18 in these terms:-

$$e \approx \sqrt{1 - \frac{(R_3^2 - 2.\Delta c.R_3)}{(2.\Delta a.R_3 + R_3^2)}}$$

Equation 2.19

or even:-

$$e_{theo} \approx \sqrt{1 - \frac{(R_3^2 - 4.\Delta a.R_3)}{(2.\Delta a.R_3 + R_3^2)}}$$
Equation 2.20

Note that none of e; its approximation; or e$_{theo}$ are *exactly* identical in value.

The Linear Strain Treatment of the Flattening, f

Allied to Eccentricity is the even simpler parameter, Flattening, f, defined as:-

$$f = \frac{a - c}{a} = 1 - \frac{c}{a} = 1 - \lambda = 1 - s$$
Equation 1.2

It follows that:-

$$f = 1 - s = 1 - \lambda \approx 2 - \frac{\Delta c}{\Delta a}$$
Equation 2.21

Or:-

$$s \approx \frac{\Delta c}{\Delta a} - 1$$
Equation 2.22

Therefore:-

$$e_{app2} \approx \sqrt{1 - \left(\frac{\Delta c}{\Delta a} - 1\right)^2}$$
Equation 2.23

Or in complex form:-

$$e_{app2} \approx \sqrt{\left(\frac{\Delta c}{\Delta a}\right)^2 - 2\left(\frac{\Delta c}{\Delta a}\right)}$$
Equation 2.24

The parameter e^2 is always real and therefore it is convenient further to study eccentricities as their squares or as approximations of their squares.

So note that:-

$$e^2 \approx 2\left(\frac{\Delta c}{\Delta a}\right) - \left(\frac{\Delta c}{\Delta a}\right)^2 = \frac{\Delta c}{\Delta a}\left(2 - \frac{\Delta c}{\Delta a}\right)$$
Equation 2.25

The Linear Treatment of Volumetric Strain

For the case of an oblate biaxial ellipsoid we have seen that:-

$$V = \frac{4}{3}\pi a^2 c$$
Equation 1.13

By substitution the infinitesimally-incremented form, V_Δ, may be written:-

$$
\begin{aligned}
V_\Delta &= \frac{4}{3}\pi(\varepsilon_a)^2(\varepsilon_c) \\
&= \frac{4}{3}\pi(\Delta a + R_3)^2(R_3 - \Delta c) \\
&= \frac{4}{3}\pi(\Delta a^2 + 2.\Delta a.R_3 + R_3^2)(R_3 - \Delta c) \\
&= \frac{4}{3}\pi\left(R_3\Delta a^2 + 2.\Delta a.R_3{}^2 + R_3^3 - \Delta c.\Delta a^2 - 2R_3.\Delta a\Delta c - \Delta c R_3^3\right) \\
&= \frac{4}{3}\pi\left[R_3(1 + \Delta a^2 - 2\Delta a\Delta c) + 2.\Delta a.R_3{}^2 + R_3^3 - \Delta c.\Delta a^2 - \Delta c R_3^2\right] \\
&= \frac{4}{3}\pi\left[R_3(\Delta a^2 - 2\Delta a\Delta c) + R_3^2(2.\Delta a - \Delta c) + R_3^3 - \Delta c.\Delta a^2\right]
\end{aligned}
$$
Equation 2.26

The Percentage Specific Defect relative to the R_3 sphere was computed to be:-

$$PSD\left(V_\Delta, \frac{4}{3}\pi R_3^3\right) = -0.000000000000266$$
Equation 2.27

I ascribe this discrepancy to numerical errors.

The Linear Treatment of Surface Area Strain

We have seen that the surface area of the general ellipsoid is given by:-

$$S = 2\pi \left\{ c^2 + ab \left[ET(\varphi, k). \sqrt{1 - s^2} + FT(\varphi, k). \frac{s^2}{\sqrt{1 - s^2}} \right] \right\}$$

Equation 1.37

With regard to s we may develop its infinitesimal form as:-

$$s = \frac{c}{a} = \frac{R_3 - \Delta c}{\Delta a + R_3}$$

Equation 2.28

whilst:-

$$s^2 = \left(\frac{R_3 - \Delta c}{\Delta a + R_3} \right)^2 = \frac{(R_3 - \Delta c)^2}{(\Delta a + R_3)^2} = \frac{R_3^2 - 2\Delta c R_3 + \Delta c^2}{\Delta a^2 + 2\Delta a R_3 + R_3^2} = \frac{R_3(R_3 - 2\Delta c) + \Delta c^2}{R_3(R_3 + 2\Delta a) + \Delta a^2}$$

Equation 2.29

and:-

$$e = \sqrt{1 - s^2} = \sqrt{1 - \frac{(R_3^2 - 2.\Delta c.R_3 + \Delta c^2)}{(\Delta a^2 + 2.\Delta a.R_3 + R_3^2)}}$$

Equation 2.30

Therefore:-

$$\varphi = \cos^{-1} s = \cos^{-1} \left(\frac{R_3 - \Delta c}{\Delta a + R_3} \right)$$

Equation 2.31

whilst:-

$$k = \sqrt{\frac{a^2(b^2 - c^2)}{b^2(a^2 - c^2)}}$$

Equation 1.5

in view of the fact that we are discussing an oblate biaxial where a = b:-

$$k = \sqrt{\frac{(b^2 - c^2)}{(a^2 - c^2)}}$$

Equation 2.32

Clearly the quotient and its root within Equation 2.32 are unity. This necessary value of k may be expressed in infinitesimal terms as:-

$$k = \sqrt{\frac{(\Delta a^2 + 2\Delta aR_3 + R_3^2 - c^2) - (R_3^2 - 2\Delta cR_3 + \Delta c^2)}{(\Delta a^2 + 2\Delta aR_3 + R_3^2 - c^2) - (R_3^2 - 2\Delta cR_3 + \Delta c^2)}} = 1$$

Equation 2.33

Now from Equation 1.35 and 1.36 we know that:-

$$ET(\varphi, 1) = \sin \varphi$$

Equation 2.34a

$$FT(\varphi, 1) = Log_n(\tan \varphi + \sec \varphi)$$

Equation 2.34b

These two elliptic integrals may be expanded in infinitesimal terms as:-

$$ET(\varphi, 1) = \sin \varphi = \sin(\cos^{-1} s) = \sqrt{1 - \left(\frac{R_3 - \Delta c}{\Delta a + R_3}\right)^2} = \sqrt{1 - \frac{R_3(R_3 - 2\Delta c) + \Delta c^2}{\Delta a^2 + 2.\Delta a. R_3 + R_3^2}}$$

Equation 2.35

which reduces to:-

$$FT(\varphi, 1) = \sqrt{1 - \frac{R_3(R_3 - 2\Delta c) + \Delta c^2}{R_3(R_3 + 2\Delta a) + \Delta a^2}}$$

Equation 2.36

Meanwhile:-

$$FT(\varphi, 1) = \ln(\tan(\varphi) + \sec(\varphi)) = \ln(\tan(acos(\varphi)) + \sec(acos(\varphi)))$$

Equation 2.37

or by elimination of inverse trigonometric functions:-

$$FT(\varphi, 1) = \ln\left(\frac{\sqrt{1 - s^2}}{s} + \frac{1}{s}\right) = \ln\left[\frac{1}{s}\left(1 + \sqrt{1 - s^2}\right)\right]$$

Equation 2.38

Appropriate substitutions for s and s^2 followed by simplification then yields:-

$$FT(\varphi, 1) = \ln\left[\frac{\Delta a + R_3}{R_3 - \Delta c}\left(1 + \sqrt{1 - \frac{R_3(R_3 - 2\Delta c) + \Delta c^2}{R_3(R_3 + 2\Delta a) + \Delta a^2}}\right)\right]$$

Equation 2.39

For the Planet Earth these integrals have the respective values:-

ET(φ,1)	0.081819190928843
FT(φ,1)	0.08200250404661

Figure 2.1 presents the Planet Earth Surface Area Ellipsoid in its fully-elaborated form:-

The Figure 2.1 S_{lab} has a value of $5.10065621721672 \times 10^{14}$ square meters for Planet Earth, but is obviously repetitive and most inconvenient in its fully-elaborated form.

In order to achieve more manageable algebra define these local variables:-

$$t = R_3(R_3 - 2\Delta c) + \Delta c^2$$

Equation 2.40

$$u = R_3(R_3 + 2\Delta a) + \Delta a^2$$

Equation 2.41

$$v = \frac{t}{u}$$

Equation 2.42

The infinitesimally-adjusted Surface Area S then becomes:-

$$S = 2\pi\left\{t + u\left[1 - v\left\{1 - \frac{1}{e} \cdot \left[ln\left(\frac{\Delta a + R_3}{R_3 - \Delta c}\right) \cdot (1 + e)\right]\right\}\right]\right\}$$

Equation 2.43

If we compare the outcome of the general formula for a triaxial ellipsoid as given in Wikipedia with the result given by S_{lab} and its Equation 2.43 equivalent we obtain a Percentage Specific Defect of:-

$$PSD(S_{\text{triwiki}}, S_{lab}) = 0.000000000000380$$

Equation 2.44

I ascribe the small discrepancy to numerical error.

$$S_{iab} := 2 \cdot \pi \cdot \left(R_3 \cdot (R_3 - 2 \cdot \Delta c) + \Delta c^2 + \left(R_3 \cdot (R_3 + 2 \cdot \Delta a) + \Delta a^2 \right) \cdot \left(\sqrt{1 - \frac{R_3 \cdot (R_3 - 2 \cdot \Delta c) + \Delta c^2}{R_3 \cdot (R_3 + 2 \cdot \Delta a) + \Delta a^2}} \cdot \sqrt{1 - \frac{R_3 \cdot (R_3 - 2 \cdot \Delta c) + \Delta c^2}{R_3 \cdot (R_3 + 2 \cdot \Delta a) + \Delta a^2}} + \ln\left(\frac{\Delta a + R_3}{R_3 - \Delta c} \cdot \left(1 + \sqrt{1 - \frac{R_3 \cdot (R_3 - 2 \cdot \Delta c) + \Delta c^2}{R_3 \cdot (R_3 + 2 \cdot \Delta a) + \Delta a^2}} \right) \right) \cdot \frac{\frac{R_3 \cdot (R_3 - 2 \cdot \Delta c) + \Delta c^2}{R_3 \cdot (R_3 + 2 \cdot \Delta a) + \Delta a^2}}{\sqrt{1 - \frac{R_3 \cdot (R_3 - 2 \cdot \Delta c) + \Delta c^2}{R_3 \cdot (R_3 + 2 \cdot \Delta a) + \Delta a^2}}} \right) \right)$$

Figure 2.1
The Infinitesimal Expression for Oblate Spheroid Surface Area

Strain Components

Amongst the several virtues of the mathematical infinitesimal is that it can be resolved into even small components that define the respective contributions of different physical forces to a relevant distortion.

By way of illustration:-

$$\Delta M = -\Delta m_{gravitation} + \Delta m_{centrifugal} - \Delta m_{cohesion} \pm \varepsilon = \Delta M_{measured}$$

Equation 2.45

where ΔM is the Summative Strain due to All Influences; $\Delta m_{gravitation}$ is the Effect of Gravity; $\Delta m_{centrifugal}$ is the Effect of Centrifugal Force (in a rotating body); $\Delta m_{cohesion}$ is the Effect of Material Bonding; and so forth until an Empirical (i.e. measured) Strain $\Delta M_{measured}$ is met within the Limit of Error, ε.

All variables are of course co-dimensional, and may act positively or negatively: The aspiration is that ε is zero.

We expect ΔM to be Δa or Δc, but it could be applied to Volume or Area.

Newton's Number

Isaac Newton developed a dimensionless number that essentially defines the ratio of the Centrifugal and the Gravitational stressor forces that oppose to shape a spinning star or planet.

We may define this number as:-

$$Ne(a, \Omega, G, M) = \frac{a^3 \cdot \Omega^2}{G \cdot M} = \frac{Centrifugal}{Gravitational}$$

Equation 3.1

where a is the Equatorial Radius, Ω is the Angular Velocity of the Rotating Body (a frequency in hertz which is the inverse of the diurnal period), G is the Universal Gravitational Constant, and M is the Mass of the Spinning Body.

Table 3.1 lists the dimensions and SI Units of these four variables:-

Name	Symbol	M	L	T	SI Unit	SI Value	Notes
Equatorial Radius	a	0	1	0	meter	6.378137000000000E+06	Earth
Angular Velocity	Ω	0	0	-1	hertz	7.292115090000000E-05	Earth: Radians per Second
Universal Gravitational Constant	G	-1	3	-2	$m^3kg^{-1}s^{-2}$	6.674080000000000E-11	General
Mass of Body	M	1	0	0	Kilogram	5.972200000000000E+24	Earth
	a^3	0	3	0		2.594666415324530E+20	
	Ω^2	0	0	-2		5.317494000000000E-09	
Centrifugal Dividend		0	3	-2		1.379712309548970E+12	
	G	-1	3	-2		6.674080000000000E-11	
	M	1	0	0		5.972200000000000E+24	
Gravitational Divisor		0	3	-2		3.985894057600000E+14	
Newton's Number	Ne()	0	0	0		3.461487660260910E-03	Earth

Table 3.1
Dimensionalities of Newton's Number and its Physical Components

Proof of the Non-Dimensionality of Newton's Number

Table 3.1 lists the mechanical dimensions of the variables of interest and it follows that:-

$$Ne(a, \Omega, G, M) = \frac{L^3 \cdot (T^{-1})^2}{(M^{-1}L^3T^{-2}) \cdot M^1} = \frac{L^3T^{-2}}{M^0L^3T^{-2}} = M^0L^0T^0$$
Equation 3.2

Flattening Revisited

Recall from Part One that:-

$$f = \frac{a-c}{a} = 1 - \frac{c}{a} = 1 - \lambda = 1 - s$$
Equation 1.2

and Eccentricity e is:-

$$e = \sqrt{1 - \lambda^2}$$
Equation 1.3

In the case of Planet Earth:-

f = 0.003352810671831
e = 0.081819190928904

It follows that:-

$$\frac{1}{f} = \frac{1}{1 - \lambda} = 298.257222932887$$
Equation 3.3

The Fundamental Equation of Inertial Fluid Oblation

The Fundamental Equation of Inertial Fluid oblation is:-

$$\alpha_j . Ne(a, \Omega, G, M) = f_j(\lambda) = f_j(s)$$
Equation 3.4

where α is a Constant (often expressed as a fraction); j is an arbitrary counter; Ne() is Newton's Number; and λ or s is the Ratio of Polar Radius to Equatorial Radius. The RHS function f(λ) is some function of Eccentricity e (and implicitly of λ) which is later to be defined.

Equation 3.4 is a generalised form of MacLaurin's Formula for the rotation of *fluid* masses as propounded in 1714 by the great Aberdonian mathematician and physicist Colin MacLaurin.

MacLaurin models say nothing about cohesion or rigidity, both of which may be surmised to be important for the terrestrial planets and their exoplanetary analogs.

α_j is a multiplier that arises in various models of the MacLaurin scheme. It is usually a product of algebraic analysis or theoretical, mechanical, thinking.

$f_j(\lambda)$ is a purely geometrical expression of elliptical flattening and may be conceptualised as defining the quadrant of the oblate ellipse that is a semi-meridian of the oblate biaxial ellipsoid.

The First Essay Deformation Equation

On Page 5: Lines 3-4 of his treatise "The Effect of Rotation on the Flattening of Celestial Bodies"[8], Wolfgang H Müller mentions a simple multiplier that involves the ratio of the Earth's Polar Acceleration due to Gravity over its Equatorial Acceleration due to Gravity (the ratio of angular acceleration to gravitational acceleration).

Both accelerations are of course geophysically accessible but can also be computed by recourse to mechanical theory.

We may define this primitive value α_4 by the equilibration of forces in the following way:-

$$F_1 = F_2$$
$$mg = \frac{GMm}{r^2}$$
$$g = \frac{GM}{r^2}$$

Equation 3.5

where g is the Local Planet Surface Acceleration due to Gravity; m is the Local Test (reference) Mass; G is the Universal Gravitational Constant; M is the Planetary Mass; and r is the Local Planetary Radius.

If the Polar Acceleration due to Gravity is g_{pol} and the Equatorial Acceleration due to Gravity is g_{equ} then it follows that:-

$$\alpha_4 = \frac{g_{pol}}{g_{equ}} = \frac{GM}{c^2} \times \frac{a^2}{GM} = \frac{a^2}{c^2} = \frac{1}{\lambda^2}$$

Equation 3.6

For the Earth:-

$$PSD\left(\alpha_4, \frac{1}{\lambda^2}\right) = -0.142947565252018$$

Equation 3.7

The Equation 3.7 discrepancy is too large to be numerical error but small enough to pledge the integrity of the MacLaurin-type model. The figure does not of course include the effects of cohesion, rigidity or any factors apart from centrifugal and gravitational inertias.

If instead of α_4 we use the preterprimitive $\alpha_0 = 1$ it is possible to state Equation 3.4 in these terms:-

$$\alpha_0 . Ne(a, \Omega, G, M) = \frac{1}{\lambda^2}$$
Equation 3.8

For the Earth:-

$$PSD(\alpha_0 . Ne(a, \Omega, G, M), 1 - \lambda) = 3.13960804812089$$
Equation 3.9

This three percent discrepancy contains not only modelling fallacies, but also actual astrophysical stressors to be explored.

Solar System Planetary Data

Table 3.2 to 3.4 reproduce the planetary a, b, c, M and I size and dynamical data drawn from a number of Web sources, more especially NASA databases. V and S are computed by the author.

In Table 3.2; a, b and c are the Principle Ellipsoid Axes in meters. a and b are of course the same and longer than c, as is the case for all ideal biaxial oblate ellipsoids, which planets are (imprecisely) assumed to be. M is the Planetary Mass in kilograms. In this and other tables the precision of figures is largely arbitrary and driven by the precision of published data. No special efforts have been made uniformly to format tabular numbers.   In Table 3.3 R_3 is computed from Equation 1.15 and V is computed using Equation 1.12. $V_{indicated}$ is calculated from:-

$$V_{indicated} = \frac{4}{3}\pi R_3^2$$
Equation 3.10

Planetary Volume V and $V_{indicated}$ should, and do, tally. The dimensionless Oblation Number V^2/S^3 is computed using V. The Planetary Moment of Inertia, I_a, is computed using (Rektorys: Page 148: Eqn (39))[4]:-

$$I_a = \frac{8}{15}\pi ab^4$$
Equation 3.11

It is a simple matter to compute the Total Mechanical Energy of Rotation (in joules) using:-

$$E = \frac{1}{2} I_a \Omega^2$$

Equation 3.12

Table 3.4 offers the parameters φ and k and the Elliptic Integrals $E(\varphi,k)$ and $F(\varphi,k)$ for the Sun, Moon and Solar Planets. It is immediately seen that for *spherical* celestial bodies φ is zero and for *all* planets k is unity.

I assumed the Wikipedia expression of triaxial ellipsoid surface area, S_{wiki}, as fiducial and checked that in every case it agreed with differently-expressed Wolfram and Rektorys outcomes. The S_{wiki} values were computed using Equation 1.17, though as we have seen it is possible to eliminate the elliptic integrals for biaxial ellipsoids.

This data will be employed in the subsequent critical evaluations.

Planet Name (Satellites italisised)	Symbol	a	b	c	M	Biax/Triax
SUN	☉	6.963420E+08	6.963420E+08	6.963420E+08	1.988500E+30	SPHERICAL
Mercury	☿	2.439700E+06	2.439700E+06	2.439700E+06	3.301100E+23	SPHERICAL
Venus	♀	6.051800E+06	6.051800E+06	6.051800E+06	4.867500E+24	SPHERICAL
Earth	⊕	6.378137E+06	6.378137E+06	6.356752E+06	5.972300E+24	BIAXIAL
Moon	☾	1.738100E+06	1.738100E+06	1.736000E+06	7.346000E+22	BIAXIAL
Mars	♂	3.396200E+06	3.396200E+06	3.376200E+06	6.417100E+23	BIAXIAL
Jupiter	♃	7.149200E+07	7.149200E+07	6.685400E+07	1.898190E+27	BIAXIAL
Saturn	♄	6.026800E+07	6.026800E+07	5.436400E+07	5.683400E+26	BIAXIAL
Uranus	♅	2.555900E+07	2.555900E+07	2.497300E+07	8.681300E+25	BIAXIAL
Neptune	♆	2.476400E+07	2.476400E+07	2.434100E+07	1.024130E+26	BIAXIAL
Pluto	♇	1.188000E+06	1.188000E+06	1.188000E+06	1.303000E+22	SPHERICAL

Table 3.2
Basic Sun, Moon and Solar Planet Size Data

Planet Name (Satellites italicised)	Symbol	R_3	$V_{indicted}$	V^2/S^3	V	I_a
SUN	☉	696342000	1.41435E+27	1.31232E-44	1.41435E+27	2.7432E+44
Mercury	☿	2439700	6.08272E+19	2.48583E-32	6.08272E+19	1.4482E+32
Venus	♀	6051800	9.28415E+20	2.64686E-34	9.28415E+20	1.3601E+34
Earth	⊕	6371000.685	1.08321E+21	2.047E-34	1.08321E+21	1.7686E+34
Moon	☽	1737399.718	2.19679E+19	1.35724E-31	2.19679E+19	2.6578E+31
Mars	♂	3389520.204	1.63119E+20	4.80251E-33	1.63119E+20	7.5703E+32
Jupiter	♃	69911308.41	1.4313E+24	1.2886E-39	1.4313E+24	3.1292E+39
Saturn	♄	58231993.02	8.2713E+23	3.22116E-39	8.2713E+23	1.3322E+39
Uranus	♅	25362154.53	6.83356E+22	2.04789E-37	6.83356E+22	1.8275E+37
Neptune	♆	24622189.48	6.25271E+22	2.37447E-37	6.25271E+22	1.5605E+37
Pluto	♇	1188000	7.02325E+18	9.0797E-31	7.02325E+18	3.9649E+30

Table 3.3
Sun, Moon and Solar Planet
Volumetric and Surface Area Properties

Planet Name (Satellites italicised)	Symbol	φ	k	φ/n	F(φ,k)	E(φ,k)	S_wiki
SUN	☉	0.000000E+00	1.000000E+00	0.000000E+00	0.000000E+00	0.000000E+00	6.093335E+18
Mercury	☿	0.000000E+00	1.000000E+00	0.000000E+00	0.000000E+00	0.000000E+00	7.479675E+13
Venus	♀	0.000000E+00	1.000000E+00	0.000000E+00	0.000000E+00	0.000000E+00	4.602343E+14
Earth	⊕	8.191136E-02	1.000000E+00	6.399325E-04	8.200311E-02	8.181979E-02	5.100656E+14
Moon	☾	4.916217E-02	1.000000E+00	3.840794E-04	4.918198E-02	4.914236E-02	3.793233E+13
Mars	♂	1.085792E-01	1.000000E+00	8.482752E-04	1.087932E-01	1.083660E-01	1.443740E+14
Jupiter	♃	3.621829E-01	1.000000E+00	2.829554E-03	3.703712E-01	3.543164E-01	6.146893E+16
Saturn	♄	4.463295E-01	1.000000E+00	3.486949E-03	4.619320E-01	4.316575E-01	4.269398E+16
Uranus	♅	2.145483E-01	1.000000E+00	1.676159E-03	2.162135E-01	2.129061E-01	8.083954E+15
Neptune	♆	1.850951E-01	1.000000E+00	1.446055E-03	1.861611E-01	1.840400E-01	7.618793E+15
Pluto	♇	0.000000E+00	1.000000E+00	0.000000E+00	0.000000E+00	0.000000E+00	1.773547E+13

Table 3.4
Sun, Moon and Solar Planet
Elliptic Outcomes

List of Trial Multipliers and their Proponents

Recollecting α_j from Equation 3.4 we may list the various Trial Multipliers as shown in Table 3.5 below:-

Trial Alphas	Proponent	Inverse
$\alpha_0 := \dfrac{1}{1}$	Plain Form	$\dfrac{1}{\alpha_0} = 1$
$\alpha_1 := \dfrac{1}{\pi}$	Wikipedia Form	$\dfrac{1}{\alpha_1} = 3.14159265358979$
$\alpha_2 := \dfrac{5}{4}$	Muller Form after Newton and Chandrasekhar (Eqn. 1-2)	$\dfrac{1}{\alpha_2} = 0.8$
$\alpha_3 := \dfrac{2}{3}$	Muller and Lofink Lecture Notes: Eqn. 2.13	$\dfrac{1}{\alpha_3} = 1.5$
$\alpha_4 := \dfrac{g_{pol}}{g_{equ}}$	Muller Recommended Multiplier	$\dfrac{1}{\alpha_4} = 0.99472552619195035$

Table 3.5
Some Proposed Values of the Inertial Strain Equation LHS Multiplier α_j

Selected Models of Fluid Oblation

Modified MacLaurin Equation

The MacLaurin Formula (1714) for planetary inertial oblation *but with the Multiplier set to unity (i.e.* $\alpha_0 = 1$ *)* can be expressed in the Wikipedia[0] form as:-

$$LHS_0 = RHS_0$$

$$Ne(a,\Omega) = \frac{2\sqrt{1-e^2}}{e^3}(3 - 2e^2)\sin^{-1}e - \frac{6}{e^2}(1 - e^2)$$
Equation 3.13

and:-

$$e^2 = 1 - \lambda^2 = 1 - \left(\frac{c}{a}\right)^2$$
Equation 3.14

I computed that for the Planet Earth:-

$$LHS_0 = a_0.Ne(a,\Omega) = 0.003461487822077$$
Equation 3.15a

$$RHS_0 = \frac{2\sqrt{1-e^2}}{e^3}(3 - 2e^2)\sin^{-1}e - \frac{6}{e^2}(1 - e^2) = 0.003573750419946$$
Equation 3.15b

Accordingly, the Percentage Specific Defect, PSD(LHS$_{mac}$,RHS$_{mac}$) is minus 3.24318916139267. This three-percent discrepancy is very alluring but as we shall see it is not very meaningful:- and is not in any case a "standard" MacLaurin form.

The Standard (1714) MacLaurin Equation

The MacLaurin Formula (1714) for planetary inertial oblation *complete with Ludolphine Multiplier* can be expressed in the Wikipedia[0] form as:-

$$LHS_1 = RHS_1$$
$$\alpha_1.Ne(a,\Omega) = \frac{2\sqrt{1-e^2}}{e^3}(3 - 2e^2)\sin^{-1}e - \frac{6}{e^2}(1 - e^2)$$
Equation 3.16

and e^2 is as of Equation 3.14.

I computed that for the Planet Earth:-

$$LHS_1 = a_1.Ne(a, \Omega) = 0.001101825794672$$
Equation 3.17a

$$RHS_1 = \frac{2\sqrt{1 - e^2}}{e^3}(3 - 2e^2)\sin^{-1} e - \frac{6}{e^2}(1 - e^2) = 0.003573750419946$$
Equation 3.17b

The only change is that we are dividing the LHS by π in the proper MacLaurin way and thus the Percentage Specific Defect, PSD(LHS$_{mac}$,RHS$_{mac}$) is minus 224.348044602613.

It is by no means preposterous that inertial oblation accounts for only 0.308310784246 of the distortive strain when we consider the fact that Newton's Number is only around 0.00346, and that things do not fly of the Earth's surface of their own accord. Even if rotational speed was much larger, say like that of a gas turbine rotor or a hard drive disk, we would hardly expect the tiny gravitational centripetation of the object to overwhelm the intermolecular cohesions of its substance: At least not for rocky planets.

The Müller[9] Form After Isaac Newton and Chandrasekhar (Eqn. 1.2 exref[9])

This submission for planetary inertial oblation has a -

$$LHS_2 = RHS_2$$
$$\alpha_2.Ne(a, \Omega) = f$$
Equation 3.18

and f is as of Equation 1.2.
I computed that for the Planet Earth:-

$$LHS_2 = a_2.Ne(a, \Omega) = 0.004326859777597$$
Equation 3.19a

$$RHS_2 = f = 0.003352810671831$$
Equation 3.19b

The Percentage Specific Defect, PSD(LHS$_2$,RHS$_2$) is plus 22.5116864384956.

The Müller[9] and Lofinck Lecture Notes ExRef Eqn. 2-13[9]

This submission for planetary inertial oblation has a -

$$LHS_2 = RHS_2$$

$$\alpha_3 . Ne(a, \Omega) = \frac{(1 - 2\lambda^2) . \cos^{-1} \lambda - 3\lambda\sqrt{1 - \lambda^2}}{(1 - \lambda^2)^{\frac{3}{2}}}$$

Equation 3.20

and λ is as implicit in Equation 3.14.
I computed that for the Planet Earth:-

$$LHS_3 = a_3 . Ne(a, \Omega) = 0.002307658548052$$

Equation 3.21a

$$RHS_3 = \frac{(1 - 2\lambda^2) . \cos^{-1} \lambda - 3\lambda\sqrt{1 - \lambda^2}}{(1 - \lambda^2)^{\frac{3}{2}}} = 0.001792886417188$$

Equation 3.21b

The Percentage Specific Defect, PSD(LHS$_3$,RHS$_3$) is plus 22.3071186722173.

The MacLaurin or Flattening RHSs and a$_4$ as Recommended by Müller

We have seen earlier in this disquisition that Müller recommends the use of this LHS multiplier for Inertial Deformation Equations:-

$$\alpha_4 = \frac{g_{pol}}{g_{equ}} = \frac{GM}{c^2} \times \frac{a^2}{GM} = \frac{a^2}{c^2} = \frac{1}{\lambda^2}$$

Equation 3.6

Unfortunately, Müller does not state a preferred RHS for the relevant Inertial Oblation Formula. So firstly I essayed The Standard MacLaurin RHS; and then Flattening. Thus this submission for planetary inertial oblation has a -

$$LHS_4 = RHS_4$$

$$\alpha_4 . Ne(a, \Omega) = \frac{2\sqrt{1 - e^2}}{e^3}(3 - 2e^2) \sin^{-1} e - \frac{6}{e^2}(1 - e^2)$$

Equation 3.22a

and e is as implicit in Equation 3.14.
I computed that for the Planet Earth:-

$$LHS_4 = a_4 . Ne(a, \Omega) = 0.003479842158387$$
Equation 3.23a

$$RHS_4 = \frac{2\sqrt{1 - e^2}}{e^3}(3 - 2e^2)\sin^{-1} e - \frac{6}{e^2}(1 - e^2) = 0.003573750419946$$
Equation 3.23b

The Percentage Specific Defect, for the case of this MacLaurin RHS (a) is $PSD(LHS_4, RHS_4)$ = -2.69863566461984.

For case (b), RHS = Flattening, these forms apply:-

$$LHS_4 = RHS_4$$
$$\alpha_4 . Ne(a, \Omega) = f$$
Equation 3.24b

and f is as of Equation 3.14.

I computed that for the Planet Earth:-

$$LHS_4 = a_4 . Ne(a, \Omega) = 0.003479842158387$$
Equation 3.25a
$$RHS_4 = f = 0.003352810671831$$
Equation 3.25b

The Percentage Specific Defect, for the case of this Flattening RHS (b) is $PSD(LHS_4, RHS_4)$ = 3.65049564820839.

We shall see that there are minor numerical discrepancies between the MATHCAD® and the EXCEL® renditions of these various numbers: It is the MATHCAD® versions presented above and the EXCEL® in Table 3.6. These discrepancies do not necessarily relate to the physics of the Sun and its satellites.

Some of the Issues Arising from Attempts to Apply
MacLaurin-Style Analytical Models

Table 3.6 shows that all of the Models are meaningless except for oblate planets, because they report 100% discrepancies. The exception is j = 3 Müller and Lofinck for which a Division-by-Zero arises when a body is spherical.

None of the models remotely satisfies the case of the Moon whose rotation is synchronous ("locked" to the Earth's gravity so that it always presents the same aspect). The percentage standard defects for the Moon are well into negative five figures for every model tested.

When we look at the means and the standard deviations for the qualifying six independently-rotating planets we discern that Müller-type models show LHS-RHS mean formulaic discrepancies of 16 to 33%. This seems very loose, but in my opinion none of these formulae allow enough latitude to accommodate the effects of cohesive body integrity which must be greater than the forces tending to tear apart the gravitational agglomerates.

I would expect a model mean discrepancy to be about one hundred percent.

Only the 1714 $j = 1$ MacLaurin original model appears with mean $\approx$ -180 and population standard deviation around 77 to offer the algebraic space to house any cohesive or rigidifying forces tending to oppose the centrifugal.

Clearly, no equilibrium model of celestial body mechanical integrity will be possible unless we add cohesive and rigidifying factors to the gravitational and the centrifugal.

Planet Name (Satellites italicised)	Symbol	Modified MacLaurin (Wikipedia)	Standard MacLaurin (Wikipedia)	Müller-Newton-Chandrasekhar	Müller and Lofink	Müller Recommended Accelarations Ratio (a) MacLaurin RHS 4a $1/\lambda2$	Müller Recommended Accelarations Ratio (b) Flattening RHS 4b $1/\lambda2$	Mean²	Population SD²
j number		0	1	2	3				
α_j		1	$1/\pi$	5/4	2/3	$1/\lambda2$	$1/\lambda2$		
SUN	☉	100.00	100.00	100.00	#DIV/0!	100.00	100.00		
Mercury	☿	100.00	100.00	100.00	#DIV/0!	100.00	100.00		
Venus	♀	100.00	100.00	100.00	#DIV/0!	100.00	100.00		
Earth	♁	-3.25	-224.36	22.51	22.30	-2.56	3.79	-30.26	87.44
Moon	☾	-16879.87	-53243.83	-12638.20	-12650.31	-16838.86	-15784.30	-72.59	116.01
Mars	♂	-36.93	-330.16	-2.82	-3.30	-35.32	-27.02	5.59	65.40
Jupiter	♃	23.58	-140.08	41.80	38.71	33.17	36.39	19.45	56.75
Saturn	♄	34.12	-106.96	49.38	45.23	46.40	48.51	-3.22	69.98
Uranus	♅	17.62	-158.82	37.90	36.76	21.35	25.89	12.61	59.07
Neptune	♆	30.40	-118.67	47.60	46.89	32.75	36.72		
Pluto	♇	100.00	100.00	100.00	#DIV/0!	100.00	100.00		
Mean[1]		10.92	-179.84	32.73	31.10	15.97	20.71		
Population SD[1]		24.54	77.09	18.15	17.32	27.40	25.39		

[1] For Oblate Planets Earth, Mars, Jupiter, Saturn, Uranus and Pluto Only

[2] For Oblate Planets Earth, Mars, Jupiter, Saturn, Uranus and Pluto Only

Table 3.6
Discrepancies between Equation LHS and RHS for
The Sun, Moon and Planets expressed as Percentage Specific Defects

Regression Studies

We wish statistically to assess the mathematical co-ordination between Planetary Flattening, λ, and Newton's Number Ne(), thought to be a mechanical determinant of body oblation.

For convenience, the required data has been gathered to Table 3.7

Planet Name (Satellites italisised)	Symbol	Nominal Composition	Newton's Number (Ne())	Observed Flattening λ
SUN	☉		0.000020888140710	0.999999998750000
Mercury	☿	TELLURIC	0.000000112604057	0.000000000000000
Venus	♀	TELLURIC	0.000000264712717	0.000000000000000
Earth	♁	TELLURIC	0.003461430632239	0.003352859933865
Moon	☾	TELLURIC	0.000007587982392	0.001208215867902
Mars	♂	TELLURIC	0.004581721797134	0.005888934691714
Jupiter	♃	GAS GIANT	0.089177685179120	0.064874391540312
Saturn	♄	GAS GIANT	0.154819984373912	0.097962434459415
Uranus	♅	ICE GIANT	0.029534907404472	0.022927344575296
Neptune	♆	ICE GIANT	0.026078316034161	0.017081246971410
Pluto	♇	TELLURIC	0.000249968864011	0.000000000000000

Table 3.7
Newton's Numbers and Observed Flattenings for Several Solar System Bodies

Appropriate linear regressions conform to the straight-line equation (first-order algebraic polynomial):-

$$y = u + v.x$$
Equation 3.26

The quality of the agreement between u and v is measured by the Coefficient of Correlation (agreement) R^2 where $R^2 = -1$ is exact inverse correlation; $R^2 = +1$ is exact positive correlation; and $R^2 = 0$ denotes utterly no relationship between u and v.

In the present instance we will employ EXCEL® to study the conformity of the Observed Planetary Flattening λ to Newton's Number Ne() for the six Solar System planets that exhibit geometrical oblation of the biaxial ellipsoid.

This will issue in three Inferred Flattenings λ_1, λ_2, λ_3 which can then be compared with the Observed Flattening values using their respective percentage specific defects, as well as the standard Correlation Coefficients, R^2.

Treating Flattening as the outcome of Newton's Number physics we may specify:-
$$\lambda_j = u + v.Ne()$$
Equation 3.27

where j is an arbitrary label of Estimated Flattening.

For these purposes we exclude the Sun, even though it has a vanishingly small inertial oblation; the Moon, which we have seen to have anomalous properties, partly due to its "locked" synchrony; and of course all spherical planets. This leaves us with the telluric (terrestrial) planets Earth and Mars; the Ice Giants Uranus and Neptune; and the Gas Giants Jupiter and Saturn; and it is these six we shall regress.

Figure 3.1 is the linear regression graph for λ_1 under the free computation of both u and v. In this condition:-

$$\lambda_1 = 0.00285057 + 0.63377614 . Ne()$$
Equation 3.28

where $R^2 = 0.99311703$.

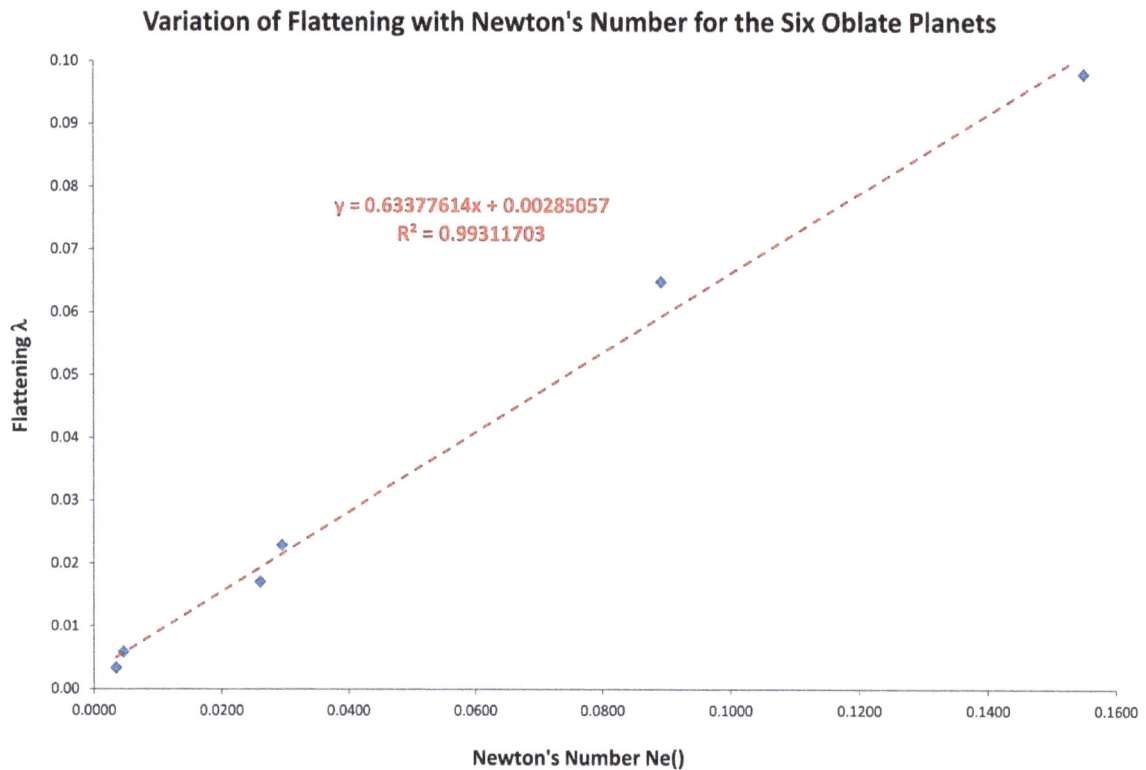

Variation of Flattening with Newton's Number for the Six Oblate Planets

$y = 0.63377614x + 0.00285057$
$R^2 = 0.99311703$

Flattening λ

Newton's Number Ne()

Figure 3.1
Fully-Variant Linear Regression of Flattening against Newton's Number
for the Six Qualifying Solar System Bodies

It is of course the case that if a celestial body is not spinning, that is to say $\Omega = 0$, then there can be no inertial distortion and accordingly Newton's Number is zero.

Therefore, we may constrain our regression to intercept the origin so that u = 0. The EXCEL-generated linear regression then becomes:-

$$\lambda_2 = +0.65994926 . Ne()$$
Equation 3.29

where $R^2 = 0.98951950$.

The quality of the regression as a statistical algorithm is thus vitiated a little but the physical fidelity of the fitment has been enhanced.

Figure 3.2 presents the resulting graph with the six participating planets named. You can see that the two telluric planets named in red have very small Newton's Numbers and Flattenings near to the origin. The ice giants occupy a lower intermediate position; and the two gas giants have notably large Flattenings and notably large Newton's Numbers.

Because of our use of dimensionless variables there is no direct and obvious relation to planetary size, but we may intuit that the rocky, telluric planets, broadly comprised of competent solid rock, are especially resistant to distortion under stress; that the intermediate bodies made of ices can flow plastically like glaciers (because ices tend crytallographically to slip along intermolecular glide planes); and that the two gas giants are more truly fluid, more truly MacLaurin-conformable, because their internal chemical cohesion is negligible.

Now regression discloses correlation, not causality, so its numerical coefficients may provisionally be modified to suggest reasonable factors of the geometry or the physics of the situation.

I propose:-

$$\lambda_3 = \frac{1}{350} + \frac{2}{\pi} . Ne()$$
Equation 3.30

Table 3.8 summarises the implications of the three model flattenings in terms of R^2 (for λ_1 and λ_2) and of PSD(λ,λ_j).

It is apparent that both the full regression and rationalised fitments have mean percentage specific defects around minus eight point five and standard deviations of plus twenty so that they are sensibly the same. Both R^2s are larger than 0.98 and this is well beyond what we could reasonably expect when correlating the geometry and physics of natural objects moving in free space.

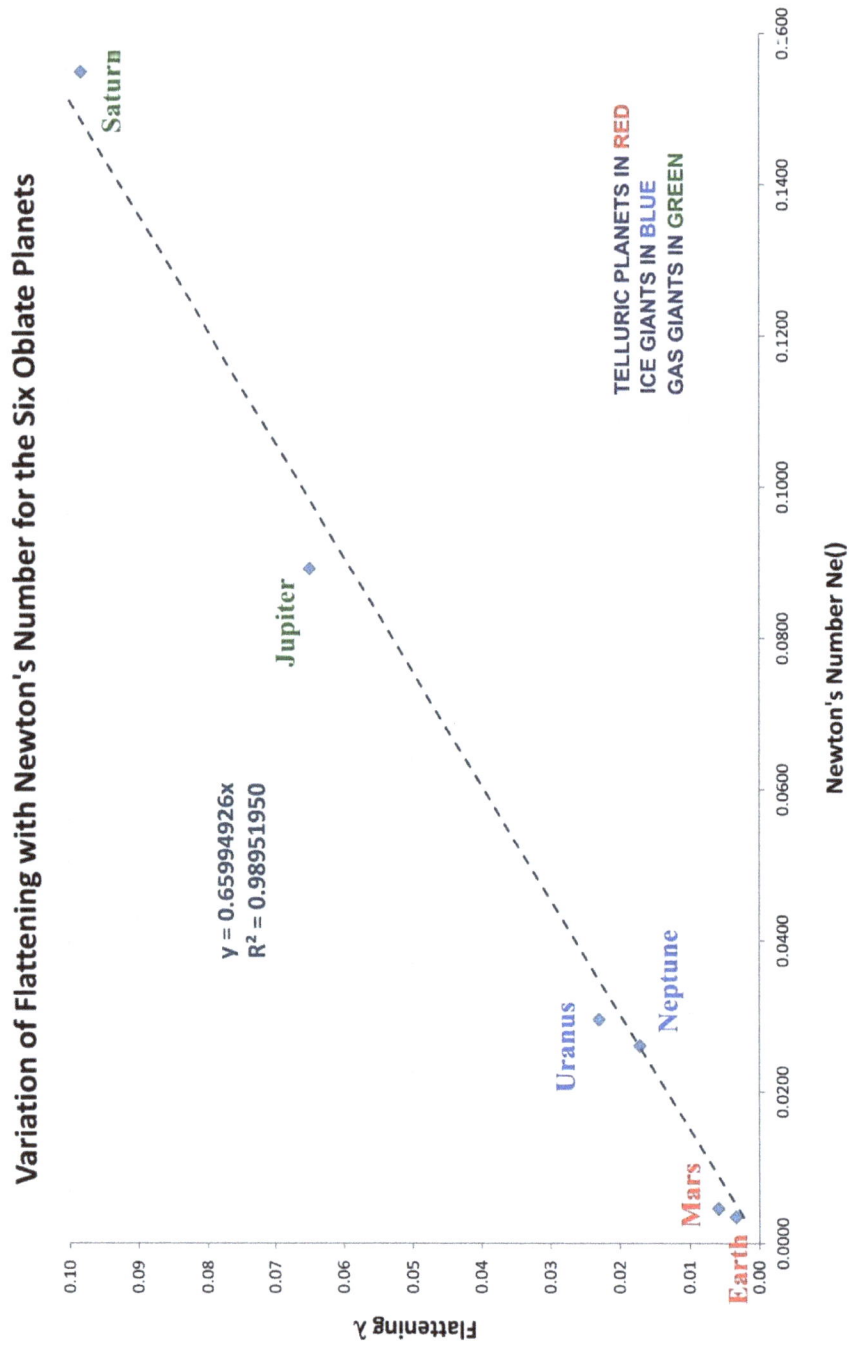

Figure 3.2
An Annotated Linear Regression of Flattening against Newton's Number
Constrained to Intercept the Origin

FLATTENING

Planet Name (Satellites italicised)	Symbol	Nominal Composition	Newton's Number (Ne())	Observed Flattening λ	Inferred Flattening λ_1	Inferred Flattening λ_2	Inferred Flattening λ_3	PSD (λ,λ_1)	PSD (λ,λ_2)	PSD (λ,λ_3)	
					0.00285057		0.00285714				Intercept u
					0.63377614	0.65994926	0.63661977				Grade v
					0.99311703	0.98951950					Coefficient R^2
Earth	♁	TELLURIC	0.00346143	0.00335286	0.00504434	0.00228437	0.00506076	-50.44893746	31.86805804	-50.93854614	
Mars	♂	TELLURIC	0.00458172	0.00588893	0.00575436	0.00302370	0.00577396	2.28528153	48.65448391	1.95242693	
Jupiter	♃	GAS GIANT	0.08917769	0.06487439	0.05936926	0.05885275	0.05962942	8.48583289	9.28200489	8.08480961	
Saturn	♄	GAS GIANT	0.15481998	0.09796243	0.10097178	0.10217333	0.10141861	-3.07194043	-4.29848409	-3.52805810	
Uranus	♅	ICE GIANT	0.02953491	0.02292734	0.02156909	0.01949154	0.02165965	5.92417042	14.98561806	5.52918671	
Neptune	♆	ICE GIANT	0.02607832	0.01708125	0.01937838	0.01721037	0.01945911	-13.44830097	-0.75590733	-13.92092455	
								-8.37898234	16.62262891	-8.80351759	Mean μ
								20.10999960	18.50555728	20.14824769	Pop. Stand. Dev. σ

Table 3.8
Models of Flattening and their Agreement with
Observed Values for Six Qualifying Planets

1530 13 February 2020
"If you think tensorial analysis is the answer you haven't really understood the question"

At this juncture I need to emphasise that we here disclose serious *internal* defects with *all* (or possibly *all but one*) of the various models examined; but only if they are expected to represent observed reality at a scientific level. (I.e the LHS of the formulae never agree with the RHS). This is entirely distinct from any *external* discrepancy due to a need for superadded cohesion or rigidity or whatever. And of course any computational imprecisions that have crept in, including simple mis-programming.

Mathematics is an Art of Man, and the fact that four or five great geniuses working for years over the space of more than three centuries cannot satisfactorily explain the Figure of a Planet is neither surprising nor reprehensible:- If it were other what labor, and what entertainment, would remain for the rest of us? God is jealous of his secrets.

1401 22 February 2020

When considering the privy properties of inaccessible things errors abound and misconceptions propagate.

On Page 30 of "Effect"[8], Müller reminds us that a three-dimensional stress tensor has six principal vectors, of which in his analysis four are non-zero. Unfortunately, those four are each very intricate. Those four involve the Poisson's Ratio of the planetary material, which is both indeterminable and probably heterogeneous; and the Mean Density, which is inferred from Newtonian Mechanics and probably heterogeneous.

Let us assume that all six of these mathematical components have an equal and separate Probability of Error, p_x, of 0.5 (i.e. 50%) then[13]:-

$$P_{Sys} = 1 - \prod_{i=1}^{n} (1 - P_i) = 1 - (1 - P_x)^n$$

Equation 4.1

where P_{Sys} is the Probability of Total System Failure (serially consequential); n is the Number of Independent Components; and P_i is the Probability of Failure of the ith. Component. P_x is the Equal Probability of a Component Failure.

Table 4.1 lists the indicated System Failure Probabilities, P_{Sys}, against P_x Component Failure Probabilities for both the Four-Component and Six-Component systems. (There are four substantive components if we consider $\sigma_{r\varphi} = \sigma_{\vartheta\varphi} = 0$ "obvious").

P_x	P_{Sys}	
	n	
	4	**6**
0	0	0
0.1	0.3439	0.468559
0.2	0.5904	0.737856
0.3	0.7599	0.882351
0.4	0.8704	0.953344
0.5	0.9375	0.984375
0.6	0.9744	0.995904
0.7	0.9919	0.999271
0.8	0.9984	0.999936
0.9	0.9999	0.999999

Table 4.1
Tensorial Analysis Chances of Rectitude for
Four or Six Component Stress Vectors

We can see that when $P_x = 0.5$, P_{Sys} is 0.9375 for n = 4, and $P_{Sys} = 0.984375$ for n = 6 components. In all humility, we can expect a chance of only one in sixty-four of being "right" in a tensorial analysis of spheroidal celestial body stresses.

Even when the probability of a component being correct is 90% the probability if six-component system failure is nearly 50%.

PART FIVE
BIBLIOGRAPHY AND REFERENCES

Compendia

0   Wikipedia contributors. (2019, August 28). Maclaurin spheroid.
         In Wikipedia, The Free Encyclopedia.
         Retrieved 15:47, March 16, 2020, from
https://en.wikipedia.org/w/index.php?title=Maclaurin_spheroid
              &oldid=912937727

1   Wikipedia contributors. (2019, November 5). Ellipsoid.
         In Wikipedia, The Free Encyclopedia.
         Retrieved 17:48, December 6, 2019, from
         https://en.wikipedia.org/w/index.php?title=Ellipsoid
              &oldid=924709651

2   Wikipedia contributors. (2019, November 1). Elliptic integral.
         In Wikipedia, The Free Encyclopedia.
         Retrieved 17:49, December 6, 2019, from
         https://en.wikipedia.org/w/index.php?title=Elliptic_integral
              &oldid=924082763

3   Wikipedia contributors. (2020, January 24). Elastic modulus.
         In Wikipedia, The Free Encyclopedia.
         Retrieved 13:27, March 16, 2020, from
https://en.wikipedia.org/w/index.php?title=Elastic_modulus
              &oldid=937277563

4   "Survey of Applicable Mathematics"   (English Edition)
         Prof. RNDr. Karel Rektorys, Dr.Sc. (editor)
         English Translation Edited by the Staff of the Department of Mathematics,
         The University of Surrey, UK
         Copyright: Karel Rektorys, 1969
         Iliffe Books Ltd of London SE1 in co-edition with SNTL Prague   1969
         592 03927 7
         pp 1369   Hardback

5   Digital Library of Mathematical Functions
         National Institute of Standards and Technology
         US Department of Commerce
         Symmetric Integrals
         https://dlmf.nist.gov/19.16#E3

6   "Physics: Parts I and II"
     Daniel Halliday and Robert Resnick
     John Wiley of New York
     January 1966
     ISBN 0-471-34524-5 (Hardback)
     pp 1214   (text)

7   Wolfram
     https://mathworld.wolfram.com/Ellipsoid.html

Published Papers and Monographs

8   "The Effect of Rotation On the Flattening of Celestial Bodies:
      A Journey through Four Centuries"
     Wolfgang H Müller
     Mathematics and Mechanics of Complex Systems
     V6   No.1   2018
     pp 1-40
     dx.doi.org/10.2140/memocs.2018.6.1
     https://msp.org/memocs/2018/6-1/p01.xhtml

9   "Lecture Notes of TICMI"
     Wolfgang H Müller and Lofinck

     IVANE JAVAKHISHVILI TBILISI STATE UNIVERSITY
     ILIA VEKUA INSTITUTE OF APPLIED MATHEMATICS
     GEORGIAN ACADEMY OF NATURAL SCIENCES
     TBILISI INTERNATIONAL CENTRE OF
     MATHEMATICS AND INFORMATICS
     L E C T U R E N O T E S
     of
     T I C M I
     Volume 15, 2014
     THE MOVEMENT OF THE EARTH:
     Tbilisi
     MODELLING THE FLATTENING PARAMETER
     Wolfgang H. Müller & Paul Lofinck
     https://www.emis.de/journals/TICMI/lnt/vol15/vol15.pdf

10   "Rotational Flattening"
     Farside
     http://farside.ph.utexas.edu/teaching/celestial/Celestialhtml/n...

11   "Geophysical Model of the Dynamical Flattening of the Earth
         in agreement with The Precision Constant"
        P DeFraigne
        Royal Observatory of Belgium
        Geophys.J.Int (1997)
        V130
        pp 47-56

12   "On the Figure of Elastic Planets I: Gravitational Collapse and
         Infinitely Many Equilibria"
        Fia Jia, Ousmane Kodio, S Jon Chapman and Alain Goriely
        Proceeding A
        The Royal Society of London
        Proc.R.Soc.A V475: 20180815
        pp18
        Accepted: 7 March 2019
        http://dx.doi.org/10.1098/rspa.2018.0815
        royalsocietypublising.org/journal/rspa

13   "Reliability of Systems with Various Element Configurations"
        pp 11
        https://ocw.mit.edu/courses/civil-and-environmental-engineering/
              1-151-probability-and-statistics-in-engineering-spring-2005/
              lecture-notes/app1_reli_final.pdf

Spheroid Contact Stresses

H   "Elastic Compression of Spheres and Cylinders at Point and Line Contact"
        MJ Puttock and EJ Thwaite
        National Standards Laboratory
        Technical Paper No. 25
        Commonwealth Scientific and Industrial Research Organisation
        (CSIRO)
        Melbourne 1969

        Division of Applied Physics
        National Standards Laboratory
        CSIRO
        University Grounds
        Chippendale
        NSW 2008

I "Tutorial on Hertz Contact Stress"
  Xiaoyin Zhu
  OPTI 521
  December 1, 2012
  https://wp.optics.arizona.edu/optomech/wp-content/uploads/sites/53/2016/10/OPTI-521-Tutorial-on-Hertz-contact-stress-Xiaoyin-Zhu.pdf
  8 pp

J "Contact Between Solid Surfaces"
  John A Williams and Rob S Dwyer-Joyce
  CRC Press LLC Copyright: 2001
  http://home.ufam.edu.br/berti/nanomateriais/8403_PDF_CH03.pdf
  pp42

K "Tutorial - Hertz Contact Stress"
  Brian Taylor
  OPTI521 Fall 2016, SID:23333406
  https://wp.optics.arizona.edu/optomech/wp-content/uploads/sites/53/2016/12/Tutorial_Taylor_Brian.pdf
  pp6

L "Shape Transitions in Soft Spheres Regulated by Elasticity"
  Craig Fogle, Amy C Rowat, Alex J Levine, Joseph Rudnick
  UCLA, Los Angeles, CA 90095
  July 9 2018
  arXiv: 1306.5864v2
  Icond-mat.softI
  16 October 2013

M "Chapter 3"
  "Contact Between Solid Surfaces"
  John A Williams and Rob S Dwyer-Joyce
  home.ufam.edu.br/berti/nanomaterials/8403_PDF_CH03.pdf
  © CRC Press LLC, 2001

N "Tutorial on Hertz Contact Stress"
  Xiaoyin Zho
  OPT1 521
  December 1, 2012
  wp.optics.arizona.edu

O "Tutorial – Hertz Contact Stress"
  Brian Taylor
  OPT 1521
  Fall 2016   SID 23333406

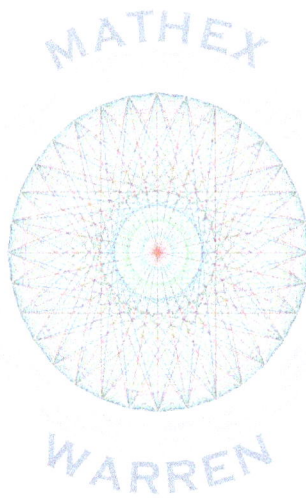

CHAPTER FIVE

The General Binomial Product

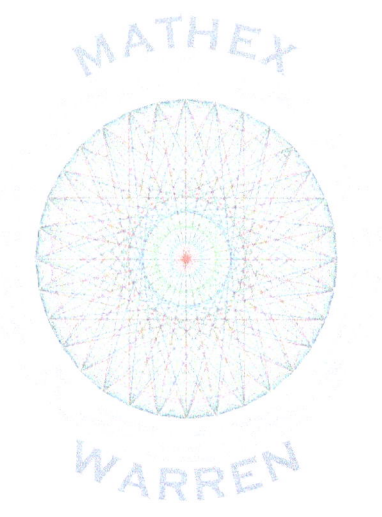

Some Observations upon
Elementary Algebraic Polynomials

by
James R Warren BSc MSc PhD PGCE

PART ZERO
INTRODUCTION

We shall consider certain developments of binomially-structured algebraic polynomials of the form[1,2]:-

$$T_n = (z + r_i)^n = (z + r_1)(z + r_2) \dots (z + r_n) = \prod_{i=0}^{n}(z + r_i)$$

$$= c_{n,0} + c_{n,0} \cdot z + c_{n,0} \cdot z^2 + \dots + c_{n,0} \cdot z^n = \sum_{i=0}^{n} c_{n,i} \cdot z^i$$

Equation 0.1

where z is some quantity (not necessarily a Zeta Function or anything); r_i is the ith Additive Co-quantity (not necessarily a Root); n is the Degree of the Polynomial; and $c_{n,i}$ is the ith. Polynomial Coefficient. Whilst this discussion will assume that z, r and c are all real numbers I hypothesise that any processes explicated here are general to the complex plane.

Furthermore, when $z \neq 0$, we may declare:-

$$R_i = \frac{r_i}{z}$$

Equation 0.2

associated with the z-Free Polynomial Coefficient, $C_{n,k}$, such that:-

$$T_n = z^n \cdot \prod_{i=0}^{n}(1 + R_i) = z^n \cdot \sum_{i=0}^{n} C_{n,i}$$
Equation 0.3

<u>Trial Data</u>

I tested elaborations using PTC® Mathcad® Prime 3.1 corroborated by MicroSoft EXCEL® tabulations running upon Windows® 7. The hardware platform was an old 64-bit HP500400 personal computer. As always my approach is conservative and parsimonious, so accordingly there are technical limitations that require to be born in mind.

In particular, it is impractical *in this context* to employ an Infinitesimal ϵ, of some such value as 10^{-14} or 10^{-15}. Do that and Prime shall shoal upon zero in the computation of (for example) the Geometric Mean.

So:-

$$\epsilon = 0.000000001 = 10^{-9}$$

Equation 0.4

and INF, the Surrogate of Summative Infinitude, had to be set to a mere 46.
The value of z was arbitrarily set to:-

$$z = \frac{7}{13} = 0.538461538461538$$

Equation 0.5

A series of eight trial r-values was similarly defined in order to facilitate the production of a fully-heterogeneous octic (degree eight) polynomial:-

$$r_1 = \frac{23}{24} \quad r_2 = \frac{14}{4} \quad r_3 = \frac{23}{12} \quad r_4 = \frac{19}{23} \quad r_5 = \frac{15}{25} \quad r_6 = \frac{5}{13} \quad r_7 = \frac{27}{10} \quad r_8 = \frac{254}{457}$$

These predicates are summarised in Table 0.1 together with selected solutions:-

Variable Symbol	Variable Subscript	Numerator	Denominator	Value	r_i Cumulative Sum	r_i Cumulative Product	Eloborated Factor Solution, EFS
z		7	13	0.538462			
r_0	0	1	1	1	1	1	
r_1	1	23	24	0.958333	1.95833333	0.95833333	1.496794872
r_2	2	14	4	3.5	5.45833333	3.35416667	6.044748521
r_3	3	23	12	1.916667	7.37500000	6.42881944	14.84063259
r_4	4	19	23	0.826087	8.20108696	5.31076389	20.25076286
r_5	5	15	25	0.6	8.80108696	3.18645833	23.05471464
r_6	6	5	13	0.384615	9.18570234	1.22556090	21.28127505
r_7	7	27	10	2.7	11.88570234	3.30901442	68.91859075
r_8	8	254	457	0.555799	12.44150103	1.83914587	75.41487266

Table 0.1
Polynomial Trial Data together with Cumulates and
Solutions at Degrees from One to Eight

The Power Mean

The Power Mean is a species of generalised algebraic average defined by the equation:-

$$PM(m,n) = \left(\frac{1}{n}\sum_{i=1}^{n} r_i^m\right)^{\frac{1}{m}}$$

Equation 0.6

The power mean LHS as stated in Equation 0.6 is not functional in that it omits the data series $r_1...r_n$ as an explicit array argument.

Notwithstanding that, the Power Mean relates to ordinary Pythagorean means in the following ways:-

Pythagorean Mean Type	Value of Argument m
Arithmetic	1
Geometric	ϵ
Harmonic	-1

Inspection of Equation 0.6 confirms that if m is zero, PM(m,n) is undefined. If however m "tends to" zero, the solution PM(m,n) becomes progressively more accurate. Therefore, the technical infinitesimal ϵ applies to actual computational mechanics, though using the value above specified (10^{-9}) on the said platform I could achieve only eight-figure accuracy in the value of the geometric mean.

Definitions of Specific Means

The Arithmetic Mean
This is defined by:-

$$AM(n,r) = \frac{1}{n}\cdot\sum_{i=1}^{n} r_i$$

Equation 0.7

for which example outcomes using the trial data include:-

AM(3,r) = 2.125
PM(1,3) = 2.125
AM(4,r) = 1.80027173913043
PM(1,4) = 1.80027173913043

The Geometric Mean

This is defined by:-

$$GM(n,r) = \left(\prod_{i=1}^{n} r_i\right)^{\frac{1}{n}}$$

Equation 0.8

for which example outcomes using the trial data include:-

GM(3,r) = 1.85941819437929
PM(ϵ,3) = 1.85941815343516
GM(4,r) = 1.51806108841865
PM(ϵ,4) = 1.5180610038641

The Harmonic Mean

This is defined by:-

$$HM(n,r) = n \cdot \left(\sum_{i=1}^{n} \frac{1}{r_i}\right)^{-1}$$

Equation 0.9

for which example outcomes using the trial data include:-

HM(3,r) = 1.62080536912752
PM(-1,3) = 1.62080536912752
HM(4,r) = 1.30656700480513
PM(-1,4) = 1.30656700480513

The Incomplete Gamma Function

The Incomplete Gamma Function is a species of false factorial often useful, and in our context it is frequently convenient in the discussion of permutation mathematics.
It is defined in a number of alternative ways, but the most basic definition is given by:-

$$\Gamma(s,t) = \int_{t}^{INF} e^{-x} . x^{s-1} . dx$$

Equation 0.10

With INF = 46 Mathcad® Prime® 3.1 manages this construct very well upon my platform, achieving some fifteen-figure accuracy. I did not bother with EXCEL® comparisons on this occasion.

PART ONE
THE LINEAR CASE

The algebraic polynomial of the first degree may be written:-

$$T_1 = (z + r_1) = T_0 \cdot (z + r_1) = 1 \times (z + r_1) = 1.49679487179487$$
Equation 1.1

The numerical value is of course special to our particular trial data.
Long multiplication confirms that the coefficients of the linear equation are given by:-

$$c_{1,0} = r_1$$
Equation 1.2a

$$c_{1,1} = 1$$
Equation 1.2b

These two sub-equations for each coefficient of the linear resolve to:-

$$c_{1,0} = \prod_{i=1}^{n} r_i = 0.958333333333333$$
Equation 1.3a

$$c_{1,1} = 1$$
Equation 1.3b

Since:-

$$T_1 = \sum_{i=0}^{n} c_{1,i} \cdot z^i = c_{1,0} + c_{1,1} \cdot z$$
Equation 1.4

Substitution yields:-

$$T_1 = \sum_{i=0}^{n} c_{1,i} \cdot z^i = \prod_{i=1}^{n} r_i + z = \sum_{i=1}^{n} r_i + z$$
Equation 1.5

Substitution of means in Equation 1.5 gives:-

$$T_1 = \sum_{i=0}^{n} c_{1,i} \cdot z^i = GM(n,r)^n + z = n.\,AM(n,r) + z$$

Equation 1.6

From which:-

$$T_1 = 1.49679487179487$$

Equation 1.7

PART TWO
THE QUADRATIC CASE

The algebraic polynomial of the second degree may be written:-

$$T_2 = (z + r_1)(z + r_2) = T_1.(z + r_2) = 6.04474852071006$$
Equation 2.1

The numerical value is of course special to our particular trial data.
Long multiplication confirms that the coefficients of the quadratic equation are given by:-

$$c_{2,0} = r_1 r_2$$
Equation 2.2a

$$c_{2,1} = r_1 + r_2$$
Equation 2.2b

$$c_{2,2} = 1$$
Equation 2.2c

These three sub-equations for each coefficient of the quadratic resolve to:-

$$c_{2,0} = \prod_{i=1}^{n} r_i = 3.35416666666667$$
Equation 2.3a

$$c_{2,1} = \sum_{i=1}^{n} r_i = 4.45833333333333$$
Equation 2.3b

$$c_{2,2} = 1$$
Equation 2.3d

Since:-

$$T_2 = \sum_{i=0}^{n} c_{2,i} \cdot z^i = c_{2,0} + c_{2,1} \cdot z + c_{2,2} \cdot z^2$$
Equation 2.4

Substitution yields:-

$$T_2 = \sum_{i=0}^{n} c_{2,i} \cdot z^i = \prod_{i=1}^{n} r_i + \sum_{i=1}^{n} r_i \cdot z + 1 \cdot z^2$$

Equation 2.5

Substitution of means in Equation 2.5 gives:-

$$T_2 = \sum_{i=0}^{n} c_{2,i} \cdot z^i = GM(n,r)^n + n.AM(n,r) \cdot z + z^2$$

Equation 2.6

From which:-

$$T_2 = 6.04474852071006$$

Equation 2.7

The algebraic polynomial of the third degree may be written:-

$$T_3 = (z + r_1)(z + r_2)(z + r_3) = T_2 \cdot (z + r_3) = 14.8406325861023$$
Equation 3.1

The numerical value is of course special to our particular trial data.
Long multiplication confirms that the coefficients of the cubic equation are given by:-

$$c_{3,0} = r_1 r_2 r_3$$
Equation 3.2a

$$c_{3,1} = r_1 r_2 + r_1 r_3 + r_2 r_3$$
Equation 3.2b

$$c_{3,2} = r_1 + r_2 + r_3$$
Equation 3.2c

$$c_{3,3} = 1$$
Equation 3.2d

These four sub-equations for each coefficient of the cubic resolve to:-

$$c_{3,0} = \prod_{i=1}^{n} r_i = 6.42881944444445$$
Equation 3.3a

$$c_{3,1} = \prod_{j=1}^{n} r_j \cdot \sum_{i=1}^{n} \frac{1}{r_i} = n \cdot \frac{GM(n,r)^n}{HM(n,r)} = 11.8993055555556$$
Equation 3.3b

$$c_{3,2} = \sum_{i=1}^{n} r_1 = 6.375$$
Equation 3.3c

$$c_{3,3} = 1$$
Equation 3.3d

Since:-

$$T_3 = \sum_{i=0}^{n} c_{3,i} \cdot z^i = c_{3,0} + c_{3,1} \cdot z + c_{3,2} \cdot z^2 + c_{3,3} \cdot z^3$$

Equation 3.4

Substitution yields:-

$$T_3 = \sum_{i=0}^{n} c_{3,i} \cdot z^i = \prod_{i=1}^{n} r_i + \prod_{j=1}^{n} r_j \cdot \sum_{i=1}^{n} \frac{1}{r_i} \cdot z + \sum_{i=1}^{n} r_i \cdot z^2 + 1 \cdot z^3$$

Equation 3.5

or:-

$$T_3 = \sum_{i=0}^{n} c_{3,i} \cdot z^i = \prod_{i=1}^{n} r_i \cdot \left(z. \sum_{i=1}^{n} r_i + 1 \right) + \sum_{i=1}^{n} r_i \cdot z^2 + z^3$$

Equation 3.6

Substitution of means in Equation 3.5 gives:-

$$T_3 = \sum_{i=0}^{n} c_{3,i} \cdot z^i = GM(n,r)^n + n \cdot \frac{GM(n,r)^n}{HM(n,r)} \cdot z + n. AM(n,r) \cdot z^2 + z^3$$

Equation 3.7

Equation 3.7 may be further condensed to:-

$$T_3 = \sum_{i=0}^{n} c_{3,i} \cdot z^i = GM(n,r)^n \cdot \left(1 + n \cdot \frac{1}{HM(n,r)} \right) \cdot z + n. AM(n,r) \cdot z^2 + z^3$$

Equation 3.8

or:-

$$T_3 = z^3 + n. \left(\frac{GM(n,r)^n}{n} + \frac{GM(n,r)^n}{HM(n,r)}. z + AM(n,r). z^2 \right)$$

Equation 3.9

or:-

$$T_3 = z^3 + n.\left[GM(n,r)^n\left(\frac{1}{n} + \frac{z}{HM(n,r)}\right) + AM(n,r).z^2\right]$$

Equation 3.10

From which.-

$$T_3 = 14.8406325861023$$

Equation 3.11

PART FOUR
THE QUARTIC CASE

The algebraic polynomial of the fourth degree may be written:-

$$T_4 = (z + r_1)(z + r_2)(z + r_3)(z + r_4) = T_3.(z + r_4) = 20.2507628599656$$
Equation 4.1

The numerical value is of course special to our particular trial data.
Long multiplication confirms that the coefficients of the quartic equation are given by:-

$$c_{4,0} = r_1 r_2 r_3 r_4$$
Equation 4.2a

$$c_{4,1} = r_1 r_2 r_3 + r_1 r_2 r_4 + r_1 r_3 r_4 + r_2 r_3 r_4$$
Equation 4.2b

$$c_{4,2} = r_1 r_2 + r_1 r_3 + r_1 r_4 + r_2 r_3 + r_2 r_4 + r_3 r_4$$
Equation 4.2c

$$c_{4,3} = r_1 + r_2 + r_3 + r_4$$
Equation 4.2d

$$c_{4,4} = 1$$
Equation 4.2e

These five sub-equations for each coefficient of the quartic resolve to:-

$$c_{4,0} = \prod_{i=1}^{n} r_i = 5.31076388888889$$
Equation 4.3a

$$c_{4,1} = \prod_{j=1}^{n} r_j \cdot \sum_{i=1}^{n} \frac{1}{r_i} = n \cdot \frac{GM(n,r)^n}{HM(n,r)} = 16.2586805555556$$
Equation 4.3b

$$c_{4,2} = \sum_{i=1}^{n-1} \sum_{j=i+1}^{n} r_i.r_j = 17.1656099033816$$
Equation 4.3c

$$c_{4,3} = \sum_{i=1}^{n} r_i = 7.20108695652174$$

Equation 4.3d

$$c_{4,4} = 1$$

Equation 4.3e

Since:-

$$T_4 = \sum_{i=0}^{n} c_{4,i} \cdot z^i = c_{4,0} + c_{4,1} \cdot z + c_{4,2} \cdot z^2 + c_{4,3} \cdot z^3 + c_{4,4} \cdot z^4$$

Equation 4.4

Substitution yields:-

$$T_4 = \sum_{i=0}^{n} c_{4,i} \cdot z^i = \prod_{i=1}^{n} r_i + \prod_{j=1}^{n} r_j \cdot \sum_{i=1}^{n} \frac{1}{r_i} \cdot z + \sum_{i=1}^{n-1} \sum_{j=i+1}^{n} r_i.r_j \cdot z^2 + \sum_{i=1}^{n} r_i \cdot z^3 + 1 \cdot z^4$$

Equation 4.5

or:-

$$T_4 = \sum_{i=0}^{n} c_{4,i} \cdot z^i = \prod_{i=1}^{n} r_i \cdot \left(z. \sum_{i=1}^{n} r_i + 1 \right) + \sum_{i=1}^{n-1} \sum_{j=i+1}^{n} r_i.r_j \cdot z^2 + \sum_{i=1}^{n} r_i \cdot z^3 + z^4$$

Equation 4.6

Substitution of means in Equation 4.5 gives:-

$$T_4 = \sum_{i=0}^{n} c_{4,i} \cdot z^i = GM(n,r)^n + n \cdot \frac{GM(n,r)^n}{HM(n,r)} \cdot z + \sum_{i=1}^{n-1} \sum_{j=i+1}^{n} r_i.r_j \cdot z^2 + n.AM(n,r) \cdot z^3 + 1 \cdot z^4$$

Equation 4.7

From which:-

$$T_4 = 20.2507628599656$$

Equation 4.8

The algebraic polynomial of the fifth degree may be written:-

$$T_5 = (z + r_1)(z + r_2)(z + r_3)(z + r_4)(z + r_5) = T_4.(z + r_5) = 23.0547146405763$$
Equation 5.1

The numerical value is of course special to our particular trial data.
Long multiplication confirms that the coefficients of the quintic equation are
given by:-

$$c_{5,0} = r_1 r_2 r_3 r_4 r_5$$
Equation 5.2a

$$c_{5,1} = r_1 r_2 r_3 r_4 + r_1 r_2 r_3 r_5 + r_1 r_2 r_4 r_5 + r_1 r_3 r_4 r_5 + r_2 r_3 r_4 r_5$$
Equation 5.2b

$$\alpha c_{5,2} = r_3 r_4 r_5 + r_2 r_4 r_5 + r_2 r_3 r_5 + r_2 r_3 r_4 + r_1 r_4 r_5$$
$$\beta c_{5,2} = r_1 r_3 r_5 + r_1 r_3 r_4 + r_1 r_2 r_5 + r_1 r_2 r_4 + r_1 r_2 r_3$$

$$c_{5,2} = \alpha c_{5,2} + \beta c_{5,2}$$
Equation 5.2c

$$\alpha c_{5,3} = r_1 r_2 + r_1 r_3 + r_1 r_4 + r_1 r_5 + r_2 r_3$$
$$\beta c_{5,3} = r_2 r_4 + r_2 r_5 + r_3 r_4 + r_3 r_5 + r_4 r_5$$

$$c_{5,3} = \alpha c_{5,3} + \beta c_{5,3}$$
Equation 5.2d

$$c_{5,4} = r_1 + r_2 + r_3 + r_4 + r_5$$
Equation 5.2e

$$c_{5,5} = 1$$
Equation 5.2f

These six sub-equations for each coefficient of the quintic resolve to:-

$$c_{5,0} = \prod_{i=1}^{n} r_i = 3.18645833333333$$
Equation 5.3a

$$c_{5,1} = \prod_{j=1}^{n} r_j \cdot \sum_{i=1}^{n} \frac{1}{r_i} = n \cdot \frac{GM(n,r)^n}{HM(n,r)} = 15.0659722222222$$

Equation 5.3b

$$c_{5,2} = \prod_{k=1}^{n} r_k \cdot \left(\sum_{j=1}^{n-1} \sum_{j=i+1}^{n-j} \frac{1}{r_j \cdot r_{i+j}} \right) = 26.5580464975845$$

Equation 5.3c

$$c_{5,3} = \sum_{i=1}^{n-1} \sum_{j=i+1}^{n} r_i \cdot r_j = 21.4862620772947$$

Equation 5.3d

$$c_{5,4} = \sum_{i=1}^{n} r_i = 7.80108695652174$$

Equation 5.3e

$$c_{5,5} = 1$$

Equation 5.3f

Since:-

$$T_5 = \sum_{i=0}^{n} c_{5,i} \cdot z^i = c_{5,0} + c_{5,1} \cdot z + c_{5,2} \cdot z^2 + c_{5,3} \cdot z^3 + c_{5,4} \cdot z^4 + c_{5,5} \cdot z^5$$

Equation 5.4

Substitution yields:-

$$T_5 = \sum_{i=0}^{n} c_{5,i} \cdot z^i = \prod_{i=1}^{n} r_i + \prod_{j=1}^{n} r_j \cdot \sum_{i=1}^{n} \frac{1}{r_i} \cdot z + \prod_{k=1}^{n} r_k \cdot \left(\sum_{j=1}^{n-1} \sum_{j=i+1}^{n-j} \frac{1}{r_j \cdot r_{i+j}} \right) \cdot z^2 + \sum_{i=1}^{n-1} \sum_{j=i+1}^{n} r_i \cdot r_j$$
$$\cdot z^3 + \sum_{i=1}^{n} r_i \cdot z^4 + 1 \cdot z^5$$

Equation 5.5

Substitution of means gives:-

$$T_5 = \sum_{i=0}^{n} c_{5,i} \cdot z^i = GM(n,r)^n + n\frac{GM(n,r)^n}{HM(n,r)} \cdot z + GM(n,r)^n \cdot \left(\sum_{j=1}^{n-1} \sum_{j=i+1}^{n-j} \frac{1}{r_j \cdot r_{i+j}}\right) \cdot z^2$$

$$+ \sum_{i=1}^{n-1} \sum_{j=i+1}^{n} r_i \cdot r_j \cdot z^3 + n.AM(n,r) \cdot z^4 + 1 \cdot z^5$$

Equation 5.6

From which:-

$$T_5 = 23.0547146405763$$
Equation 5.7

The algebraic polynomial of the sixth degree may be written:-

$$T_6 = (z + r_1)(z + r_2)(z + r_3)(z + r_4)(z + r_5)(z + r_6) = T_5.(z + r_6) = 21.2812750528396$$
Equation 6.1

The numerical value is of course special to our particular trial data.
Long multiplication confirms that the coefficients of the sextic equation are given
by:-

$$c_{6,0} = r_1 r_2 r_3 r_4 r_5 r_6$$
Equation 6.2a

$$c_{6,1} = r_1 r_2 r_3 r_4 r_5 + r_1 r_2 r_3 r_4 r_6 + r_1 r_2 r_3 r_5 r_6 + r_1 r_2 r_4 r_5 r_6 + r_1 r_3 r_4 r_5 r_6 + r_2 r_3 r_4 r_5 r_6$$
Equation 6.2b

$$\alpha c_{6,2} = r_1 r_2 r_3 r_4 + r_1 r_2 r_3 r_5 + r_1 r_2 r_3 r_6 + r_1 r_2 r_4 r_5 + r_1 r_2 r_4 r_6$$
$$\beta c_{6,2} = r_1 r_2 r_5 r_6 + r_1 r_3 r_4 r_5 + r_1 r_3 r_4 r_6 + r_1 r_3 r_5 r_6 + r_1 r_4 r_5 r_6$$
$$\gamma c_{6,2} = r_2 r_3 r_4 r_5 + r_2 r_3 r_4 r_6 + r_2 r_3 r_5 r_6 + r_2 r_4 r_5 r_6 + r_3 r_4 r_5 r_6$$

$$c_{6,2} = \alpha c_{6,2} + \beta c_{6,2} + \gamma c_{6,2}$$
Equation 6.2c

$$\alpha c_{6,3} = r_1 r_2 r_3 + r_1 r_2 r_4 + r_1 r_2 r_5 + r_1 r_2 r_6 + r_1 r_3 r_4$$
$$\beta c_{6,3} = r_1 r_3 r_5 + r_1 r_3 r_6 + r_1 r_4 r_5 + r_1 r_4 r_6 + r_1 r_5 r_6$$
$$\gamma c_{6,3} = r_2 r_3 r_4 + r_2 r_3 r_5 + r_2 r_3 r_6 + r_2 r_4 r_5 + r_2 r_4 r_6$$
$$\delta c_{6,3} = r_2 r_5 r_6 + r_3 r_4 r_5 + r_3 r_4 r_6 + r_3 r_5 r_6 + r_4 r_5 r_6$$

$$c_{6,3} = \alpha c_{6,3} + \beta c_{6,3} + \gamma c_{6,3} + \delta c_{6,3}$$
Equation 6.2d

$$\alpha c_{6,4} = r_1 r_2 + r_1 r_3 + r_1 r_4 + r_1 r_5 + r_1 r_6$$
$$\beta c_{6,4} = r_2 r_3 + r_2 r_4 + r_2 r_5 + r_2 r_6 + r_3 r_4$$
$$\gamma c_{6,4} = r_3 r_5 + r_3 r_6 + r_4 r_5 + r_4 r_6 + r_5 r_6$$

$$c_{6,4} = \alpha c_{6,4} + \beta c_{6,4} + \gamma c_{6,4}$$
Equation 6.2e

$$c_{6,5} = r_1 + r_2 + r_3 + r_4 + r_5 + r_6$$
Equation 6.2f

$$c_{6,6} = 1$$
Equation 6.2g

These seven sub-equations for each coefficient of the sextic resolve to:-

$$c_{6,0} = \prod_{i=1}^{n} r_i = 1.2255608974359$$
Equation 6.3a

$$c_{6,1} = \sum_{i=1}^{n} \left(\frac{\prod_{j=1}^{n} r_j}{r_i} \right) = \prod_{j=1}^{n} r_j \cdot \sum_{i=1}^{n} \frac{1}{r_i} = n \cdot \frac{GM(n,r)^n}{HM(n,r)} = 8.98106303418804$$
Equation 6.3b

$$c_{6,2} = \prod_{k=1}^{n} r_k \cdot \left(\sum_{i=1}^{n-1} \sum_{j=i+1}^{n} \frac{1}{r_i \cdot r_j} \right) = 25.280605490524$$
Equation 6.3c

$$c_{6,3} = \sum_{i=1}^{n-2} \sum_{j=i+1}^{n-1} \sum_{k=j+1}^{n} r_i \cdot r_j \cdot r_k = 34.8219934503902$$
Equation 6.3d

$$c_{6,4} = \sum_{i=1}^{n-1} \sum_{j=i+1}^{n} r_i \cdot r_j = 24.4866801374954$$
Equation 6.3e

$$c_{6,5} = \sum_{i=1}^{n} r_i = 8.18570234113712$$
Equation 6.3f

$$c_{6,6} = 1$$
Equation 6.3g

Since:-

$$T_6 = \sum_{i=0}^{n} c_{6,i} \cdot z^i = c_{6,0} + c_{6,1} \cdot z + c_{6,2} \cdot z^2 + c_{6,3} \cdot z^3 + c_{6,4} \cdot z^4 + c_{6,5} \cdot z^5 + c_{6,6} \cdot z^6$$
Equation 6.4

Substitution yields:-

$$T_6 = \prod_{i=1}^{n} r_i + \prod_{j=1}^{n} r_j \cdot \sum_{i=1}^{n} \frac{1}{r_i} \cdot z + \prod_{k=1}^{n} r_k \cdot \left(\sum_{i=1}^{n-1} \sum_{j=i+1}^{n} \frac{1}{r_i \cdot r_j} \right) \cdot z^2 + \sum_{i=1}^{n-2} \sum_{j=i+1}^{n-1} \sum_{k=j+1}^{n} r_i \cdot r_j \cdot r_k \cdot z^3$$

$$+ \sum_{i-1}^{n-1} \sum_{j=l+1}^{n} r_i \cdot r_j \cdot z^4 + \cdot \sum_{i=1}^{n} r_i \, z^5 + 1 \cdot z^6$$

Equation 6.5

Substitution of some means gives:-

$$T_6 = GM(n,r)^n + n \cdot \frac{GM(n,r)^n}{HM(n,r)} \cdot z + GM(n,r)^n \cdot \left(\sum_{i=1}^{n-1} \sum_{j=i+1}^{n} \frac{1}{r_i \cdot r_j} \right) \cdot z^2$$

$$+ \sum_{i=1}^{n-2} \sum_{j=i+1}^{n-1} \sum_{k=j+1}^{n} r_i \cdot r_j \cdot r_k \cdot z^3 + \sum_{i=1}^{n-1} \sum_{j=i+1}^{n} r_i \cdot r_j \cdot z^4 + n \cdot AM(n,r) \cdot z^5 + 1 \cdot z^6$$

Equation 6.6

From which:-

$$T_6 = 21.2812750528396$$

Equation 6.7

The case of the seventh degree has an unfortunate English-language connotation, for whilst the SEX of the Sextic Equation is always a happy interpolation even in the least promising of contexts, the sepsis of the Septic is only infelicitous.

A Septic Equation is an algebraic polynomial of the seventh degree ($n = 7$). A specific solution may be explicated as:-

$$T_7 = (z + r_1)(z + r_2)(z + r_3)(z + r_4)(z + r_5)(z + r_6)(z + r_7) = \prod_{i=1}^{7}(z + r_i)$$

Equation 7.1

Using our data this is the value of T_7:-

$$T_7 = 68.9185907480421$$

Equation 7.2

Amongst the several possibilities the corresponding Polynomial Coefficients $c_{7,k}$ may be elaborated as the following nested summations[3,4], (the zeroth is a product):-

$$c_{7,0} = \prod_{i=1}^{n} r_i$$

Equation 7.3a

$$c_{7,1} = \sum_{h=1}^{n-5}\sum_{i=h+1}^{n-4}\sum_{j=i+1}^{n-3}\sum_{k=j+1}^{n-2}\sum_{l=k+1}^{n-1}\sum_{m=l+1}^{n} r_h.r_i.r_j.r_k.r_l.r_m = 25.4744310897436$$

Equation 7.3b

$$c_{7,2} = \sum_{i=1}^{n-4}\sum_{j=i+1}^{n-3}\sum_{k=j+1}^{n-2}\sum_{l=k+1}^{n-1}\sum_{m=l+1}^{n} r_i.r_j.r_k.r_l.r_m = 77.2386978586028$$

Equation 7.3c

$$c_{7,3} = \sum_{j=1}^{n-3}\sum_{k=j+1}^{n-2}\sum_{l=k+1}^{n-1}\sum_{m=l+1}^{n} r_j.r_k.r_l.r_m = 119.299987806577$$

Equation 7.3d

$$c_{7,4} = \sum_{k=1}^{n-2} \sum_{l=k+1}^{n-1} \sum_{m=l+1}^{n} r_k . r_l . r_m = 100.936029821628$$
Equation 7.3e

$$c_{7,5} = \sum_{l=1}^{n-1} \sum_{m=l+1}^{n} r_l . r_m = 46.5880764585656$$
Equation 7.3f

$$c_{7,6} = \sum_{m=1}^{n} r_m = 10.8857023411371$$
Equation 7.3g

$$c_{7,7} = 1$$
Equation 7.3h

This leads directly to the summative formulation:-

$$T_7 = c_{7,0} + c_{7,1}z + c_{7,2}z^2 + c_{7,3}z^3 + c_{7,4}z^4 + c_{7,5}z^5 + c_{7,6}z^6 + c_{7,7}z^7$$
$$= \sum_{i=1}^{7} c_{7,i}.z^i$$
$$= 68.9105907480421$$
Equation 7.4

Composite Elaboration

Long multiplication, especially of polynomials, is always tedious and even more error-prone, at least in human hands. The septic elaboration is truly baleful, especially for naturally idle enthusiasts like me, and virtually impossible for those whose worktables are so crowded with gadgets, cables and other rubbish that there is hardly a patch of comfort for a sheet of A4.

Accordingly, like all good algebraists, we seek concision, knowing that some shortenings are briefer or more accurate than others.

Among the principles worth exploration is the fact that the product of two polynomials forms a third polynomial of a degree equal to the sum of the two degrees, i.e.:-

$$P(n_{m1}, z, r_1 \ldots r_{m1}) \times P(n_{m2}, z, r_{m1+1} \ldots r_{m2+m1}) = P(n_{m1+m2}, z, r_1 \ldots r_{m1+m2})$$
Equation 7.5

For example, it is possible to compute the coefficients of a dependent Septic Polynomial of Degree m1+m2 = 7 from the coefficients of an m1 = 3 Cubic whose r-Co-quantities are r_1, r_2, and r_3; and an m2 = 4 Quartic whose Co-quantities are r_4, r_5, r_6 and r_7.

The resulting n = 7 Septic evidently incorporates $r_1 \ldots r_7$ according to this scheme:-

$$P(n_3, z, r_1 \ldots r_3) \times P(n_4, z, r_4 \ldots r_7) = P(n_7, z, r_1 \ldots r_7)$$
Equation 7.6

Equation 7.6 summarises our intent, but I should point-out in passing that the Polynomial Product Equation is general enough to be useful for any adequately-long series of co-quantities. For instance, we could "over-square" a cubic by re-employing the co-quantities of a series of four r's r_1, r_2, r_3 and r_4 as in the product $P(n_3,z,r_1\ldots r_3) \times P(n_4,z,r_1\ldots r_4)$. Clearly, the resulting Septic would be a different animal to that whom you and I require.

Firstly, for our exercise of Composite Elaboration we need to define a general Cubic:-

$$T_3 = F + Gz + Hz^2 + Iz^3$$
Equation 7.7

Now from Part Three we know that:-

$$T_3 = \prod_{i=1}^{m1} r_i + \left(\prod_{i=1}^{m1} r_i \cdot \sum_{i=1}^{m1} \frac{1}{r_i} \right) z + \sum_{i=1}^{m1} r_i z^2 + 1 \cdot z^3$$
Equation 7.8

From which we may abstract:-

$$F = \prod_{i=1}^{m1} r_i = 6.42881944444445$$
Equation 7.9a

$$G = \prod_{i=1}^{m1} r_i \cdot \sum_{i=1}^{m1} \frac{1}{r_i} = 11.8993055555556$$
Equation 7.9b

$$H = \sum_{i=1}^{m1} r_i = 6.375$$
Equation 7.9c

$$I = 1$$
Equation 7.9d

Secondly, a General Quartic may be written as:-

$$T_4 = A + Bz + Cz^2 + Dz^3 + Ez^4$$
Equation 7.10

whilst from Part Four we know that:-

$$T_4 = \prod_{i=m1+1}^{m1+m,2} r_i + \left(\prod_{i=m1+1}^{m1+m2} r_i \cdot \sum_{i=m1+1}^{m1+m2} \frac{1}{r_i}\right)z + \left(\sum_{i=m1+1}^{m1+m2-1} \sum_{j=i+1}^{m1+m2} r_i.r_j\right)z^2 + \left(\sum_{i=m1+1}^{m1+m2} r_i\right) \cdot z^3 + z^4$$

Equation 7.11

Accordingly, we may abstract of the Quartic:-

$$A = \prod_{i=m1+1}^{m1+m2} r_i = 0.514715719063545$$

Equation 7.11a

$$B = \prod_{i=m1+1}^{m1+m2} r_i \cdot \sum_{i=m1+1}^{m1+m2} \frac{1}{r_i} = 3.00983277591973$$

Equation 7.11b

$$C = \sum_{i=m1+1}^{m1+m2-1} \sum_{j=i+1}^{m1+m2} r_i.r_j = 5.93304347826087$$

Equation 7.11c

$$D = \sum_{i=m1+1}^{m1+m2} r_i = 4.51070234113712$$

Equation 7.11d

$$E = 1$$
Equation 7.11e

The Septic Equation T_7 can be expressed as the product of Equations 7.7 and 7.10 to give:-

$$T_7 = T_3.T_4 = (F + Gz + Hz^2 + Iz^3).(A + Bz + Cz^2 + Dz^3 + Ez^4)$$
Equation 7.12

Long multiplication of this expression supplies:-

$$T_7 = (AF) + (BF + AG)z + (CF + BG + AH)z^2 + (DF + CG + BH + AI)z^3$$
$$= (EF + DG + CH + BI)z^4 + (EH + DH + CI)z^5 + (EH + DI)z^6 + (EI)z^7$$
Equation 7.13

This sum of products in the co-efficient form resolves to:-

$$c_{7,0} = AF$$
Equation 7.14a

$$c_{7,1} = BF + AG = 25.4744310897436$$
Equation 7.14b

$$c_{7,2} = CF + BG + AH = 77.2386978586028$$
Equation 7.14c

$$c_{7,3} = DF + CG + BH + AI = 119.299987806577$$
Equation 7.14d

$$c_{7,4} = EF + DG + CH + BI = 100.936029821628$$
Equation 7.14e

$$c_{7,5} = EG + DH + CI = 46.5880764585656$$
Equation 7.14f

$$c_{7,6} = EH + DI = 10.8857023411371$$
Equation 7.14g

$$c_{7,7} = EI = 1$$
Equation 7.14h

And the set of Equations 7.14* may be expanded as:-

$$c_{7,0} = AF = \left(\prod_{i=m1+1}^{m1+m2} r_i\right) \cdot \left(\prod_{i=1}^{m1} r_i\right) = 3.30901442307692$$
Equation 7.14a

$$c_{7,1} = BF + AG = \left(\prod_{i=m1+1}^{m1+m2} r_i \cdot \sum_{i=m1+1}^{m1+m2} \frac{1}{r_i}\right)\left(\prod_{i=1}^{m1} r_i\right) + \prod_{i=m1+1}^{m1+m2} r_i \cdot \left(\prod_{i=1}^{m1} r_i \cdot \sum_{i=1}^{m1} \frac{1}{r_i}\right)$$
$$= 25.4744310897436$$
Equation 7.14b

$$c_{7,2} = CF + BG + AH$$

$$= \left(\sum_{i=m1+1}^{m1+m2-1} \sum_{j=i+1}^{m1+m2} r_i . r_j \right) . \left(\prod_{i=1}^{m1} r_i \right) + \left(\prod_{i=m1+1}^{m1+m2} r_i . \sum_{i=m1+1}^{m1+m2} \frac{1}{r_i} \right) . \left(\prod_{i=1}^{m1} r_i . \sum_{i=1}^{m1} \frac{1}{r_i} \right)$$

$$+ \prod_{i=m1+1}^{m1+m2} r_i . \sum_{i=1}^{m1} r_i = 77.2386978586028$$

Equation 7.14c

$$c_{7,3} = DF + CG + BH + AI$$

$$= \left(\sum_{i=m1+1}^{m1+m2} r_i \right) . \left(\prod_{i=1}^{m1} r_i \right) + \left(\sum_{i=m1+1}^{m1+m2-1} \sum_{j=i+1}^{m1+m2} r_i . r_j \right) . \left(\prod_{i=1}^{m1} r_i . \sum_{i=1}^{m1} \frac{1}{r_i} \right)$$

$$+ \left(\prod_{i=m1+1}^{m1+m2} r_i . \sum_{i=m1+1}^{m1+m2} \frac{1}{r_i} \right) . \sum_{i=1}^{m1} r_i + \prod_{i=m1+1}^{m1+m2} r_i . 1 = 119.299987806577$$

Equation 7.14d

$$c_{7,4} = EF + DG + CH + BI$$

$$= 1. \left(\prod_{i=1}^{m1} r_i \right) + \left(\sum_{i=m1+1}^{m1+m2} r_i \right) . \left(\prod_{i=1}^{m1} r_i . \sum_{i=1}^{m1} \frac{1}{r_i} \right) + \left(\sum_{i=m1+1}^{m1+m2-1} \sum_{j=i+1}^{m1+m2} r_i . r_j \right) . \sum_{i=1}^{m1} r_i$$

$$+ \left(\prod_{i=m1+1}^{m1+m2} r_i . \sum_{i=m1+1}^{m1+m2} \frac{1}{r_i} \right) . 1 = 100.936029821628$$

Equation 7.14e

$$c_{7,5} = EG + DH + CI = 1. \left(\prod_{i=1}^{m1} r_i . \sum_{i=1}^{m1} \frac{1}{r_i} \right) + \left(\sum_{i=m1+1}^{m1+m2} r_i \right) . \sum_{i=1}^{m1} r_i + \left(\sum_{i=m1+1}^{m1+m2-1} \sum_{j=i+1}^{m1+m2} r_i . r_j \right) . 1$$

$$= 46.5880764585656$$

Equation 7.14f

$$c_{7,6} = EH + DI = 1. \sum_{i=1}^{m1} r_i + \left(\sum_{i=m1+1}^{m1+m2} r_i \right) . 1 = 10.8857023411371$$

Equation 7.14g

$$c_{7,7} = 1 \times 1 = 1$$

Equation 7.14h

The set of Equations 7.14* is susceptible of extensive simplifications but these will be deferred for the time being.

Tables 7.1, 7.2 and 7.3 summarise the numerical intermediates and outcomes of these septic coefficient computations.

	Coefficient	Value
QUARTIC COMPONENT	**A**	0.514715719063545
	B	3.009832775919730
	C	5.933043478260870
	D	4.510702341137120
	E	1.000000000000000
CUBIC COMPONENT	**F**	6.428819444444450
	G	11.899305555555600
	H	6.375000000000000
	I	1.000000000000000

Table 7.1
Quartic and Cubic Coefficient Values for
Composite Composition of
The Septic Polynomial Example

	A	B	C	D	E
F	3.30901442	19.34967147	38.14246528	28.99849092	6.42881944
G	6.12475962	35.81491987	70.59909722	53.67422543	11.89930556
H	3.28131271	19.18768395	37.82315217	28.75572742	6.37500000
I	0.51471572	3.00983278	5.93304348	4.51070234	1.00000000

Table 7.2
Input Coefficient Products Values for
Composite Composition of
The Septic Polynomial Example

k	z^k	Coefficient Product Code	Coefficient Product Values	ΣCPV	$z^k \cdot \Sigma CPV$
0	1.00000000	AF	3.30901442	3.30901442	3.30901442
1	0.53846154	BF AG	19.34967147 6.12475962	25.47443109	13.71700136
2	0.28994083	CF BG AH	38.14246528 35.81491987 3.28131271	77.23869786	22.39465204
3	0.15612198	DF CG BH AI	28.99849092 70.59909722 19.18768395 0.51471572	119.29998781	18.62535085
4	0.08406568	EF DG CH BI	6.42881944 53.67422543 37.82315217 3.00983278	100.93602982	8.48525638
5	0.04526614	EG DH CI	11.89930556 28.75572742 5.93304348	46.58807646	2.10886228
6	0.02437407	EH DI	6.37500000 4.51070234	10.88570234	0.26532891
7	0.01312450	EI	1.00000000	1.00000000	0.01312450
				68.91859075	Sum

Table 7.3
Numerical Inputs and Outputs for
Composite Composition of
The Septic Polynomial Example

Treatment of the Septic as a Polynomial Ring[5]

At least at numerical levels, considerable labor (man or machine) may be spared by the resolution of the septic equation as a polynomial ring.

Recall from Equation 7.12 that the septic equation can be construed as the product of a cubic and a quartic that partition the septic's co-quantities (and the resulting coefficients) between them:-

$$T_7 = T_3.T_4 = (F + Gz + Hz^2 + Iz^3).(A + Bz + Cz^2 + Dz^3 + Ez^4)$$
Equation 7.12

Ring theory allows us to define any polynomial coefficient of degree n = m1+m2 using:-

$$c_{n,k} = \sum_{j=0}^{k} c_{m1,j}.c_{m2,k-j}$$
Equation 7.15

whence:-

$$T_7 = \sum_{k=0}^{n} z^k.\left(\sum_{j=0}^{k} c_{m1,j}.c_{m2,k-j}\right) = 68.9185907480421$$
Equation 7.16

whilst:-

$$U_7 = \sum_{i=0}^{k}\left(\sum_{j=0}^{k} c_{m1,j}.c_{m2,k-j}\right) = 8$$
Equation 7.17

The directness of the ring-structure algorithm permits not only economy but also numerical accuracy.

Table 7.4 illustrates the diagonalising structure of the ring computational system and its coefficient products, whilst Table 7.5 presents the numerical outcomes in summative terms.

z   0.53846154

MATRIX A

P(3)	P(4) c_{4,0}	c_{4,1}	c_{4,2}	c_{4,3}	c_{4,4}	i
	0	1	2	3	4	
$c_{3,0}$	0	0	1	2	3	4
$c_{3,1}$	1	1	2	3	4	5
$c_{3,2}$	2	2	3	4	5	6
$c_{3,3}$	3	3	4	5	6	7
j						

MATRIX B

P(3)	P(4) c_{4,0}	c_{4,1}	c_{4,2}	c_{4,3}	c_{4,4}	i
		0.514716	3.009833	5.933043	4.510702	1
$c_{3,0}$	6.428819	3.309014	19.34967	38.14247	28.99849	6.428819
$c_{3,1}$	11.89931	6.12476	35.81492	70.5991	53.67423	11.89931
$c_{3,2}$	6.375	3.281313	19.18768	37.82315	28.75573	6.375
$c_{3,3}$	1	0.514716	3.009833	5.933043	4.510702	1
$c_{3,4}$	0	0	0	0	0	0
j						

Table 7.4
The Ring Structure of
Coefficient Products for the
Example Septic Produced from a Cubic and a Quartic

Table 7.5

Coefficient Product Summations
for the
Example Septic Produced from a Cubic and a Quartic

k		Actual $c_{7,k}$	Actual $c_{7,k}*z^k$						$\Sigma c_{3,j}c_{4,j}$	$z^k \cdot \Sigma c_{3,j}c_{4,j}$
0	$c_{7,0}$	3.30901442	3.30901442	$c_{3,0}*c_{4,0}$					3.30901442	3.30901442
1	$c_{7,1}$	25.47443109	13.71700136	$c_{3,1}*c_{4,0}$	$c_{3,0}*c_{4,1}$				25.47443109	13.71700136
2	$c_{7,2}$	77.23869786	22.39465204	$c_{3,2}*c_{4,0}$	$c_{3,1}*c_{4,1}$	$c_{3,0}*c_{4,2}$			77.23869786	22.39465204
3	$c_{7,3}$	119.29998781	18.62535085	$c_{3,3}*c_{4,0}$	$c_{3,2}*c_{4,1}$	$c_{3,1}*c_{4,2}$	$c_{3,0}*c_{4,3}$		119.29998781	18.62535085
4	$c_{7,4}$	100.93602982	8.48525638	$c_{3,4}*c_{4,0}$	$c_{3,3}*c_{4,1}$	$c_{3,2}*c_{4,2}$	$c_{3,1}*c_{4,3}$	$c_{3,0}*c_{4,4}$	100.93602982	8.48525638
5	$c_{7,5}$	46.58807646	2.10886228		$c_{3,4}*c_{4,1}$	$c_{3,3}*c_{4,2}$	$c_{3,2}*c_{4,3}$	$c_{3,2}*c_{4,4}$	46.58807646	2.10886228
6	$c_{7,6}$	10.88570234	0.26532891			$c_{3,4}*c_{4,2}$	$c_{3,3}*c_{4,3}$	$c_{3,2}*c_{4,4}$	10.88570234	0.26532891
7	$c_{7,7}$	1.00000000	0.01312450				$c_{3,4}*c_{4,3}$	$c_{3,3}*c_{4,4}$	1.00000000	0.01312450
8	$c_{7,8}$	0.00000000	0.00000000					$c_{3,4}*c_{4,4}$	0.00000000	0.00000000
			68.91859075						384.73193980	68.91859075 Sums

PART EIGHT
THE OCTIC CASE

An Octic Equation is an algebraic polynomial of the eighth degree (n = 8). A specific solution may be explicated as:-

$$T_8 = (z + r_1)(z + r_2)(z + r_3)(z + r_4)(z + r_5)(z + r_6)(z + r_7)(z + r_8) = \prod_{i=1}^{8}(z + r_i)$$

Equation 8.1

Using our data this is the value of T_8:-

$$T_8 = 75.4148726566272$$
Equation 8.2

Amongst the several possibilities the corresponding Polynomial Coefficients $c_{8,k}$ may be elaborated as the following nested summations (the zeroth is a product):-

$$c_{8,0} = \prod_{i=1}^{n} r_i = 1.83914587190709$$

Equation 8.3a

$$c_{8,1} = \sum_{g=1}^{n-6}\sum_{h=1}^{n-5}\sum_{i=h+1}^{n-4}\sum_{j=i+1}^{n-3}\sum_{k=j+1}^{n-2}\sum_{l=k+1}^{n-1}\sum_{m=l+1}^{n} r_g.r_h.r_i.r_j.r_k.r_l.r_m = 17.4676697771138$$
Equation 8.3b

$$c_{8,2} = \sum_{h=1}^{n-5}\sum_{i=h+1}^{n-4}\sum_{j=i+1}^{n-3}\sum_{k=j+1}^{n-2}\sum_{l=k+1}^{n-1}\sum_{m=l+1}^{n} r_h.r_i.r_j.r_k.r_l.r_m = 68.4035979520742$$
Equation 8.3c

$$c_{8,3} = \sum_{i=1}^{n-4}\sum_{j=i+1}^{n-3}\sum_{k=j+1}^{n-2}\sum_{l=k+1}^{n-1}\sum_{m=l+1}^{n} r_i.r_j.r_k.r_l.r_m = 143.545474451318$$
Equation 8.3d

$$c_{8,4} = \sum_{j=1}^{n-3}\sum_{k=j+1}^{n-2}\sum_{l=k+1}^{n-1}\sum_{m=l+1}^{n} r_j.r_k.r_l.r_m = 175.400100661487$$
Equation 8.3e

$$c_{8,5} = \sum_{k=1}^{n-2} \sum_{l=k+1}^{n-1} \sum_{m=l+1}^{n} r_k.r_l.r_m = 126.829621551334$$

Equation 8.3f

$$c_{8,6} = \sum_{l=1}^{n-1} \sum_{m=l+1}^{n} r_l.r_m = 52.6383355278191$$

Equation 8.3g

$$c_{8,7} = \sum_{m=1}^{n} r_m = 11.4415010282268$$

Equation 8.3h

$$c_{8,8} = 1$$

Equation 8.3i

This leads directly to the summative formulation:-

$$T_8 = c_{8,0} + c_{8,1}z + c_{8,2}z^2 + c_{8,3}z^3 + c_{8,4}z^4 + c_{8,5}z^5 + c_{8,6}z^6 + c_{8,7}z^7 + c_{8,8}z^8$$

$$= \sum_{i=1}^{8} c_{8,i}.z^i$$

$$= 75.4148726566272$$

Equation 8.4

Composite Elaboration

Allow that whilst n = 8; m1 = 1, m2 = 4, m3 = 5 and m4 = 8. These m* auxiliary bounds shall facilitate the partition of the octic polynomial into two component quartics.

Long multiplication, especially of polynomials, is always tedious and even more error-prone, at least in human hands. The octic elaboration is truly baleful, especially for naturally idle enthusiasts like me, and virtually impossible for those whose worktables are so crowded with gadgets, cables and other rubbish that there is hardly a patch of comfort for a sheet of A4.

Accordingly, like all good algebraists, we seek concision, knowing that some shortenings are briefer or more accurate than others.

Among the principles worth exploration is the fact that the product of two polynomials forms a third polynomial of a degree equal to the sum of the two degrees, i.e.:-

$$P(n_{m2}, z, r_{m1} \ldots r_{m2}) \times P(n_{m4}, z, r_{m3} \ldots r_{m4}) = P(n_{m2+m4}, z, r_1 \ldots r_{m2+m4})$$

Equation 8.5

For example, it is possible to compute the coefficients of a dependent Octic Polynomial of Degree m2+m4 = 8 from the coefficients of an m2 = 4 Quartic whose r-Co-quantities are r_1, r_2, r_3 and r_4; and an m4 = 4 Quartic whose Co-quantities are r_5, r_6, r_7 and r_8.

The resulting n = 8 Octic evidently incorporates $r_1...r_8$ according to this scheme:-

$$P(n_{4a}, z, r_1 ... r_4) \times P(n_{4b}, z, r_5 ... r_8) = P(n_8, z, r_1 ... r_8)$$
Equation 8.6

Equation 8.6 summarises our intent, but I should point-out in passing that the Polynomial Product Equation is general enough to be useful for any adequately-long series of co-quantities. For instance, we could "square" a quartic by re-employing the co-quantities of a series of four r's r_1, r_2, r_3 and r_4 as in the product $P(n_4,z,r_1...r_4) \times P(n_4,z,r_1...r_4)$. Clearly, the resulting Octic would be a different animal to that whom you and I require.

Firstly, for our exercise of Composite Elaboration we need to define a general Quartic:-

$$T_{4a} = A + Bz + Cz^2 + Dz^3 + Ez^4$$
Equation 8.7

Now from Part Four we know that:-

$$T_{4a} = \prod_{i=m1}^{m2} r_i + \left(\prod_{i=m1}^{m2} r_i \cdot \sum_{i=m1}^{m2} \frac{1}{r_i} \right) z + \left(\sum_{i=m1}^{m2-1} \sum_{j=i+1}^{m2} r_i \cdot r_j \right) z^2 + \sum_{i=m1}^{m2} r_i \cdot z^3 + 1. z^4$$
Equation 8.8

From which we may abstract:-

$$A = \prod_{i=m1}^{m2} r_i = 5.31076388888889$$
Equation 8.9a

$$B = \prod_{i=m1}^{m2} r_i \cdot \sum_{i=m1}^{m2} \frac{1}{r_i} = 16.2586805555556$$
Equation 8.9b

$$C = \sum_{i=m1}^{m2-1} \sum_{j=i+1}^{m2} r_i \cdot r_j = 17.1656099033816$$
Equation 8.9c

$$D = \sum_{i=m1}^{m2} r_i = 7.20108695652174$$
Equation 8.9d

$$E = 1$$
Equation 8.9e

The Second Quartic may be written as:-

$$T_{4b} = F + Gz + Hz^2 + Iz^3 + Jz^4$$
Equation 8.10

whilst again from Part Four we know that:-

$$T_{4b} = \prod_{i=m3}^{m4} r_i + \left(\prod_{i=m3}^{m4} r_i \cdot \sum_{i=m3}^{m4} \frac{1}{r_i} \right) z + \left(\sum_{i=m3}^{m4} \sum_{j=i+1}^{m4} r_i.r_j \right) z^2 + \left(\sum_{i=m3}^{m4} r_i \right) \cdot z^3 + z^4$$
Equation 8.11

Accordingly, we may abstract of the Second Quartic:-

$$F = \prod_{i=m3}^{m4} r_i = 0.346305335802054$$
Equation 8.12a

$$G = \prod_{i=m3}^{m4} r_i \cdot \sum_{i=m3}^{m4} \frac{1}{r_i} = 2.22890759131459$$
Equation 8.12b

$$H = \sum_{i=m3}^{m4} \sum_{j=i+1}^{m4} r_i.r_j = 4.93713516243057$$
Equation 8.12c

$$I = \sum_{i=m3}^{m4} r_i = 4.2404140717051$$
Equation 8.12d

$$J = 1$$
Equation 8.12e

The Octic Equation T_8 can be expressed as the product of Equations 8.7 and 8.10 to give:-

$$T_8 = T_{4a}.T_{4b} = (A + Bz + Cz^2 + Dz^3 + Ez^4).(F + Gz + Hz^2 + Iz^3 + Jz^4)$$
Equation 8.13

Long multiplication of this expression supplies:-

$$\begin{aligned}T_8 &= (AF) + (BF + AG)z + (CF + BG + AH)z^2 + (DF + CG + BH + AI)z^3\\ &= (EF + DG + CH + BI + AJ)z^4 + (EH + DH + CI + BJ)z^5 + (EH + DI + CJ)z^6\\ &\quad + (EI + DJ)z^7 + (EJ)\end{aligned}$$
Equation 8.14

This sum of products in the co-efficient form resolves to:-

$$c_{8,0} = AF = 1.83914587190709$$
Equation 8.15a

$$c_{8,1} = BF + AG = 17.4676697771139$$
Equation 8.15b

$$c_{8,2} = CF + BG + AH = 68.4035979520743$$
Equation 8.15c

$$c_{8,3} = DF + CG + BH + AI = 143.545474451318$$
Equation 8.15d

$$c_{8,4} = EF + DG + CH + BI + AJ = 175.400100661487$$
Equation 8.15e

$$c_{8,5} = EG + DH + CI + BJ = 126.829621551334$$
Equation 8.15f

$$c_{8,6} = EH + DI + CJ = 52.638335527819$$
Equation 8.15g

$$c_{8,7} = EI + DJ = 11.4415010282268$$
Equation 8.15h

$$c_{8,8} = EJ = 1$$
Equation 8.15i

And the set of Equations 8.15* may be expanded as:-

$$c_{8,0} = AF = \prod_{i=m1}^{m2} r_i \cdot \prod_{i=m3}^{m4} r_i = 1.83914587190709$$

Equation 8.16a

$$c_{8,1} = BF + AG = \left(\prod_{i=m1}^{m2} r_i \cdot \sum_{i=m1}^{m2} \frac{1}{r_i} \right) \cdot \prod_{i=m3}^{m4} r_i + \prod_{i=m1}^{m2} r_i \cdot \left(\prod_{i=m3}^{m4} r_i \cdot \sum_{i=m3}^{m4} \frac{1}{r_i} \right)$$
$$= 17.4676697771139$$

Equation 8.16b

$$c_{8,2} = CF + BG + AH$$
$$= \left(\sum_{i=m1}^{m2-1} \sum_{j=i+1}^{m2} r_i \cdot r_j \right) \cdot \prod_{i=m3}^{m4} r_i + \left(\prod_{i=m1}^{m2} r_i \cdot \sum_{i=m1}^{m2} \frac{1}{r_i} \right) \cdot \left(\prod_{i=m3}^{m4} r_i \cdot \sum_{i=m3}^{m4} \frac{1}{r_i} \right)$$
$$+ \prod_{i=m1}^{m2} r_i \cdot \left(\sum_{i=m3}^{m4} \sum_{j=i+1}^{m4} r_i \cdot r_j \right) = 68.4035979520743$$

Equation 8.16c

$$c_{8,3} = DF + CG + BH + AI$$
$$= \left(\sum_{i=m1}^{m2} r_i \right) \cdot \prod_{i=m3}^{m4} r_i + \left(\sum_{i=m1}^{m2-1} \sum_{j=i+1}^{m2} r_i \cdot r_j \right) \cdot \left(\prod_{i=m3}^{m4} r_i \cdot \sum_{i=m3}^{m4} \frac{1}{r_i} \right)$$
$$+ \left(\prod_{i=m1}^{m2} r_i \cdot \sum_{i=m1}^{m2} \frac{1}{r_i} \right) \cdot \left(\sum_{i=m3}^{m4} \sum_{j=i+1}^{m4} r_i \cdot r_j \right) + \prod_{i=m1}^{m2} r_i \cdot \left(\sum_{i=m3}^{m4} r_i \right)$$
$$= 143.545474451318$$

Equation 8.16d

$$c_{8,4} = EF + DG + CH + BI + AJ$$
$$= 1. \prod_{i=m3}^{m4} r_i + \left(\sum_{i=m1}^{m2} r_i \right) \cdot \left(\prod_{i=m3}^{m4} r_i \cdot \sum_{i=m3}^{m4} \frac{1}{r_i} \right)$$
$$+ \left(\sum_{i=m1}^{m2-1} \sum_{j=i+1}^{m2} r_i \cdot r_j \right) \cdot \left(\sum_{i=m3}^{m4} \sum_{j=i+1}^{m4} r_i \cdot r_j \right) + \left(\prod_{i=m1}^{m2} r_i \cdot \sum_{i=m1}^{m2} \frac{1}{r_i} \right) \cdot \left(\sum_{i=m3}^{m4} r_i \right)$$
$$+ \prod_{i=m1}^{m2} r_i . 1 = 175.400100661487$$

Equation 8.16e

$$c_{8,5} = EG + DH + CI + BJ$$

$$= 1.\left(\prod_{i=m3}^{m4} r_i \cdot \sum_{i=m3}^{m4} \frac{1}{r_i}\right) + \left(\sum_{i=m1}^{m2} r_i\right)\cdot\left(\sum_{i=m3}^{m4}\sum_{j=i+1}^{m4} r_i\cdot r_j\right)$$

$$+ \left(\sum_{i=m1}^{m2-1}\sum_{j=i+1}^{m2} r_i\cdot r_j\right)\cdot\left(\sum_{i=m3}^{m4} r_i\right) + \left(\prod_{i=m1}^{m2} r_i \cdot \sum_{i=m1}^{m2} \frac{1}{r_i}\right).1 = 126.829621551334$$

Equation 8.16f

$$c_{8,6} = EH + DI + CJ = 1.\left(\sum_{i=m3}^{m4}\sum_{j=i+1}^{m4} r_i\cdot r_j\right) + \left(\sum_{i=m1}^{m2} r_i\right)\cdot\left(\sum_{i=m3}^{m4} r_i\right) + \left(\sum_{i=m1}^{m2-1}\sum_{j=i+1}^{m2} r_i\cdot r_j\right).1$$

$$= 52.638335527819$$

Equation 8.16g

$$c_{8,7} = EI + DJ = 1.\left(\sum_{i=m3}^{m4} r_i\right) + \left(\sum_{i=m1}^{m2} r_i\right).1 = 11.4415010282268$$

Equation 8.16h

$$c_{8,8} = EJ = 1.1 = 1$$

Equation 8.16i

The set of Equation 8.16* is susceptible of extensive simplifications but these will be deferred for the time being.

Treatment of the Octic as a Polynomial Ring

At least at numerical levels, considerable labor (man or machine) may be spared by the resolution of the octic equation as a polynomial ring.

At a technical level, if you are using Mathcad® Prime® 3.1 or, I dare say, many other tools, you will find that you need to zeroise the coefficient array elements implicated. I achieved this by setting:-

n=8
i=0...n: j=1...n
c1$_{i,j}$=0: c2$_{i,j}$=0

Pseudocode Block 8.1

This was not necessary for the septic ring of Part Seven. Also, you may need explicitly to declare c1$_{i,j}$ and c2$_{i,j}$ as real in several programming languages.

Recall from Equation 8.13 that the octic equation can be construed as the product of two quartics that partition the octic's co-quantities (and the resulting coefficients) between them:-

$$T_8 = T_{4a}.T_{4b} = (A + Bz + Cz^2 + Dz^3 + Ez^4).(F + Gz + Hz^2 + Iz^3 + Jz^4)$$
Equation 8.13

Ring theory allows us to define any polynomial coefficient of degree n = m3+m4 using:-

$$c_{n,k} = \sum_{j=0}^{k} c_{m2,j} \cdot c_{m4,k-j}$$
Equation 8.17

whence:-

$$T_8 = \sum_{k=0}^{n} z^k \cdot \left(\sum_{j=0}^{k} c1_{m2,j} \cdot c2_{m4,k-j} \right) = 75.4148726566272$$
Equation 8.18

whilst:-

$$U_8 = \sum_{i=0}^{k} \left(\sum_{j=0}^{k} c1_{m2,j} \cdot c2_{m4,k-j} \right) = 9$$
Equation 8.19

The directness of the ring-structure algorithm permits not only economy but also numerical accuracy.

Because the r-value is variable in our product scheme the simple process of classical binomial expansion is only of value when we are able to calculate some Representative Co-quantity r_{rep} that is a constant average for each $(z+r)$ multiplicand.

Let us propose that there is some species of that unique average value of the Co-quantities $r_1...r_n$ that when substituted into the General Binomial Expansion:-

$$B_n = (z + r_i)^n = P(n,z,r) = \sum_{k=0}^{n} BE(k,n).z^{n-k}.r_{rep}^k = \sum_{k=0}^{n} \frac{n!}{k!\,(n-k)!}.z^{n-k}.r_{rep}^k$$

Equation 9.1

will yield up the Summative Value T_n defined by:-

$$T_n = \sum_{i=0}^{n} c_{n,i}.z^i$$

Equation 9.2

B_n is the Binomial Expansion of the Polynomial of Degree n; z is an Arbitrary Constant; r_i is the ith. Coefficient Co-quantity whose Representative Co-quantity is r_{rep}; P(n,z,r) is the Polynomial generated by n, z and r-values; k is the Serial of the Polynomial Term that transits from 0 to n; and BE(k,n) is the Combination count of the number of k objects abstractable of population n.

B_n = P(n,z,r) constitutes a Local Solution of the Polynomial of Degree n and the stated data (i.e. the elaborated factor solution EFS).

A Heuristic Approach

My initial approach to the Representative Co-quantity r_{rep} was merely heuristic.

My scheme for the estimation of r_{rep} involved the calculation of the arithmetic mid-point between two successively-guessed points r_{less} and r_{more} using:-

$$r_{rep} = r_{less} + \frac{(r_{more} - r_{less})}{2}$$

Equation 9.3

The (signed) Error of Estimation, EE, is then:-

$$EE = B_n - T_n$$

Equation 9.4

The ideal value of EE is of course zero.

The estimation of this average results is a symmetrisation of the polynomial and therefore of a loss of information, as does naturally the calculation of any average.

Table 9.1 presents the values of r_{rep} and their absolute errors for the six specimen polynomials elaborated.

Degree n	Heuristic Representative Co-quantity r_{rep}	Product Estimate $B_n=(z+r_{rep})^n$	Estimation Error B_n-T_n
0	not available		
1	not available		
2	1.920127775	6.0446614	-8.710844E-05
3	1.918985	14.840627	-6.067466E-06
4	1.58287875	20.250762	-1.277754E-06
5	1.33459955	23.054712	-2.238285E-06
6	1.126227115	21.281298	2.310290E-05
7	1.292410633	68.961415	4.282432E-02
8	1.178188	75.4142	-6.726034E-04

Table 9.1
Heuristic
Values of Representative Co-quantities and their Errors
for Seven Trial Polynomials of Degree 2 to 8

An Analytic Approach

Dis-satisfied with trial-and-error, I also essayed the use of Regula Falsi and Newton-Raphson iteration to refine estimates of r_{rep}. In both cases, however, the required functional series gradually diverged to give results consistently inferior to guess-work.

Notwithstanding the above, I later realised that there was a simple exact analytic way of determining the appropriate r_{rep} for any binomial product *where T_n is known*.

Because:-

$$T_n = \left(z + r_{rep}\right)^n$$
Equation 9.5

it is elementary that:-

$$r_{rep} = T_n^{\frac{1}{n}} - z$$

Equation 9.6

Now in this analytic scenario the Error of Estimation EE is everywhere zero so that we may write:-

$$T_n = B_n = \sum_{k=0}^{n} \frac{n!}{k!\,(n-k)!} \cdot z^{n-k} \cdot r_{rep}^k$$

Equation 9.7

Table 9.2 shows the relationship between the heuristic and analytic r_{rep}'s for the given data and expresses the error of the guestimates in terms of Specific Defect.

Degree n	Heuristic Representative Co-quantity r_{rep}	Product Estimate $B_n=(z+r_{rep})^n$	Estimation Error B_n-T_n	Analytic r_{rep} as $(T_n)^{(1/n)}-z$	Specific Defect PSD (Analytic r_{rep} / Heuristic r_{rep})
0	not available				
1	not available			0.95833333	
2	1.92012778	6.04466141	-0.00008711	1.92014549	0.00092259
3	1.91898500	14.84062652	-0.00000607	1.91898533	0.00001745
4	1.58287875	20.25076158	-0.00000128	1.58287878	0.00000211
5	1.33459955	23.05471240	-0.00000224	1.33459959	0.00000273
6	1.12622712	21.28129816	0.00002310	1.12622681	-0.00002674
7	1.29241063	68.96141507	0.04282432	1.29224817	-0.01257228
8	1.17818800	75.41420005	-0.00067260	1.17818991	0.00016244
Σ			0.04207813	11.31160742	-0.01149171
μ	1.47905955		0.00601116	1.41395093	-0.00164167
σ			0.08424337	19.74473602	-0.02390600

Table 9.2
Heuristic and Analytic
Values of Representative Co-quantities and their Specific Defect
for Seven Trial Polynomials of Degree 2 to 8

Failed Paradigms of t and T Estimation

I thought that it should be possible to approximate the values of individual Polynomial Terms $t_{n,k}$ using some product of Combination Values, Permutation Values and the Representative Co-quantity r_{rep}.

Table 9.3 presents the actual Polynomial Term Coefficients $c_{n,k}$ and Terms for Summation $t_{n,k}$ for the cited test data.

Table Nine presents the failed $c_{n,k}$ estimation formulae.

Some of the failed formulae, you may think, are more logical than others. None successfully approximated the actual terms.

$c_{n,k}$ Actual — Term

Degree	0	1	2	3	4	5	6	7	8	
1	0.95833333									1.958333333
2	3.354166667	4.45833333								8.8125
3	6.428819444	11.89930556	6.37500000	1						25.703125
4	5.310763889	16.25868056	17.16560990	7.20108696	1					46.9361413
5	3.186458333	15.06597222	26.55804650	21.48626208	7.80108696	1				75.09782609
6	1.225560901	8.98106303	25.28060549	34.82193345	24.48668014	8.18570234	1			103.9816054
7	3.309014421	25.47443109	77.23869786	119.29998781	100.93602982	46.58807646	10.88570234	1		384.7319398
8	1.839145871	17.467766978	68.40359795	143.54547445	175.40010066	126.82962155	52.63833553	11.44150103	1	598.5654468

$t_{n,k}$ — z^k

Degree	0	1	2	3	4	5	6	7	8	
1	1.00000000	0.53846154	0.28994083	0.15612198	0.08406568	0.04526614	0.02437407	0.01312450	0.00706704	1.496794872
2	0.958333333	2.400641026	0.289940828	0	0	0	0	0	0	6.044748521
3	6.428819444	6.407318376	1.848372781	0.156121985	0	0	0	0	0	14.84063259
4	5.310763889	8.754674145	4.977011155	1.124247986	0.084065684	0	0	0	0	20.25076286
5	3.186458333	8.112446581	7.700262002	3.354477876	0.655803711	0.045266138	0	0	0	23.05471464
6	1.225560897	4.835957018	7.329879698	5.436478723	2.058489514	0.370535128	0.024374074	0	0	21.28127505
7	3.309014423	13.71700136	22.39465204	18.62535085	8.485256385	2.108862276	0.265328915	0.013124501	0	68.91859075
8	1.839145872	9.405668342	19.83299586	22.41060434	14.74512943	5.741087091	1.283010688	0.150163996	0.0070670390	75.41487266

Table 9.3
The Actual Polynomial Coefficients $c_{n,k}$ and Term $t_{n,k}$
for the z and r_i Test Fractions introduced in Part Zero

Term Estimate Type 1	$E_1 = (1/1) \cdot {}^nC_r \cdot (n-k) \cdot \mu(r_{rep})$
Term Estimate Type 2	$E_2 = r_{rep} * ((n-k)/\Pi(i=1,n))$
Term Estimate Type 3	$E_3 = {}^nC_k \cdot r_{rep}^{k-1}$
Term Estimate Type 4	$E_4 = {}^nC_k \cdot {}^nP_k \cdot r_{rep}$
Term Estimate Type 5	$E_5 = $ Actual ${}^nC_k \cdot$ Actual ${}^nP_k \cdot r_{rep}$
Term Estimate Type 6	$E_6 = $ Actual ${}^nC_k \cdot$ Actual ${}^nP_k \cdot r_{rep}^{1/(2n)}$
Term Estimate Type 7	$E_7 = $ Actual ${}^nC_k \cdot k! \cdot r_{rep}$
Term Estimate Type 8	$E_8 = $ Actual ${}^nC_k \cdot$ Actual ${}^nP_k \cdot r_{rep}^{1/(n-1)}$

Table 9.4
The Failed Estimation Formulae

Knowledge of Polynomial Numeric Value, T_n

Equation 9.6 implies a knowledge of the value of the polynomial from which, together with the input Constant and Co-quantities, the polynomial's coefficients may be reconstructed.

Among some of the important scenarios in which that summative value, T_n, might be available we could list:-

A   Mixture or Compounding

T_n has an actual real-world weight or magnitude, because the sum of the Co-quantities, r_i, is finite and each r_i is known:-

$$T_n = \sum_{i=1}^{n} r_i = W$$

Equation 9.8

where W is a measured or known Natural-World Quantity.

B   Definitional Completion

The set of r_i constitute an exhaustive constitution *by definition*. Perhaps the most famous example of this condition is the axiom that the sum of probabilities of a specified set of outcomes *must* always total unity:-

$$T_n = \sum_{i=1}^{n} r_i = 1$$

Equation 9.9

C   Theoretical Limit

The exact value of T_n is neither known nor exactly determinable but may be approached by generating and assimilating a very large number of analytically-determined terms. The assimilation strategy may be summation, multiplication or something.

That is:-

$$\lim_{n \to \infty} T_n = \sum_{i=1}^{\infty} r_i$$

Equation 9.10

where the summation sign indicates a general assimilative operator.

A straight-forward example is the Fourier Series:-

$$f(x) = a_0 + \sum_{n=1}^{\infty} \left(a_n \cos \frac{n\pi x}{L} + b_n \sin \frac{n\pi x}{L} \right)$$

Equation 9.11

conveniently provided by WORD®, but there are an infinite number of other examples, some simpler, some more complicated.

From the point of view of the construction of $R_i = r_i/z$, it is notable that these three set species A, B and C can always be configured to eliminate redundant r_i's that have values of zero.

The Normalisation of the General Binomial Product

Certain gains of efficiency and formulaic generalisation might greatly assist computerisation of this product process.

This may be done by reducing the Constant z to unity and dividing each value of the Co-quantity r_i by z. The principles are illustrated below for the Cubic ($n = 3$) case:-

$$T_3 = (z - r_1)(z - r_2)(z - r_3) = \left[z \left(\frac{z}{z} + \frac{r_1}{z} \right) \right] \times \left[z \left(\frac{z}{z} + \frac{r_2}{z} \right) \right] \times \left[z \left(\frac{z}{z} + \frac{r_3}{z} \right) \right]$$

Equation 9.12

Because as we previously remarked:-

$$R_i = \frac{r_i}{z}$$

Equation 9.13

Cancellation and the gathering of constants allows us to write the simple generalisation:-

$$T_n = z^n \prod_{i=1}^{n} (1 + R_i)$$
Equation 9.14

It is now obvious that the product on the RHS of Equation 9.14 is a classic combinatorial Binomial Expansion.

Now assume that the Representative Normalised Value of the Co-quantities, R_{rep}, is a surrogate for the collective of r_i's and is constant for any particular set of r_i's. Then we may write:-

$$T_n = z^n \prod_{i=1}^{n} \left(1 + R_{rep}\right) = z^n \cdot \left(1 + R_{rep}\right)^n$$
Equation 9.15

Clearly:-

$$\sqrt[n]{T_n} = z \cdot \left(1 + R_{rep}\right)$$
Equation 9.16

and the elementary transposition introduced in Equation 9.6 may be developed to give:-

$$R_{rep} = \frac{\sqrt[n]{T_n}}{z} - 1$$
Equation 9.17

For the given z, r_1, r_2 and r_3 data the resulting $R_{rep} = 3.56382990767533$ and:-

$$T_3 = z^n \cdot \prod_{i=0}^{n} \left(1 + R_{rep}\right) = z^n \cdot \left(1 + R_{rep}\right)^n = 14.8406325861023$$
Equation 9.18

It is possible to detect several patterns of formulaic conformation in the general binomial expansion defined in Equation 0.1 of the Introduction. Specifically, there are discernable patterns in the development of Polynomial Coefficients $c_{n,k}$ as functions of Polynomial Degree n and Polynomial Coefficient Term Position k as represented by:-

$$P(n) = \sum_{i=1}^{n} c_{n,k}.z^i$$
Equation 10.1

where P(n) is the Polynomial of Degree n, given the Constant z and the series of independent Co-quantities $r_1 \ldots r_n$.

Some of the discernable patterns were studied for the test data applied to P(1,k) until P(6,k). Notwithstanding these suggestive patterns I have been unable to elicit a general formula for P(n,k) except for Nested Summation which will be discussed in detail in its own subsection. It is happily the case, however, that simple, general, formulae involving Representative Co-quantities can be established.

<u>Reciprocal Relations</u>

When dealing with nested summations of the types that summarise permuted variables it is sometimes succinct to define coefficients in terms of summed reciprocals $1/r_i$ rather than the direct Co-quantity value r_i. Whether this device assists computational economies is moot.

For example the k = 1 coefficient of the cubic is defined by:-

$$c_{3,1} = r_1.r_3 + r_2.r_3 + r_1.r_2$$
Equation 10.2

which may conveniently be re-expressed as:-

$$c_{3,1} = \frac{r_1.r_2.r_3}{r_2} + \frac{r_1.r_2.r_3}{r_1} + \frac{r_1.r_2.r_3}{r_3}$$
Equation 10.3

from which minor rearrangement gives:-

$$c_{3,1} = r_1.r_2.r_3 \left(\frac{1}{r_1} + \frac{1}{r_2} + \frac{1}{r_3} \right)$$
Equation 10.4

from which it obviously follows that:-

$$c_{n,k} = \prod_{i=1}^{n} r_i \sum_{j=1}^{n} \frac{1}{r_j}$$

Equation 10.5

From the point of view of Mathcad® elaborations it is entirely happy to use i as the counter in place of j in the summation.

Note that Equation 10.5 is identical to the special nested summation:-

$$c_{3,1} = \sum_{i=1}^{n-1} \sum_{m=l+1}^{n} r_l . r_m$$

Equation 10.6

and the method is immediately extensible to the special equation:-

$$c_{3,2} = \prod_{k=1}^{n} r_k . \left(\sum_{i=1}^{n-1} \sum_{j=i+1}^{n} \frac{1}{r_i . r_j} \right) = \sum_{i=1}^{n} r_i$$

Equation 10.7

and indeed to the quartic and other higher degrees, as for example in the identity:-

$$c_{4,1} = \prod_{i=1}^{n} r_i \sum_{i=1}^{n} \frac{1}{r_i} = \sum_{k=1}^{n-2} \sum_{l=k+1}^{n-1} \sum_{m=l+1}^{n} r_k . r_l . r_m$$

Equation 10.8

Note, however, that Equation 10.8 is not general for n other than four.

Type 1 Coefficient Equations

Table 10.1 presents some of the apparent properties of the row, column and diagonal vectors of the matrix $c_{n,k}$ for the maxima n = k = 6. It also presents a convenient color code for the EXCEL® arrangement of the relevant coefficient equations.

Subsequently, Table 10.2 presents the equations themselves including their outcome values for the given z value and the attendant r Co-quantity data.

Color Code	Leading Product	Leading Sum	k-Wise Variation	Formula Conformation
RED	Yes	No	Same Formula Structure for Each k	Product of All r_i
ORANGE	Yes	No	Same Formula Structure for Each k	Product of One Nested Sum of Reciprocals
YELLOW	Yes	No	Same Formula Structure for Each k	Product of Two Nested Sums of Reciprocals
GREEN	No	No	Same Formula Structure for Each k	Coefficient is Unity
BLUE	No	Yes	Different Formula Structure for Each k	Sum of All r_i
INDIGO	No	Yes	Different Formula Structure for Each k	Two Nested Sums
VIOLET	No	Yes	Different Formula Structure for Each k	Three Nested Sums

Table 10.1
Selected Apparent Properties of the Polynomial Coefficient Equations

Degree	Term 0	Term 1	Term 2
1	$c_{1,0} := \prod_{i=1}^{n} r_i = 0.9583333333333333$	$c_{1,1} := 1$	
2	$c_{2,0} := \prod_{i=1}^{n} r_i = 3.3541666666666667$	$c_{2,1} := \sum_{i=1}^{n} r_i = 4.458333333333333$	$c_{2,2} := 1$
3	$c_{3,0} := \prod_{i=1}^{n} r_i = 6.428819444444445$	$c_{3,1} := \prod_{i=1}^{n} r_i \cdot \sum_{i=1}^{n} \frac{1}{r_i} = 11.899305555555556$	$c_{3,2} := \sum_{i=1}^{n} r_i = 6.375$
4	$c_{4,0} := \prod_{i=1}^{n} r_i = 5.310763888888889$	$c_{4,1} := \prod_{i=1}^{n} r_i \cdot \sum_{i=1}^{n} \frac{1}{r_i} = 16.258680555555556$	$c_{4,2} := \prod_{i=1}^{n} r_i \cdot \left(\sum_{j=1}^{n-1} \sum_{i=1}^{n-j} \frac{1}{r_j \cdot r_{i+j}} \right) = 17.165609903816$
5	$c_{5,0} := \prod_{i=1}^{n} r_i = 3.1864583333333333$	$c_{5,1} := \prod_{i=1}^{n} r_i \cdot \sum_{i=1}^{n} \frac{1}{r_i} = 15.065972222222222$	$c_{5,2} := \prod_{i=1}^{n} r_i \cdot \left(\sum_{j=1}^{n-1} \sum_{i=1}^{n-j} \frac{1}{r_j \cdot r_{i+j}} \right) = 26.558046464975845$
6	$c_{6,0} := \prod_{i=1}^{n} r_i = 1.2255608974359$	$c_{6,1} := \prod_{j=1}^{n} r_j \cdot \sum_{i=1}^{n} \frac{1}{r_i} = 8.98106303418804$	$c_{6,2} := \prod_{i=1}^{n} r_i \cdot \left(\sum_{j=1}^{n-1} \sum_{i=1}^{n-j} \frac{1}{r_j \cdot r_{i+j}} \right) = 25.280605490524$

Table 10.2a
Tabulation of Polynomial Coefficient Equations of Type 1

Degree	Term 3	Term 4	Term 5	Term 6
1				
2				
3	$c_{3,3} := 1$			
4	$c_{4,3} := \sum_{i=1}^{n} r_i = 7.2010869565652174$	$c_{4,4} := 1$		
5	$c_{5,3} := \sum_{i=1}^{n-1} \sum_{j=i+1}^{n} r_i \cdot r_j = 21.4862620772947$	$c_{5,4} := \sum_{i=1}^{n} r_i = 7.8010869565652174$	$c_{5,5} := 1$	
6	$c_{6,3} := \sum_{i=1}^{n-2} \sum_{j=i+1}^{n-1} \sum_{k=j+1}^{n} r_i \cdot r_j \cdot r_k = 34.8219934503902$	$c_{6,4} := \sum_{i=1}^{n-1} \sum_{j=i+1}^{n} r_i \cdot r_j = 24.4866801374954$	$c_{6,5} := \sum_{i=1}^{n} r_i = 8.1857023413712$	$c_{6,6} := 1$

Table 10.2b
Tabulation of Polynomial Coefficient Equations of Type 1

<u>Partial Functions (Type 2)</u>

As you have doubtless discerned from Table 10.2 certain formula phrases occur so frequently that it could be appropriate to define them as partial functions, at least as a matter of computational convenience or economy.

In the spirit of the foregoing we could define:-

$$PP(m,n) = \prod_{i=m}^{n} r_i$$

Equation 10.9

$$PR(m,n) = \prod_{i=m}^{n} \frac{1}{r_i}$$

Equation 10.10

$$SS(m,n) = \sum_{i=m}^{n} r_i$$

Equation 10.11

$$RR(m,n) = \sum_{i=m}^{n} \frac{1}{r_i}$$

Equation 10.12

$$NS(m1,m2) = \sum_{i=m1}^{m2-1} \sum_{j=i+1}^{m2} r_i . r_j$$

Equation 10.13

$$NR(m1,m2) = \sum_{j=m1}^{m2-1} \sum_{i=1}^{m2-j} \frac{1}{r_j . r_{i+1}}$$

Equation 10.14

$$SR(m1,m2,m3,m4) = \sum_{i=m1}^{m2} r_i . \sum_{j=m3}^{m4} \frac{1}{r_j}$$

Equation 10.15

$$PRS(m1,m2,m3,m4) = \prod_{i=m1}^{m2} r_i . \sum_{j=m3}^{m4} \frac{1}{r_j}$$

Equation 10.16

Table 10.3 presents the coefficient formulae cast in partial terms:-

Degree	Term		
	0	**1**	**2**
1	$c_{1,0} := PP(1,1) = 0.958333333333333$	$c_{1,1} := 1$	
2	$c_{2,0} := PP(1,n) = 3.35416666666667$	$c_{2,1} := SS(1,n) = 4.45833333333333$	$c_{2,2} := 1$
3	$c_{3,0} := PP(1,n) = 6.42881944444445$	$c_{3,1} := PRS(1,n,1,n) = 11.8993055555556$	$c_{3,2} := SS(1,n) = 6.375$
4	$c_{4,0} := PP(1,n) = 5.31076388888889$	$c_{4,1} := PRS(1,n,1,n) = 16.2586805555556$	$c_{4,2} := NS(1,n) = 17.1656099033816$
5	$c_{5,0} := PP(1,n) = 3.18645833333333$	$c_{5,1} := PP(1,n)\cdot RR(1,n) = 15.0659722222222$	$c_{5,2} := PP(1,n)\cdot NR(1,n) = 26.5580464975845$
6	$c_{6,0} := PP(1,n) = 1.22556089743359$	$c_{6,1} := PRS(1,n,1,n) = 8.98106303418804$	$c_{6,2} := PP(1,n)\cdot NR(1,n) = 25.280605490524$

Table 10.3a
Tabulation of Polynomial Coefficient Equations of Type 2 (Partial Functions)

Term

Degree	3	4	5	6
1				
2				
3	$c_{3,3} := 1$			
4	$c_{4,3} := SS(1,n) = 7.20108695652174$	$c_{4,4} := 1$		
5	$c_{5,3} := NS(1,n) = 21.48626620772947$	$c_{5,4} := SS(1,n) = 7.80108695652174$	$c_{5,5} := 1$	
6	$c_{6,3} := \sum_{i=1}^{n-2} r_i \cdot \left(\sum_{k=i+1}^{n-1} r_k \left(SS(1,n) - SS(1,k) \right) \right) = 34.8219934503902$	$c_{6,4} := NS(1,n) = 24.4866801374954$	$c_{6,5} := SS(1,n) = 8.18570234113712$	$c_{6,6} := 1$

Table 10.3b
Tabulation of Polynomial Coefficient Equations of Type 2 (Partial Functions)

Pythagorean Means (Type 3)

Table 10.4 presents the polynomial coefficient equations re-designed to incorporate Pythagorean means as defined in the Introduction.

Nested Summations (Type 4)

The expression of coefficients as nested summations furnishes an elegant and possibly the most economical avenue of generalisation, at least for machine estimations of the numerical values. Having said this, and at the expense of apparent contradiction, the instantiation of an algorithm, especially in a procedural language, is likely to be intricate, and that intricacy may well vitiate both economy and reliability.

In my opinion the nested summation lacks the analytic suggestivity of the partial and mean simplifications as developed for the first six polynomial degrees.

Table 10.5 presents the coefficient formulae for the first six polynomial equations expressed as nested summations.

I was unable generally to condense these ever-lengthening expressions using the notation available within Mathcad® Prime® 3.1.

Binomial Normalisation (Type 5)

In my opinion, classical binomial expansion yields the only really satisfactory generalisation for the coefficients of polynomial equations. They combine simplicity with obvious extensibility to all (n,r) and are likely to be numerically economical.

The method depends firstly upon:-

$$T_n = (z + r_i)^n = (z + r_1)(z + r_2) \ldots (z + r_n) = \prod_{i=0}^{n}(z + r_i) = z^n . \prod_{l=0}^{n}(1 + R_i)$$

$$= z^n . \left(1 + R_{rep}\right)^n = z^n . S_n = c_{n,0} + c_{n,0} . z + c_{n,0} . z^2 + \cdots + c_{n,0} . z^n = \sum_{i=0}^{n} c_{n,i} . z^i$$

Equation 10.17

in which S_n, The Binomial Summation Auxiliary, is defined by:-

$$S_n = \left(1 + R_{rep}\right)^n$$

Equation 10.18

And secondly upon:-

$$R_{rep} = \frac{\sqrt[n]{T_n}}{z} - 1$$

Equation 10.19

Implicit in the Binomial Normalisation avenue is knowledge of T_n, and accordingly the finitude of n, or at least a convergent series from which tolerably-precise S and T totals may be established.

Equation 10.18 is amenable to classical Binomial Expansion ("Pascal's Triangle") and is accordingly expressible as:-

$$S_n = \sum_{k=0}^{n} \binom{n}{k} \cdot 1^{n-k} \cdot R_{rep}^k = \sum_{k=0}^{n} \left(\frac{n!}{k!\,(n-k)!} \right) \cdot 1^{n-k} \cdot R_{rep}^k = \sum_{k=0}^{n} \left(\frac{n!}{k!\,(n-k)!} \right) \cdot R_{rep}^k$$

Equation 10.20

Equation 10.20 may be recast as:-

$$S_n = n! \sum_{k=0}^{n} \left(\frac{R_{rep}^k}{k!\,(n-k)!} \right)$$

Equation 10.21

which is not to preclude further economies.

The Normalisation Polynomial Coefficients, $C_{n,k}$, that conform to the transformation R_{rep} are analogous to, but not identical with, the Coefficients $c_{n,k}$.

Hence we may define:-

$$S_n = \sum_{k=0}^{n} \binom{n}{k} \cdot R_{rep}{}^k = \sum_{k=0}^{n} C_{n,k} \cdot R_{rep}{}^k$$

Equation 10.22

Equation 10.22 obviously confirms that the coefficients of the summation S_n and the Binomial Coefficients are identical.

For the case of the Cubic Polynomial in which of course n = 3 we may write:-

$$C_{3,0} = \prod_{i=1}^{3} R_i = 41.1781816893424$$

Equation 10.23a

$$C_{3,1} = R_1 R_2 + R_1 R_3 + R_2 R_3 = 41.0404620181406$$

Equation 10.23b

$$C_{3,2} = \sum_{i=1}^{3} R_i = 11.8392857142857$$

Equation 10.23c

$$C_{3,3} = 1$$

Equation 10.23d

Accordingly:-

$$S_n = S_3 = \sum_{i=0}^{n} C_{n,i} = 95.0579294217687$$

Equation 10.24

$$T_n = T_3 = z^n . S_n = z^3 . \sum_{i=0}^{3} C_{n,i} = 14.8406325861023$$

Equation 10.25

An Elaboration for the Sextic Polynomial

In terms of The Binomial Theorem[6]:-

$$S_n = \left(1 + R_{rep}\right)^n =$$
$$= 1 + n . R_{rep} + \frac{n(n-1)R_{rep}^2}{2!} + \frac{n(n-1)(n-2)R_{rep}^3}{3!} +$$
$$+ \frac{n(n-1)(n-2)(n-3)R_{rep}^4}{4!} + \frac{n(n-1)(n-2)(n-3)(n-4)R_{rep}^5}{5!}$$
$$+ \frac{n(n-1)(n-2)(n-3)(n-4)(n-5)R_{rep}^6}{6!}$$

Equation 10.26

Or:-

$$S_n = \left(1 + R_{rep}\right)^n =$$
$$= \frac{R_{rep}^0}{0!} + \frac{n . R_{rep}^1}{1!} + \frac{n(n-1)R_{rep}^2}{2!} + \frac{n(n-1)(n-2)R_{rep}^3}{3!} +$$
$$+ \frac{n(n-1)(n-2)(n-3)R_{rep}^4}{4!} + \frac{n(n-1)(n-2)(n-3)(n-4)R_{rep}^5}{5!}$$
$$+ \frac{n(n-1)(n-2)(n-3)(n-4)(n-5)R_{rep}^6}{6!}$$

Equation 10.27

Which may be condensed to:-

$$S_n = \sum_{k=0}^{n} \left(\frac{n!}{k!\,(n-k)!} \right) . R_{rep}^k$$

Equation 10.28

Which expands as:-

$$S_n = \left(\frac{n!}{0!\,(6-0)!} \right) . R_{rep}^0 + \left(\frac{n!}{1!\,(6-1)!} \right) . R_{rep}^1 + \left(\frac{n!}{2!\,(6-2)!} \right) . R_{rep}^2 + \left(\frac{n!}{3!\,(6-3)!} \right) . R_{rep}^3$$
$$+ \left(\frac{n!}{4!\,(6-4)!} \right) . R_{rep}^4 + \left(\frac{n!}{5!\,(6-5)!} \right) . R_{rep}^5 + \left(\frac{n!}{6!\,(6-6)!} \right) . R_{rep}^6$$

Equation 10.29

then gathering n!:-

$$S_n = n! \left[\left(\frac{R_{rep}^0}{0!\,(6-0)!} \right) + \left(\frac{R_{rep}^1}{1!\,(6-1)!} \right) + \left(\frac{R_{rep}^2}{2!\,(6-2)!} \right) + \left(\frac{R_{rep}^3}{3!\,(6-3)!} \right) + \left(\frac{R_{rep}^4}{4!\,(6-4)!} \right) \right.$$
$$\left. + \left(\frac{R_{rep}^5}{5!\,(6-5)!} \right) + \left(\frac{R_{rep}^6}{6!\,(6-6)!} \right) \right]$$

Equation 10.30

Equation 10.29 allows us to specify:-

$$S_n = \left(\frac{6.5.4.3.2.1}{1.6.5.4.3.2.1} \right) . R_{rep}^0 + \left(\frac{6.5.4.3.2.1}{1.5.4.3.2.1.} \right) . R_{rep}^1 + \left(\frac{6.5.4.3.2.1}{2.1.4.3.2.1.} \right) . R_{rep}^2 + \left(\frac{6.5.4.3.2.1}{3.2.1.3.2.1.} \right) . R_{rep}^3$$
$$+ \left(\frac{6.5.4.3.2.1}{4.3.2.1.2.1.} \right) . R_{rep}^4 + \left(\frac{6.5.4.3.2.1}{5.4.3.2.1.1} \right) . R_{rep}^5 + \left(\frac{6.5.4.3.2.1}{6.5.4.3.2.1.1.} \right) . R_{rep}^6$$

Equation 10.31

From which cancellation gives:-

$$S_n = \left(\frac{1}{1.1} \right) . R_{rep}^0 + \left(\frac{6}{1} \right) . R_{rep}^1 + \left(\frac{6.5}{2.1} \right) . R_{rep}^2 + \left(\frac{6.5.4}{3.2.1} \right) . R_{rep}^3$$
$$+ \left(\frac{6.5}{2.1.} \right) . R_{rep}^4 + \left(\frac{6}{1} \right) . R_{rep}^5 + \left(\frac{1}{1.1} \right) . R_{rep}^6$$

Equation 10.32

$$S_n = (1). R_{rep}^0 + (6). R_{rep}^1 + (15). R_{rep}^2 + (20). R_{rep}^3$$
$$+ (15). R_{rep}^4 + (6). R_{rep}^5 + (1). R_{rep}^6$$

Equation 10.33

Equation 10.33 is equivalent to:-

$$S_n = \binom{6}{0}.R_{rep}^0 + \binom{6}{1}.R_{rep}^1 + \binom{6}{2}.R_{rep}^2 + \binom{6}{3}.R_{rep}^3 + \binom{6}{4}.R_{rep}^4 + \binom{6}{5}.R_{rep}^5 + \binom{6}{6}.R_{rep}^6$$
Equation 10.33

Or in general summary:-

$$S_n = \sum_{k=0}^{n} \binom{n}{k}.R_{rep}^k = \sum_{k=0}^{n} \left(\frac{n!}{k!\,(n-k)!}\right).R_{rep}^k$$
Equation 10.34

whilst Equation 10.30 is summarised as:-

$$S_n = n! \sum_{k=0}^{n} \frac{R_{rep}^k}{k!\,(n-k)!}$$
Equation 10.35

Accordingly the numerical value of the specified sextic polynomial is given by:-

$$T_6 = z^6.S_6 = 21.2812750528396$$
Equation 10.36

Table 10.6 shows the $C_{n,k}$ Polynomial Equation Normalisation Coefficients for the Algebraic Polynomials of Degrees 1 to 6 extended to the $n = 7$ and $n = 8$ Septic and Octal cases.

Degree	0	1	2
1	$c_{1,0} := GM(n,r)^n = 0.958333333333333$	$c_{1,1} := 1$	
2	$c_{2,0} := GM(n,r)^n = 3.35416666666667$	$c_{2,1} := n \cdot \dfrac{GM(n,r)^n}{HM(n,r)} = 4.4583333333333$	$c_{2,2} := 1$
3	$c_{3,0} := GM(n,r)^n = 6.42881944444444$	$c_{3,1} := n \cdot \dfrac{GM(n,r)^n}{HM(n,r)} = 11.8993055555556$	$c_{3,2} := \prod_{h=1}^{n} r_h \cdot \left(\sum_{i=1}^{n-2} \dfrac{1}{r_i} \cdot \left(\dfrac{n}{HM(n,r)} - \dfrac{i}{HM(i,r)} \right) + \dfrac{1}{r_{n-1} \cdot r_n} \right) = 6.375$
4	$c_{4,0} := GM(n,r)^n = 5.31076388888889$	$c_{4,1} := n \cdot \dfrac{GM(n,r)^n}{HM(n,r)} = 16.2586805555556$	$c_{4,2} := \prod_{h=1}^{n} r_h \cdot \left(\sum_{i=1}^{n-2} \dfrac{1}{r_i} \cdot \left(\dfrac{n}{HM(n,r)} - \dfrac{i}{HM(i,r)} \right) + \dfrac{1}{r_{n-1} \cdot r_n} \right) = 17.1650099033816$
5	$c_{5,0} := GM(n,r)^n = 3.18645833333333$	$c_{5,1} := \dfrac{n \cdot GM(n,r)^n}{HM(n,r)} = 15.0659722222222$	$c_{5,2} := \prod_{h=1}^{n} r_h \cdot \left(\sum_{i=1}^{n-2} \dfrac{1}{r_i} \cdot \left(\dfrac{n}{HM(n,r)} - \dfrac{i}{HM(i,r)} \right) + \dfrac{1}{r_{n-1} \cdot r_n} \right) = 26.5580464975845$
6	$c_{6,0} := GM(n,r)^n = 1.2255608974359$	$c_{6,1} := \dfrac{n \cdot GM(n,r)^n}{HM(n,r)} = 8.98106303418804$	$c_{6,2} := \prod_{h=1}^{n} r_h \cdot \left(\sum_{i=1}^{n-2} \dfrac{1}{r_i} \cdot \left(\dfrac{n}{HM(n,r)} - \dfrac{i}{HM(i,r)} \right) + \dfrac{1}{r_{n-1} \cdot r_n} \right) = 25.2806050490524$

Table 10.4a
Tabulation of Polynomial Coefficient Equations of Type 3 (Pythagorean Means)

Degree	Term 3	Term 4	Term 5	Term 6
1				
2				
3	$c_{3,3} := 1$			
4	$c_{4,3} := n \cdot AM(n,r) = 7.2010869565174$	$c_{4,4} := 1$		
5	$c_{5,3} := \sum_{k=1}^{n-1} r_k \cdot (n \cdot AM(n,r) - k \cdot AM(k,r)) = 21.4862620772947$	$c_{5,4} := n \cdot AM(n,r) = 7.8010869565174$	$c_{5,5} := 1$	
6	$c_{6,3} := \sum_{i=1}^{n-2} r_i \cdot \left(\sum_{k=i+1}^{n-1} r_k \cdot (n \cdot AM(n,r) - k \cdot AM(k,r)) \right) = 34.8219934503902$	$c_{6,4} := \sum_{j=1}^{n-1} r_j \cdot (n \cdot AM(n,r) - j \cdot AM(j,r)) = 24.4866801374953$	$c_{6,5} := n \cdot AM(n,r) = 8.1857023411712$	$c_{6,6} := 1$

Table 10.4b
Tabulation of Polynomial Coefficient Equations of Type 3 (Pythagorean Means)

Degree	Term 0	Term 1	Term 2
1	$c_{1,0} := \sum_{i=1}^{n} r_i = 0.9583333333333333$	$c_{1,1} := 1$	
2	$c_{2,0} := \sum_{l=1}^{n-1}\sum_{m=l+1}^{n} r_l \cdot r_m = 3.354166666666667$	$c_{2,1} := \sum_{i=1}^{n} r_i = 4.458333333333333$	$c_{2,2} := 1$
3	$c_{3,0} := \sum_{k=1}^{n-2}\sum_{l=k+1}^{n-1}\sum_{m=l+1}^{n} r_k \cdot r_l \cdot r_m = 6.428819444444445$	$c_{3,1} := \sum_{l=1}^{n-1}\sum_{m=l+1}^{n} r_l \cdot r_m = 11.899305555555556$	$c_{3,2} := \sum_{m=1}^{n} r_m = 6.375$
4	$c_{4,0} := \prod_{i=1}^{n} r_i = 5.310763888888889$	$c_{4,1} := \sum_{k=1}^{n-2}\sum_{l=k+1}^{n-1}\sum_{m=l+1}^{n} r_k \cdot r_l \cdot r_m = 16.258680555555556$	$c_{4,2} := \sum_{l=1}^{n-1}\sum_{m=l+1}^{n} r_l \cdot r_m = 17.165009033816$
5	$c_{5,0} := \prod_{i=1}^{n} r_i = 3.186458333333333$	$c_{5,1} := \sum_{j=1}^{n-3}\sum_{k=j+1}^{n-2}\sum_{l=k+1}^{n-1}\sum_{m=l+1}^{n} r_j \cdot r_k \cdot r_l \cdot r_m = 15.065972222222222$	$c_{5,2} := \sum_{k=1}^{n-2}\sum_{l=k+1}^{n-1}\sum_{m=l+1}^{n} r_k \cdot r_l \cdot r_m = 26.5580464975845$
6	$c_{6,0} := \prod_{i=1}^{n} r_i = 1.2255608974359$	$c_{6,1} := \sum_{i=1}^{n-4}\sum_{j=i+1}^{n-3}\sum_{k=j+1}^{n-2}\sum_{l=k+1}^{n-1}\sum_{m=l+1}^{n} r_i \cdot r_j \cdot r_k \cdot r_l \cdot r_m = 8.98106303418803$	$c_{6,2} := \sum_{g=1}^{n-3}\sum_{k=g+1}^{n-2}\sum_{j=k+1}^{n-1}\left(r_g \cdot r_k \cdot r_j\right) = 25.2806054490524$

Table 10.5a
Tabulation of Polynomial Coefficient Equations of Type 4 (Nested Summations)

Degree	Term 3	Term 4	Term 5	Term 6
1				
2				
3	$c_{3,3} := 1$			
4	$c_{4,3} := \sum_{m=1}^{n} r_m = 7.2010869565652174$	$c_{4,4} := 1$		
5	$c_{5,3} := \sum_{i=1}^{n-1} \sum_{j=i+1}^{n} r_i \cdot r_j = 21.4862620772947$	$c_{5,4} := \sum_{i=1}^{n} r_i = 7.80108695652174$	$c_{5,5} := 1$	
6	$c_{6,3} := \sum_{i=1}^{n-2} \sum_{j=i+1}^{n-1} \sum_{k=j+1}^{n} r_i \cdot r_j \cdot r_k = 34.8219934503902$	$c_{6,4} := \sum_{i=1}^{n-1} \sum_{j=i+1}^{n} r_i \cdot r_j = 24.4866801374954$	$c_{6,5} := \sum_{i=1}^{n} r_i = 8.18570234113712$	$c_{6,6} := 1$

Table 10.5b
Tabulation of Polynomial Coefficient Equations of Type 4 (Nested Summations)

Degree \ Term	0	1	2	3	4	5	6	7	8
1	$1 \cdot R_{rep}^{0}$	$1 \cdot R_{rep}^{1}$							
2	$1 \cdot R_{rep}^{0}$	$2 \cdot R_{rep}^{1}$	$1 \cdot R_{rep}^{2}$						
3	$1 \cdot R_{rep}^{0}$	$3 \cdot R_{rep}^{1}$	$3 \cdot R_{rep}^{2}$	$1 \cdot R_{rep}^{3}$					
4	$1 \cdot R_{rep}^{0}$	$4 \cdot R_{rep}^{1}$	$6 \cdot R_{rep}^{2}$	$4 \cdot R_{rep}^{3}$	$1 \cdot R_{rep}^{4}$				
5	$1 \cdot R_{rep}^{0}$	$5 \cdot R_{rep}^{1}$	$10 \cdot R_{rep}^{2}$	$10 \cdot R_{rep}^{3}$	$5 \cdot R_{rep}^{4}$	$1 \cdot R_{rep}^{5}$			
6	$1 \cdot R_{rep}^{0}$	$6 \cdot R_{rep}^{1}$	$15 \cdot R_{rep}^{2}$	$20 \cdot R_{rep}^{3}$	$15 \cdot R_{rep}^{4}$	$6 \cdot R_{rep}^{5}$	$1 \cdot R_{rep}^{6}$		
7	$1 \cdot R_{rep}^{0}$	$7 \cdot R_{rep}^{1}$	$21 \cdot R_{rep}^{2}$	$35 \cdot R_{rep}^{3}$	$35 \cdot R_{rep}^{4}$	$21 \cdot R_{rep}^{5}$	$7 \cdot R_{rep}^{6}$	$1 \cdot R_{rep}^{7}$	
8	$1 \cdot R_{rep}^{0}$	$8 \cdot R_{rep}^{1}$	$28 \cdot R_{rep}^{2}$	$56 \cdot R_{rep}^{3}$	$70 \cdot R_{rep}^{4}$	$56 \cdot R_{rep}^{5}$	$28 \cdot R_{rep}^{6}$	$8 \cdot R_{rep}^{7}$	$1 \cdot R_{rep}^{8}$

Table 10.6
Tabulation of Polynomial Coefficient Equations of Type 5 (Binomial Normalisation)

References

3   Wikipedia contributors. (2020, September 5).
Polynomial.
In Wikipedia, The Free Encyclopedia. Retrieved 09:06, September 17, 2020, from
https://en.wikipedia.org/w/index.php?title=Polynomial&oldid=976911232

2   "Euler and the Fundamental Theorem of Algebra"
William Dunham
pp 12
Pages   282-293
The College Mathematics Journal
Vol. 22 No. 4 September 1991
https://www.maa.org/sites/default/files/pdf/upload_library/
        22/Polya/07468342.di020748.02p0019l.pdf

3   "A Note on Nested Sums"
Steve Butler and Pavel Karasik
Journal of Integer Sequences, Vol. 13 (2010), Article 10.4.4
pp 8
https://cs.uwaterloo.ca/journals/JIS/VOL13/Butler/butler7.pdf

4   "Summation – How to write product as a nested sums"
Elaqqad
Mathematics StackExchange
pp 2
https://math.stackexchange.com/questions/1234031/how-to-write-product-as-a-nested-sums

5   Wikipedia contributors. (2020, September 3).
Polynomial ring.
In Wikipedia, The Free Encyclopedia. Retrieved 09:09, September 17, 2020, from
https://en.wikipedia.org/w/index.php?title=Polynomial_ring&oldid=976587896

6   "A Course of Mathematics for Engineers and Scientists"
"Volume 1"
Brian H Chirgwin and Charles Plumpton
Pergamon Press of Oxford   1961
Library of Congress Card No.: 60-13094
Page No. 186 Equation 5.13
pp 326

7 Wikipedia contributors. (2020, June 21).
 Binomial theorem.
 In Wikipedia, The Free Encyclopedia. Retrieved 09:00, September 17, 2020, from
https://en.wikipedia.org/w/index.php?title=Binomial_theorem&oldid=963685641

8 Wikipedia contributors. (2020, September 13).
 Combinatorics.
 In Wikipedia, The Free Encyclopedia. Retrieved 09:04, September 17, 2020, rom
https://en.wikipedia.org/w/index.php?title=Combinatorics&oldid=978125411

9 Wikipedia contributors. (2020, September 1).
 Sums of powers.
 In Wikipedia, The Free Encyclopedia. Retrieved 09:13, September 17, 2020, from
https://en.wikipedia.org/w/index.php?title=Sums_of_powers&oldid=976176606

10 Wikipedia contributors. (2020, August 25).
 Vieta's formulas.
 In Wikipedia, The Free Encyclopedia. Retrieved 09:15, September 17, 2020, from
https://en.wikipedia.org/w/index.php?title=Vieta%27s_formulas&oldid=974866347

11 "The Summation of Series"
 Harold Thayer Davis
 Dover Publications, Inc. of Mineola, New York 2015
 after
 The Principia Press of Trinity University of San Antonio 1962
 ISBN-13: 978-0-486-78968-2 (pbk)
 pp 140 (plus extra seven catalog pages)

12 "5.3. Generalized Permutations and Combinations"
 pp 5
 https://sites.math.northwestern.edu/~mlerma/courses/cs310-05s/notes/dm-gcomb

Explorations of the Products of
Binarion Complex Numbers

by
James R Warren BSc MSc PhD PGCE

PART ONE
REVIEW OF RELATIONS

The Complex Domain (system) $\mathbb{C}$ is an extension of the common Real Domain of numbers $\mathbb{R}$. The Complex Domain was developed to facilitate intricate analyses that were possible but intractable in the Real Domain. Complex Numbers can also be computationally convenient.

The key to the definition and manipulation of complex numbers is the so-called Imaginary Square Root of Minus One, i. In order to avoid confusion I shall herein adopt only the letters j, k, l ... as array subscripts (where subscripts are convenient).

Thus:-

$$i = \sqrt{-1}$$
Equation 1.1

Therefore, a particular Complex Number, z_j, may be defined as:-

$$z_j = a_j + b_j.i$$
Equation 1.2

where a_j is the Real Part of the Complex Number formally denoted $\mathfrak{R}_j$ and b_j is the Imaginary Part or $\mathfrak{I}_j$. There is nothing metaphysically special about either a or b: All numbers are imaginary, it is just that the Real Part is an ordinary common quantity and the Imaginary Part an ancillary calculated to render the determination of negative roots practicable. Historically, this was done in order to simplify the finding of cube and quartic roots of algebraic polynomial equations. But the availability of the complex number concept permits the simplification of many other mathematical structures.

If a is zero the number is wholly complex, the square root of a negative number; and if b is zero z and a are equivalent common ("real") numbers.

Useful complex numbers usually boast both finite a and b.

We should note straight away that:-

$$i^2 = -1$$
Equation 1.3

and accordingly:-

$$\left(a_j + b_j\right)\left(a_j - b_j\right) = a_j^2 - b_j^2$$
Equation 1.4

The RHS of Equation 1.4 is a positive or negative common quantity: A Real number.

The multiplier (a_j-b_j) is termed the Complex Conjugate of the Complex Number z_j, conventionally denoted $\bar{z}$, so that:-

$$z_j \bar{z}_j = a_j^2 - b_j^2$$
Equation 1.5

At this point it is convenient to introduce our own invented quantity, the Component Dividend, Q_j, defined as the quotient:-

$$Q_j = \frac{a_j}{b_j}$$
Equation 1.6

We will find employment for Q_j later.

The Integer Powers of the Imaginary Unit

Returning to mainstream complex number doctrine we may further note that for the Natural Numbers $\mathbb{N} = 0, 1, 2, 3, 4... = j$ the (integer) exponents of i may take only one of four outcomes:-

$$i^j = \begin{cases} 0,4,8,12\ ... \Rightarrow & 4n + 0 \Rightarrow & i^j = 1 \\ 2,6,10,14\ ... \Rightarrow & 4n + 2 \Rightarrow & i^j = -1 \\ 1,5,9,13\ ... \Rightarrow & 4n + 1 \Rightarrow & i^j = i \\ 3,7,11,15\ ... \Rightarrow & 4n + 3 \Rightarrow & i^j = -i \end{cases}$$

Selection Rule 1.1

From the above selection rule we may educe that even exponents of i furnish real results and odd exponents complex results.

Complex Modulus and Complex Argument (Phase)

It is possible to resolve the components of a complex number $\mathfrak{R}_j + \mathfrak{I}_j \times i$ in the following way:-

$$\mathfrak{R}_j = \frac{z_j + \bar{z}}{2}$$
Equation 1.7

$$\Im_j = \frac{z_j - \bar{z}}{2i}$$
Equation 1.8

Thus the Modulus of the Complex Number, r_j, is given by:-

$$r_j = \sqrt{\Re_j{}^2 + \Im_j{}^2}$$
Equation 1.9

and the Complex Argument, otherwise known as the Phase, φ_j, is given by[1]:-

$$\varphi_j = atan2(\Im_j, \Re_j)$$
Equation 1.10

which may otherwise be specified using[2,3]:-

$$\varphi_j = atan2(\Re_j, \Im_j) = 2.\,atan\left(\frac{b_j}{\sqrt{a_j{}^2 + b_j{}^2} + a_j}\right) = 2.\,atan\left(\frac{b_j}{r_j + a_j}\right) = 2.\,atan\left(\frac{r_j - a_j}{b_j}\right)$$

$$= atan\left(\frac{b_j}{a_j}\right)$$

Equation 1.11

Note that a and $\Re$ symbols are interchangeable, as are b and $\Im$.

Two Exemplary Cubics

Further to conceptualise the mechanics of complex arithmetic we will examine two simple triple multiplications which of course generate cubic polynomials.

A $\Re = 4$ $\Im = 3$

$$(a + bi)^3 =$$
$$= (4 + 3i)^3$$
$$= (4 + 3i)(4 + 3i)(4 + 3i)$$

By long multiplication:-

$$64 + 144i + 108i^2 + 27i^3 = -44 + 117i$$

Because $i^2 = -1$:-

$$64 + 144i - 108 + 27i^3 = -44 + 117i$$

Because $i^3 = -i$:-

$$64 + 144i - 108 - 27i = -44 + 117i$$

Therefore:-

$$-44 + 117i = -44 + 117i$$

B   Permit the following adventitious data:-

$a_1 = 714821$	$b_1 = 40228$
$a_2 = 587611$	$b_2 = 953263$
$a_3 = 196169$	$b_3 = 981283$

Mathcad® renders the required cubic value solution as:-

$$\text{Soln} = -616978876047348000 + 512853864578919000i$$

By Definition:-

$$\left(a_j + b_j i\right)^3 = (a_1 + b_1 i)(a_2 + b_2 i)(a_3 + b_3 i)$$

By long multiplication:-

$$\left(a_j + b_j i\right)^3 = a_1 a_2 a_3 + i(b_1 a_2 a_3 + b_2 a_1 a_3 + b_3 a_1 a_2) + i^2(a_1 b_2 b_3 + a_2 b_1 b_3 + a_3 b_1 b_2) + b_1 b_2 b_3 i^3$$

By isolating term coefficients:-

$a_1 a_2 a_3$	$i^0 = 1$
$b_1 a_2 a_3 + b_2 a_1 a_3 + b_3 a_1 a_2$	$i^1 = i$
$a_1 b_2 b_3 + a_2 b_1 b_3 + a_3 b_1 b_2$	$i^2 = -1$
$b_1 b_2 b_3$	$i^3 = -i$

Gathering Reals:-

$$(a_1 a_2 a_3) - (a_1 b_2 b_3 + a_2 b_1 b_3 + a_3 b_1 b_2) = -616978876047348000$$

Gathering Imaginaries:-

$$(b_1 a_2 a_3 + b_2 a_1 a_3 + b_3 a_1 a_2) - (b_1 b_2 b_3) = 512853864578919000$$

Gathering components:-

$$Soln = \left[\left((a_1 a_2 a_3) - (a_1 b_2 b_3 + a_2 b_1 b_3 + a_3 b_1 b_2)\right) \right.$$
$$\left. + \left((b_1 a_2 a_3 + b_2 a_1 a_3 + b_3 a_1 a_2) - (b_1 b_2 b_3)\right)i\right]$$

De Moivre's Formula[4] (1722AD)

A more sophisticated approach to the exponentiation of complex numbers is offered by De Moivre's Formula.

$$\lambda_k = z_k^n = (a_k + b_k i)^n = r_k^n.[\cos n\varphi_k + i.\sin n\varphi_k]$$
Equation 1.12

or alternatively:-

$$\lambda_k = z_k^n = (a_k + b_k i)^n = \{r_k.[\cos \varphi_k + i.\sin \varphi_k]\}^n$$
Equation 1.13

where n is the Degree of the Exponent (e.g. n = 3 for a cubic expansion). *Now for the special case that n = 1 only* we may assert that:-

$$P_k = \prod_{k=1}^{k} z_k = \prod_{k=1}^{k} \lambda_k$$
Equation 1.14

where P_k is the Product of the 1 to k Complex Numbers as far as the Upper Limit k only. It is clear that this product relies upon the k family of data-derived r_j and φ_j only. This manipulation may or may not offer a pathway to the industrial economisation of the multiplication of series of complex numbers.

Simplification of the Special De Moivre Formula

As we have seen:-

$$\lambda_k = r_k.[\cos \varphi_k + i.\sin \varphi_k]$$
Equation 1.13

If there is indeed an imaginary component then it is possible, by substitution of the appropriate identities, to recast this as:-

$$\lambda_k = r_k.\left[\cos\left\{2.atan\left(\frac{r_j - a_j}{b_j}\right)\right\} + i.\sin\left\{2.atan\left(\frac{r_j - a_j}{b_j}\right)\right\}\right]$$
Equation 1.15

(See Equation 1.11 and read $k \equiv j$).

This is the same as:-

$$\lambda_k = r_k.\left[\left\{\frac{1-\left(\frac{r_k-a_k}{b_k}\right)^2}{1+\left(\frac{r_k-a_k}{b_k}\right)^2}\right\}+i.\left\{\frac{2\left(\frac{r_k-a_k}{b_k}\right)}{1+\left(\frac{r_k-a_k}{b_k}\right)^2}\right\}\right]$$
Equation 1.16

Allow that:-

$$t_k = \frac{r_k - a_k}{b_k}$$
Equation 1.17

then:-

$$\lambda_k = r_k.\left[\left\{\frac{1-t_k^2}{1+t_k^2}\right\}+i.\left\{\frac{2t_k}{1+t_k^2}\right\}\right]$$
Equation 1.18

Gathering yields:-

$$\lambda_k = r_k\left\{\frac{(1-t_k^2)+i.2t_k}{1+t_k^2}\right\}$$
Equation 1.19

<u>Euler's Theorem[4] (1749AD)</u>

Euler's Theorem states that:-

$$z_j = r_j.e^{i.\varphi_j}$$
Equation 1.20

where e is the Napierian Base or the base of natural logarithms (sometimes called Euler's Number or something, especially on The Continent). The Napierian Base is a transcendental real number with a value near to 2.718281828459045.
Euler's Theorem and De Moivre's Formula are intimately related.

PART TWO
THE EMPLOYMENT OF THE COMPONENT DIVIDEND Q

The Component Dividend Q_j is defined as the quotient:-

$$Q_j = \frac{a_j}{b_j}$$

Equation 1.6

Therefore we can establish that the Product of the Complex Numbers between one and k, P_k, is given by:-

$$P_k = b_1.(Q_1 + i) \times b_2.(Q_2 + i) \times b_3.(Q_3 + i) \times ... \times b_k.(Q_k + i) = \prod_{j=1}^{k} b_j . \prod_{j=1}^{k} (Q_j + i)$$

Equation 2.1

Q_j is invariably real.

Q in its Relation to the Complex Modulus r

We have seen that:-

$$r_j = \sqrt{\mathfrak{R}_j{}^2 + \mathfrak{I}_j{}^2} = \sqrt{a_j{}^2 + b_j{}^2}$$

Equation 1.9

And when all b are unity it is numerically the case that:-

$$r_j = \sqrt{\mathfrak{R}_j{}^2 + \mathfrak{I}_j{}^2} = \sqrt{a_j{}^2 + b_j{}^2} = \sqrt{a_j{}^2 + 1} \approx a_j + 1$$

Equation 2.2

The Derivation of Q_{REP}

Q_{REP} is the Representative Value of Q_j that is a moiety that shall be general to all the binarion multipliers of a complex product such that:-

$$P_k = b_1.(Q_{REP} + i) \times b_2.(Q_{REP} + i) \times b_3.(Q_{REP} + i) \times ... \times b_k.(Q_{REP} + i)$$

$$= \prod_{j=1}^{k} b_j . \prod_{j=1}^{k} (Q_{REP} + i) = \prod_{j=1}^{k} b_j . (Q_{REP} + i)^k$$

Equation 2.3

Q_{REP} is not any mean or average of standard type.

First, allow that, for Polynomial Degree n equivalent to series limit k:-

$$B_n = \prod_{j=1}^{n} b_j$$

Equation 2.4

Then by Equation 2.1:-

$$P_n = \prod_{j=1}^{n} b_j \cdot \prod_{j=1}^{n} (Q_j + i)$$

$$= B_n \cdot \prod_{j=1}^{n} (Q_j + i)$$

Equation 2.5

By transposition, and substitution of Q_j by Q_{REP}:-

$$\frac{P_n}{B_n} = \prod_{j=1}^{n} (Q_{REP} + i)$$

$$= (Q_{REP} + i)^n$$

Equation 2.6

It can be demonstrated that:-

$$P_n = \prod_{j=1}^{k} b_j \cdot \prod_{j=1}^{k} (Q_j + i) = \prod_{j=1}^{k} b_j \cdot \prod_{j=1}^{k} (Q_{REP} + i) = B_n \cdot (Q_{REP} + i)^n$$

Equation 2.7

Transposition for Q_{REP} and substitution of Equation 2.1 in Equation 2.8 then allows:-

$$Q_{REP} = \sqrt[n]{\frac{\prod_{j=1}^{k} b_j \cdot \prod_{j=1}^{k} (Q_j + i)}{B_n}}$$

$$= \sqrt[n]{\frac{\prod_{j=1}^{k} b_j \cdot \prod_{j=1}^{k} (Q_j + i)}{\prod_{j=1}^{k} b_j}}$$

Equation 2.8

Cancellation of B_n then allows:-

$$Q_{REP} = \sqrt[n]{\prod_{j=1}^{n}(Q_j + i)}$$

Equation 2.9

From which it is also apparent that:-

$$(Q_{REP} + i)^n = \prod_{j=1}^{n}(Q_j + i)$$

Equation 2.10

Q_{REP} is invariably complex.

The Selection Rule for the powers of the imaginary unit i follows this pattern:-

$$i^j = \begin{cases} 0,4,8,12 \ldots \Rightarrow & 4n+0 \Rightarrow & i^j = 1 \\ 2,6,10,14 \ldots \Rightarrow & 4n+2 \Rightarrow & i^j = -1 \\ 1,5,9,13 \ldots \Rightarrow & 4n+1 \Rightarrow & i^j = i \\ 3,7,11,15 \ldots \Rightarrow & 4n+3 \Rightarrow & i^j = -i \end{cases}$$

Selection Rule 1.1

Trial and inspection allow us to define these alternating functions:-

Even and Odd j Alternators

$$Ev_j = \frac{(-1)^j + 1}{2}$$
Equation 3.1

$$Ev_j = \begin{cases} 0 & j \text{ is Odd} \\ 1 & j \text{ is Even} \end{cases}$$
Selection Rule 3.1

$$Ev_j = \frac{(-1)^{j-1} + 1}{2}$$
Equation 3.2

$$Od_j = \begin{cases} 1 & j \text{ is Odd} \\ 0 & j \text{ is Even} \end{cases}$$
Selection Rule 3.2

Cycle Function S_j

$$S_j = mod(j - 1,4)$$
Equation 3.3

$$S_j = \begin{cases} mod(j-1,4) = 0 & S_j & = & 0 \\ mod(j-1,4) = 1 & S_j & = & 1 \\ mod(j-1,4) = 2 & S_j & = & 2 \\ mod(j-1,4) = 3 & S_j & = & 3 \end{cases}$$
Selection Rule 3.3

<u>Intermediary Functions X_j and Y_j</u>

$$X_j = \frac{\left[(-1)^j - (-1)^{j+1}\right]}{2} = (-1)^j$$
Equation 3.4

$$X_j = \begin{cases} -1 & for\ j\ is\ Odd \\ +1 & for\ j\ is\ Even \end{cases}$$
Selection Rule 3.4

$$Y_j = S_j . X_j$$
Equation 3.5

$$Y_j = \begin{cases} mod(j-1,4) = 0 & S_j & = & 0 \\ mod(j-1,4) = 1 & S_j & = & 1 \\ mod(j-1,4) = 2 & S_j & = & -2 \\ mod(j-1,4) = 3 & S_j & = & 3 \end{cases}$$
Selection Rule 3.5

<u>Selector Function W</u>

The further auxiliaries U_j and V_j respectively represent the Real and Complex components of the Power Selector Function W_j, whose simple four-choice output may be real (in two cases) or complex, but always unitary.

U_j the Real Part is given by:-

$$U_j = Ev_j\left[S_j . \frac{\left[(-1)^j - (-1)^{j+1}\right]}{2} - 2\right] = Ev_j\left[S_j . (-1)^j - 2\right]$$
Equation 3.6

$$U_j = \begin{cases} mod(j-1,4) = 0 & S_j & = & 0 \\ mod(j-1,4) = 1 & S_j & = & -1 \\ mod(j-1,4) = 2 & S_j & = & 0 \\ mod(j-1,4) = 3 & S_j & = & 1 \end{cases}$$
Selection Rule 3.6

Please note that attempts to simplify U_j are dangerous, and that in particular:-

$$U_j \neq Ev_j\left\{2S_j\left[(-1)^j - (-1)^{j+1}\right] - 1\right\}$$

V_j the Complex Part is given by:-

$$V_j = Od_j\left[S_j.\frac{[(-1)^j - (-1)^{j+1}]}{2} + 1\right].i = Od_j[S_j.(-1)^j + 1].i$$

Equation 3.7

$$V_j = \begin{cases} mod(j-1,4) = 0 & S_j & = & i \\ mod(j-1,4) = 1 & S_j & = & 0 \\ mod(j-1,4) = 2 & S_j & = & -i \\ mod(j-1,4) = 3 & S_j & = & 0 \end{cases}$$

Selection Rule 3.7

Accordingly:-

$$W_j = U_j + V_j$$

Equation 3.8

$$W_j = \begin{cases} mod(j-1,4) = 0 & S_j & = & i \\ mod(j-1,4) = 1 & S_j & = & -1 \\ mod(j-1,4) = 2 & S_j & = & -i \\ mod(j-1,4) = 3 & S_j & = & 1 \end{cases}$$

Selection Rule 3.8

W_j may alternatively be elaborated as:-

$$\begin{aligned} W_j &= Ev_j[(-1)^j - 2] + Od_j[(-1)^j + 1].i \\ &= Ev_j[Y_j - 2] + Od_j[Y_j + 1].i \end{aligned}$$

Equation 3.9

Table 3.1 elaborates these alternators for j = 1 to 16 (four complete cycles of W_j).

j	Ev_j	Od_j	S_j	X_j	Y_j	U_j	V_j	W_j
1	0	1	0	-1	0	0	i	i
2	1	0	1	1	1	-1	0	-1
3	0	1	2	-1	-2	0	-i	-i
4	1	0	3	1	3	1	0	1
5	0	1	0	-1	0	0	i	i
6	1	0	1	1	1	-1	0	-1
7	0	1	2	-1	-2	0	-i	-i
8	1	0	3	1	3	1	0	1
9	0	1	0	-1	0	0	i	i
10	1	0	1	1	1	-1	0	-1
11	0	1	2	-1	-2	0	-i	-i
12	1	0	3	1	3	1	0	1
13	0	1	0	-1	0	0	i	i
14	1	0	1	1	1	-1	0	-1
15	0	1	2	-1	-2	0	-i	-i
16	1	0	3	1	3	1	0	1

Table 3.1
The Values of the Described Alternators
For j = 1 to 16

The special equation for the Complex Product of a Sextic Equation (n = 6) is:-

$$P_6 = \prod_{j=1}^{6} z_j = (a_1 + b_1 i)\,(a_2 + b_2 i)(a_3 + b_3 i)(a_4 + b_4 i)(a_5 + b_5 i)(a_6 + b_6 i)$$

$$= \prod_{j=1}^{6} = b_1 b_2 b_3 b_4 b_5 b_6 (Q_1 + i)\,(Q_2 + i)(Q_3 + i)(Q_4 + i)(Q_5 + i)(Q_6 + i)$$

$$P_6 = \prod_{j=1}^{6} z_j = \prod_{j=1}^{6} b_j \prod_{j=1}^{6} (Q_j + i)$$

Equation 4.1

which may readily be generalised as:-

$$P_6 = \prod_{j=1}^{n} z_j = (a_1 + b_1 i)\,(a_2 + b_2 i)(a_3 + b_3 i) \dots (a_n + b_n i)$$

$$= \prod_{j=1}^{n} = b_1 b_2 b_3 \dots b_n (Q_1 + i)\,(Q_2 + i)(Q_3 + i) \dots (Q_n + i)$$

$$P_n = \prod_{j=1}^{n} z_j = \prod_{j=1}^{n} b_j \prod_{j=1}^{n} (Q_j + i)$$

Equation 4.2

or:-

$$P_n = B_n \prod_{j=1}^{n} (Q_j + i)$$

Equation 4.3

Furthermore:-

$$P_n = B_n \cdot \left[Q_n \prod_{k=1}^{n-1} (Q_k + i) + i \cdot \prod_{k=1}^{n-1} (Q_k + i) \right] = \prod_{j=1}^{n} b_j \cdot \left[Q_n \prod_{k=1}^{n-1} (Q_k + i) + i \cdot \prod_{k=1}^{n-1} (Q_k + i) \right]$$

Equation 4.4

It is obvious from Equation 4.3 that:-

$$P_n = B_n \prod_{j=1}^{n}(Q_j + i) = B_n.(Q_n + i).\prod_{j=1}^{n-1}(Q_j + i)$$

Equation 4.5

It is perhaps more surprisingly the case that:-

$$P_n = B_n \prod_{j=1}^{n}(Q_j + i) = B_n.(Q_n + i).\sum_{j=0}^{n-1} c_{n-1,j}.i^j$$

Equation 4.6

where $c_{n,j}$ is the jth. coefficient of the Degree-n Binomial Equation[5] (which is itself an algebraic polynomial).

The summative side of Equation 4.6 is a corollary of the well-known Equation of the Binomial Expansion:-

$$(x + y)^n = \sum_{k=1}^{n} {}^nC_k.x^k.y^{n-k} = \frac{n!}{k!(n-k)!}x^k.y^{n-k}$$

Equation 4.7

where ${}^nC_k = n!/(k!(n-k)!)$ is the Number of Combinations of n Objects taken k at a time.

The evident implication is that:-

$$P_n = B_n \prod_{j=1}^{n}(Q_j + i) = B_n.(Q_n + i).\prod_{j=1}^{n-1}(Q_j + i) = B_n.(Q_n + i).\sum_{j=0}^{n-1} c_{n-1,j}.i^j$$

Equation 4.8

from which we may anticipate that:

$$P_n = B_n \prod_{j=1}^{n}(Q_j + i) = B_n(Q_n + i).\sum_{j=0}^{n} c_{n-1,j}.i^j$$

Equation 4.9

Now we know that Q_{REP}, the Representative Value of the Set of Q_k for k = 1 to n, is given by:-

$$Q_{REP} = \sqrt[n]{\frac{P_n}{B_n}} - i$$
Equation 4.10

From which we are able to construct:-

$$P_n = (x + y)^n = B_n.\left[(Q_{REP} + i)^n\right] = B_n.\sum_{k=1}^{n} {}^nC_k.Q_{REP}{}^k.i^{n-k} \equiv B_n.\sum_{k=1}^{n} {}^nC_k.Q_{REP}{}^{n-k}.i^k$$
Equation 4.11

or:-

$$P_n = (x + y)^n = B_n.\sum_{k=1}^{n} {}^nC_k.Q_{REP}{}^{n-k}.i^k = B_n.\sum_{k=1}^{n} c_{n,k}.i^k$$
Equation 4.12

A third remarkable implication is that:-

$$\sum_{k=0}^{n} c_{n,k}.i^k = \sum_{k=0}^{n} \binom{n}{k} Q_{REP}{}^{n-k}.i^k$$
Equation 4.13

We can see that Equation 4.8 and its homologs furnish a method of bootstrapping solutions for products of degree n provided that we have knowledge of the value of Product n-1, P_{n-1}, or else its summative terms coefficients ${}^{n-1}C_r$.

<u>Permutations and their Grouping</u>

To assist solution and analysis, it is also possible to arrange a binarion product as a (summative) series of polynomial terms each of which are products of suitable permutations with an exponent of the imaginary unit such that:-

$$P_n = \prod_{j=1}^{n} z_j = \prod_{j=1}^{n} b_j \prod_{j=1}^{n} (Q_j + i)$$
Equation 4.3

Whilst it would be tedious and unclear to elaborate Equation 4.3 for the sextic or some other higher product, I shall demonstrate the action for the cubic case:-

$$P_3 = \prod_{j=1}^{3} z_j = (a_1 + b_1 i)(a_2 + b_2 i)(a_3 + b_3 i)$$

$$= \prod_{j=1}^{3} = b_1 b_2 b_3 (Q_1 + i)(Q_2 + i)(Q_3 + i)$$

$$P_3 = \prod_{j=1}^{3} z_j = \prod_{j=1}^{3} b_j \prod_{j=1}^{3}(Q_j + i)$$

Equation 4.14

Hence:-

$$P_3 = \prod_{j=1}^{3} b_j \cdot (Q_1 Q_2 Q_3 + Q_1 Q_2 i + Q_2 Q_3 i + Q_3 Q_1 i + Q_1 i^2 + Q_2 i^2 + Q_3 i^2 + i^3)$$

Equation 4.15

Equation 4.15 offers an early hint of the cyclical, though involved, character of binomial permutations.

Equation 4.15 may be re-arranged as a conventional algebraic polynomial thus:-

$$P_3 = \prod_{j=1}^{3} b_j \cdot [Q_1 Q_2 Q_3 + i.(Q_1 Q_2 + Q_2 Q_3 + Q_3 Q_1) + i^2.(Q_1 + Q_2 + Q_3) + i^3]$$

Equation 4.16

Noting that $i^2 = -1$ and that $i^3 = -i$, we may rewrite Equation 4.16 as:-

$$P_3 = \prod_{j=1}^{3} b_j \cdot [Q_1 Q_2 Q_3 + i.(Q_1 Q_2 + Q_2 Q_3 + Q_3 Q_1 - 1) - (Q_1 + Q_2 + Q_3)]$$

Equation 4.17

From which it is clear enough that:-

$$P_3 = B_3 . \{[Q_1 Q_2 Q_3 - (Q_1 + Q_2 + Q_3)] + [Q_1 Q_2 + Q_2 Q_3 + Q_3 Q_1 - 1].i\}$$
Equation 4.18

Trial Data with Derived Inputs Q_j and z_j

$a_1 := 714821$ $a_2 := 587611$ $a_3 := 196169$ $a_4 := 181284$ $a_5 := 351758$

$a_6 := 274867$ $a_7 := 818505$ $a_8 := 2094$

$b_1 := 40228$ $b_2 := 953263$ $b_3 := 981283$ $b_4 := 278900$ $b_5 := 638156$

$b_6 := 173879$ $b_7 := 111527$ $b_8 := 872630$

$$
j = \begin{bmatrix} 1 \\ 2 \\ 3 \\ 4 \\ 5 \\ 6 \\ 7 \\ 8 \end{bmatrix}
\quad
i^j = \begin{bmatrix} 1i \\ -1 \\ -1i \\ 1 \\ 1i \\ -1 \\ -1i \\ 1 \end{bmatrix}
\quad
a_j = \begin{bmatrix} 714821 \\ 587611 \\ 196169 \\ 181284 \\ 351758 \\ 274867 \\ 818505 \\ 2094 \end{bmatrix}
\quad
b_j = \begin{bmatrix} 40228 \\ 953263 \\ 981283 \\ 278900 \\ 638156 \\ 173879 \\ 111527 \\ 872630 \end{bmatrix}
$$

$$
Q_j = \begin{bmatrix} 17.76924 \\ 0.616421 \\ 0.199911 \\ 0.649996 \\ 0.55121 \\ 1.580795 \\ 7.339075 \\ 0.0024 \end{bmatrix}
\quad
z_j = \begin{bmatrix} 714821 + 40228i \\ 587611 + 953263i \\ 196169 + 981283i \\ 181284 + 278900i \\ 351758 + 638156i \\ 274867 + 173879i \\ 818505 + 111527i \\ 2094 + 872630i \end{bmatrix}
$$

The Linear Case

$$C_{1,0} := Q_1 = 1.776924 \cdot 10$$

$$C_{1,1} := 1$$

The Quadratic Case

$$C_{2,0} := Q_1 \cdot Q_2 = 10.953327$$

$$C_{2,1} := Q_1 + Q_2 = 18.385661$$

$$C_{2,2} := 1$$

$$C_{2,0} := \prod_{j=1}^{n} Q_j = 10.953327$$

$$C_{2,1} := \prod_{j=1}^{n} Q_j \cdot \sum_{j=1}^{n} \frac{1}{Q_j} = 18.385661$$

$$C_{2,2} := 1$$

The Cubic Case

$$C_{3,0} := Q_1 \cdot Q_2 \cdot Q_3 = 2.189688$$

$$C_{3,1} := Q_1 \cdot Q_2 + Q_2 \cdot Q_3 + Q_3 \cdot Q_1 = 14.628818$$

$$C_{3,2} := Q_1 + Q_2 + Q_3 = 18.585572$$

$$C_{3,3} := 1 = 1$$

$$C_{3,0} := \prod_{j=1}^{n} Q_j = 2.189688$$

$$C_{3,1} := \prod_{j=1}^{n} Q_j \cdot \sum_{j=1}^{n} \frac{1}{Q_j} = 14.628818$$

$$C_{3,2} := \sum_{j=1}^{n} Q_j = 18.585572$$

$$C_{3,2} := \prod_{j=1}^{n} Q_j \cdot \left(\sum_{k=1}^{n-1} \sum_{j=1}^{n-k} \frac{1}{Q_k \cdot Q_{k+j}} \right) = 18.585572$$

$$C_{3,3} := 1 = 1$$

The Quartic Case

$$C_{4,0} := Q_1 \cdot Q_2 \cdot Q_3 \cdot Q_4 = 1.423289$$

$$C_{4,1} := Q_1 \cdot Q_2 \cdot Q_3 + Q_1 \cdot Q_2 \cdot Q_4 + Q_1 \cdot Q_3 \cdot Q_4 + Q_2 \cdot Q_3 \cdot Q_4 = 11.698367$$

$$C_{4,2} := Q_1 \cdot Q_2 + Q_2 \cdot Q_3 + Q_3 \cdot Q_4 + Q_4 \cdot Q_1 + Q_1 \cdot Q_3 + Q_2 \cdot Q_4 = 26.709373$$

$$C_{4,3} := Q_1 + Q_2 + Q_3 + Q_4 = 19.235568$$

$$C_{4,4} := 1 = 1$$

$$C_{4,0} := \prod_{j=1}^{n} Q_j = 1.423289$$

$$C_{4,1} := \prod_{j=1}^{n} Q_j \cdot \sum_{j=1}^{n} \frac{1}{Q_j} = 11.698367$$

$$C_{4,1} := \sum_{k=1}^{n-2} \sum_{l=k+1}^{n-1} \sum_{m=l+1}^{n} Q_k \cdot Q_l \cdot Q_m = 11.698367$$

$$C_{4,2} := \prod_{j=1}^{n} Q_j \cdot \left(\sum_{k=1}^{n-1} \sum_{j=1}^{n-k} \frac{1}{Q_k \cdot Q_{k+j}} \right) = 26.709373$$

$$C_{4,2} := \sum_{l=1}^{n-1} \sum_{m=l+1}^{n} Q_l \cdot Q_m = 26.709373$$

$$C_{4,3} := \sum_{j=1}^{n} Q_j$$

$$C_{4,4} := 1 = 1$$

The Quintic Case

$$C_{5,0} := Q_1 \cdot Q_2 \cdot Q_3 \cdot Q_4 \cdot Q_5 = 0.784531$$

$$C_{5,1} := Q_1 \cdot Q_2 \cdot Q_3 \cdot Q_4 + Q_1 \cdot Q_2 \cdot Q_3 \cdot Q_5 + Q_1 \cdot Q_2 \cdot Q_4 \cdot Q_5 + Q_1 \cdot Q_3 \cdot Q_4 \cdot Q_5 + Q_2 \cdot Q_3 \cdot Q_4 \cdot Q_5 = 7.871546$$

$$C_{5,2} := Q_2 \cdot Q_5 \cdot Q_3 + Q_2 \cdot Q_4 \cdot Q_5 + Q_2 \cdot Q_3 \cdot Q_4 + Q_1 \cdot Q_3 \cdot Q_4 + Q_1 \cdot Q_4 \cdot Q_5 + Q_1 \cdot Q_2 \cdot Q_3 \cdot Q_4 + Q_1 \cdot Q_2 \cdot Q_4 + Q_5 \cdot Q_1 \cdot Q_2 \cdot Q_4 + Q_1 \cdot Q_2 \cdot Q_3 = 26.420841$$

$$C_{5,3} := Q_1 \cdot Q_2 + Q_1 \cdot Q_3 + Q_1 \cdot Q_4 + Q_1 \cdot Q_5 + Q_2 \cdot Q_3 + Q_2 \cdot Q_4 + Q_2 \cdot Q_5 + Q_3 \cdot Q_4 + Q_3 \cdot Q_5 + Q_4 \cdot Q_5 = 37.312211$$

$$C_{5,4} := Q_1 + Q_2 + Q_3 + Q_4 + Q_5 = 19.786778$$

$$C_{5,5} := 1 = 1$$

$$C_{5,0} := \prod_{j=1}^{n} Q_j = 0.784531$$

$$C_{5,1} := \prod_{j=1}^{n} Q_j \cdot \sum_{j=1}^{n} \frac{1}{Q_j} = 7.871546$$

$$C_{5,2} := \sum_{k=1}^{n-2} \sum_{l=k+1}^{n-1} \sum_{m=l+1}^{n} Q_k \cdot Q_l \cdot Q_m = 26.420841$$

$$C_{5,2} := \prod_{j=1}^{n} Q_j \cdot \left(\sum_{k=1}^{n-1} \sum_{j=1}^{n-k} \frac{1}{Q_k \cdot Q_{k+j}} \right) = 26.420841$$

$$C_{5,3} := \sum_{l=1}^{n-1} \sum_{m=l+1}^{n} Q_l \cdot Q_m = 37.312211$$

$$C_{5,4} := \sum_{j=1}^{n} Q_j = 19.786778$$

$$C_{5,5} := 1$$

The Sextic Case

$$C_{6,0} := Q_1 \cdot Q_2 \cdot Q_3 \cdot Q_4 \cdot Q_5 \cdot Q_6 = 1.240183$$

$$\alpha_{6,1} := Q_1 \cdot Q_2 \cdot Q_3 \cdot Q_4 \cdot Q_5 + Q_1 \cdot Q_2 \cdot Q_3 \cdot Q_4 \cdot Q_6 + Q_1 \cdot Q_2 \cdot Q_3 \cdot Q_5 \cdot Q_6$$

$$\beta_{6,1} := Q_1 \cdot Q_2 \cdot Q_4 \cdot Q_5 \cdot Q_6 + Q_1 \cdot Q_3 \cdot Q_4 \cdot Q_5 \cdot Q_6 + Q_2 \cdot Q_3 \cdot Q_4 \cdot Q_5 \cdot Q_6$$

$$\gamma_{6,1} := 0$$

$$C_{6,1} := \alpha_{6,1} + \beta_{6,1} + \gamma_{6,1} = 13.22783$$

$$\alpha_{6,2} := Q_1 \cdot Q_2 \cdot Q_3 \cdot Q_4 + Q_1 \cdot Q_2 \cdot Q_3 \cdot Q_5 + Q_1 \cdot Q_2 \cdot Q_3 \cdot Q_6 + Q_1 \cdot Q_2 \cdot Q_4 \cdot Q_5 + Q_1 \cdot Q_2 \cdot Q_4 \cdot Q_6$$

$$\beta_{6,2} := Q_1 \cdot Q_2 \cdot Q_5 \cdot Q_6 + Q_1 \cdot Q_3 \cdot Q_4 \cdot Q_5 + Q_1 \cdot Q_3 \cdot Q_4 \cdot Q_6 + Q_1 \cdot Q_3 \cdot Q_5 \cdot Q_6 + Q_1 \cdot Q_4 \cdot Q_5 \cdot Q_6$$

$$\gamma_{6,2} := Q_2 \cdot Q_3 \cdot Q_4 \cdot Q_5 + Q_2 \cdot Q_3 \cdot Q_4 \cdot Q_6 + Q_2 \cdot Q_3 \cdot Q_5 \cdot Q_6 + Q_2 \cdot Q_4 \cdot Q_5 \cdot Q_6 + Q_3 \cdot Q_4 \cdot Q_5 \cdot Q_6$$

$$C_{6,2} := \alpha_{6,2} + \beta_{6,2} + \gamma_{6,2} = 49.637472$$

$$\alpha_{6,3} := Q_1 \cdot Q_2 \cdot Q_3 + Q_1 \cdot Q_2 \cdot Q_4 + Q_1 \cdot Q_2 \cdot Q_5 + Q_1 \cdot Q_2 \cdot Q_6 + Q_1 \cdot Q_3 \cdot Q_4$$

$$\beta_{6,3} := Q_1 \cdot Q_3 \cdot Q_5 + Q_1 \cdot Q_3 \cdot Q_6 + Q_1 \cdot Q_4 \cdot Q_5 + Q_1 \cdot Q_4 \cdot Q_6 + Q_1 \cdot Q_5 \cdot Q_6$$

$$\gamma_{6,3} := Q_2 \cdot Q_3 \cdot Q_4 + Q_2 \cdot Q_3 \cdot Q_5 + Q_2 \cdot Q_3 \cdot Q_6 + Q_2 \cdot Q_4 \cdot Q_5 + Q_2 \cdot Q_4 \cdot Q_6$$

$$\delta_{6,3} := Q_2 \cdot Q_5 \cdot Q_6 + Q_3 \cdot Q_4 \cdot Q_5 + Q_3 \cdot Q_4 \cdot Q_6 + Q_3 \cdot Q_5 \cdot Q_6 + Q_4 \cdot Q_5 \cdot Q_6$$

$$C_{6,3} := \alpha_{6,3} + \beta_{6,3} + \gamma_{6,3} + \delta_{6,3} = 85.403786$$

$$\alpha_{6,4} := Q_1 \cdot Q_2 + Q_1 \cdot Q_3 + Q_1 \cdot Q_4 + Q_1 \cdot Q_5 + Q_1 \cdot Q_6$$

$$\beta_{6,4} := Q_2 \cdot Q_3 + Q_2 \cdot Q_4 + Q_2 \cdot Q_5 + Q_2 \cdot Q_6 + Q_3 \cdot Q_4$$

$$\gamma_{6,4} := Q_3 \cdot Q_5 + Q_3 \cdot Q_6 + Q_4 \cdot Q_5 + Q_4 \cdot Q_6 + Q_5 \cdot Q_6$$

$$C_{6,4} := \alpha_{6,4} + \beta_{6,4} + \gamma_{6,4} = 68.591045$$

$$C_{6,5} := Q_1 + Q_2 + Q_3 + Q_4 + Q_5 + Q_6 = 21.367573$$

$$C_{6,6} := 1 = 1$$

The Septic Case

Complex Coefficients were not elaborated for the Septic Case.

PART SIX

CALCULATION OF THE POLYNOMIAL CO-EFFICIENTS C_{n,k}

CALCULATION OF THE POLYNOMIAL CO-EFFICIENTS $C_{n,k}$

General Observations

As we have seen it is possible to elaborate the requisite Co-Efficients $C_{n,k}$ as series of permutations of Q-values though that exercise, always tedious, is impracticable for Degrees much greater than five.

Instead we shall evolve procedures to compute the co-efficients in terms of the polynomial's Representative Q-Dividend defined as:-

$$Q_{REP} = \sqrt[n]{\prod_{j=1}^{n}(Q_j + i)} - i$$

Equation 2.9

Equation 2.9 should conveniently be framed as the more general Representative Q-Dividend, R_n, for use as an arrayed parameter. I.e.:-

$$R_n = (Q_{REP})_n = \left(\sqrt[n]{\prod_{j=1}^{n}(Q_j + i)} - i \right)_n$$

Equation 6.1

We have seen that:-

$$P_k = \prod_{k=1}^{k} z_k = \prod_{k=1}^{k} \lambda_k$$

Equation 1.14

and that:-

$$P_k = b_1.(Q_1 + i) \times b_2.(Q_2 + i) \times b_3.(Q_3 + i) \times ... \times b_k.(Q_k + i) = \prod_{j=1}^{k} b_j . \prod_{j=1}^{k}(Q_j + i)$$

Equation 2.1

It is further the case that:-

$$Z_n = B_n.(R_n + i)^n = B_n. \prod_{j=1}^{n}(Q_j + i) = B_n.\psi_n$$

Equation 6.2

from which it is readily ascertained that:-

$$\psi_n = (R_n + i)^n = \prod_{j=1}^{n}(Q_j + i)$$

Equation 6.3

Polynomial Co-efficient Calculations

We have already seen how:-

$$P_n = \prod_{j=1}^{k} b_j . \prod_{j=1}^{k}(Q_j + i) = \prod_{j=1}^{k} b_j . \prod_{j=1}^{k}(Q_{REP} + i) = B_n.(Q_{REP} + i)^n$$

Equation 2.7

where P_n is a Product of Arbitrary Binomial Sums.

It may further be demonstrated that for at least one to eight binomial complex multiplicands:-

$$T_n = P_n = B_n. \sum_{k=0}^{n} \binom{n}{k}. R_n^{n-k} . i^k$$

Equation 6.4

where:-

$$\binom{n}{k} = {^n}C_k = \frac{n!}{k!\,(n-k)!}$$

Equation 6.5

From Equation 6.4 it is manifest that Co-efficient $C_{n,k}$ is given by:-:-

$$C_{n,k} = \binom{n}{k}. R_n^{n-k}$$

Equation 6.6

Certain Properties of the Coefficient $C_{n,k}$ and the Representative Q-Dividend, R_j

We have seen at an earlier stage that binomial products are capable of being bootstrapped from the subjacent product according to:-

$$P_n = B_n. \left[Q_n \prod_{k=1}^{n-1}(Q_k + i) + i. \prod_{k=1}^{n-1}(Q_k + i) \right]$$

Equation 4.4a

from which it is possible to isolate:-

$$\prod_{k=1}^{n}(Q_k + i) = (Q_n + i)\prod_{k=1}^{n-1}(Q_k + i) = \sum_{j=0}^{n} C_{n,j} \cdot i^j$$
Equation 6.7

Because:-

$$B_n \prod_{k=1}^{n}(Q_k + i) = B_n \sum_{j=0}^{n} C_{n,j} \cdot i^j$$
Equation 6.8

Further we may state that:-

$$\prod_{k=1}^{n}(Q_k + i) = (Q_n + i)\prod_{k=1}^{n-1}(Q_k + i) = \sum_{j=0}^{n} C_{n,j} \cdot i^j = (R_j + i)^n$$
Equation 6.9

and that:-

$$C_{n,k} = \binom{n}{k} \cdot R_n{}^{n-k} = \binom{n}{k} \cdot \left(\sqrt[n]{\prod_{j=1}^{n}(Q_j + i)} - i \right)^{n-k} = \binom{n}{k} \cdot \left(\psi^{\frac{1}{n}} - i \right)^{n-k}$$
Equation 6.10

From which the logical (but untested) implication is that:-

$$R_n = \psi^{\frac{1}{n}} - i$$
Equation 6.11

Development for the Quartic Case

The logic of Equation 6.3 may conveniently be demonstrated for the Quartic, which is neither too simple nor too involved to vitiate clarity, but similar elaborations may be performed for any binomial product, and indeed I worked checks for all kinds of expansions from the Linear to the Octic, using random values of Real a and Imaginary b on Mathcad®.

Firstly use long multiplication or whatever to establish:-

$$P_4 = \prod_{j=1}^{4} b_j \cdot [Q_1 Q_2 Q_3 Q_4 + i \cdot (Q_1 Q_2 Q_3 + Q_1 Q_2 Q_4 + Q_1 Q_3 Q_4 + Q_2 Q_3 Q_4)$$
$$+ i^2 \cdot (Q_1 Q_2 + Q_2 Q_3 + Q_3 Q_4 + Q_4 Q_1 + Q_1 Q_3 + Q_2 Q_4)$$
$$+ i^3 \cdot (Q_1 + Q_2 + Q_3 + Q_4) + i^4]$$

Equation 6.12
after
Equation 4.16

Note that the product P₄ is tantamount to:-

$$P_4 = \prod_{j=1}^{4} b_j \cdot \left[i^0 \cdot (C_{4,0}) + i^1 \cdot (C_{4,1}) + i^2 \cdot (C_{4,2}) + i^3 \cdot (C_{4,3}) + i^4 \cdot (C_{4,4}) \right] = B_4 \cdot \sum_{k=0}^{4} i^j \cdot (C_{4,k})$$

Equation 6.13

Substitute the Representative Q-Value, R₄, for all the Q in Equation 6.12:-

$$P_4 = \prod_{j=1}^{4} b_j \cdot [R_4 R_4 R_4 R_4 + i \cdot (R_4 R_4 R_4 + R_4 R_4 R_4 + R_4 R_4 R_4 + R_4 R_4 R_4)$$
$$+ i^2 \cdot (R_4 R_4 + R_4 R_4 + R_4 R_4 + R_4 R_4 + R_4 R_4 + R_4 R_4) + i^3 \cdot (R_4 + R_4 + R_4 + R_4)$$
$$+ i^4]$$

Equation 6.14

Simplifying:-

$$P_4 = \prod_{j=1}^{4} b_j \cdot [1 \cdot R_4^4 + i \cdot (4 \cdot R_4^3) + i^2 \cdot (6 \cdot R_4^2) + i^3 \cdot (4 \cdot R_4^1) + i^4 \cdot (1 \cdot R_4^0)]$$

Equation 6.15

Substitution and summation then gives:-

$$P_4 = \prod_{j=1}^{4} b_j \cdot \left[i^0 \cdot (C_{4,0}) + i^1 \cdot (C_{4,1}) + i^2 \cdot (C_{4,2}) + i^3 \cdot (C_{4,3}) + i^4 \cdot (C_{4,4}) \right] = B_4 \cdot \sum_{k=0}^{4} i^j \cdot \binom{4}{k} \cdot R_n^{n-k}$$

Equation 6.16

<u>General Observations</u>

The foregoing methods are applicable to situations in which:-

$$P_k = (a_1 + b_1) \times (a_2 + b_2) \times (a_3 + b_3) \times \ldots \times (a_n + b_n) = \prod_{j=1}^{n}(a_j + b_j)$$

$$= b_1.(Q_1 + 1) \times b_2.(Q_2 + 1) \times b_3.(Q_3 + 1) \times \ldots \times b_k.(Q_k + 1) = \prod_{j=1}^{k} b_j . \prod_{j=1}^{k}(Q_j + 1)$$

Equation 7.1

This is a circumstance in which:-

$$Q_j = \frac{a_j}{b_j}$$

Equation 7.2

Whilst by definition, and for computational purposes:-

$$i \equiv +1$$

Equation 7.3

It evidently follows that:-

$$i^j = +1$$

Equation 7.4

for any j being a Natural Number $\mathbb{N} = 0, 1, 2, 3, 4\ldots = j$.

As we have seen it is possible to elaborate the requisite Co-Efficients $C_{n,k}$ as series of permutations of Q-values though that exercise, always tedious, is impracticable for Degrees much greater than five.

Instead we shall evolve procedures to compute the co-efficients in terms of the polynomial's Representative Q-Dividend defined as:-

$$Q_{REP} = \sqrt[n]{\prod_{j=1}^{n}(Q_j + 1)} - 1$$

Equation 7.5

Equation 7.5 should conveniently be framed as the more general Representative Q-Dividend, R_n, for use as an arrayed parameter. I.e.:-

$$R_n = (Q_{REP})_n = \left(\sqrt[n]{\prod_{j=1}^{n}(Q_j + 1)} - 1 \right)_n$$

Equation 7.6

We have seen that:-

$$P_k = \prod_{k=1}^{k} z_k = \prod_{k=1}^{k} \lambda_k$$

Equation 1.14

and that:-

$$P_k = b_1.(Q_1 + i) \times b_2.(Q_2 + i) \times b_3.(Q_3 + i) \times \ldots \times b_k.(Q_k + i) = \prod_{j=1}^{k} b_j . \prod_{j=1}^{k}(Q_j + i)$$

Equation 2.1

It is further the case that:-

$$Z_n = B_n.(R_n + 1)^n = B_n. \prod_{j=1}^{n}(Q_j + 1) = B_n.\psi_n$$

Equation 7.7

from which it is readily ascertained that:-

$$\psi_n = (R_n + 1)^n = \prod_{j=1}^{n}(Q_j + 1)$$

Equation 7.8

Polynomial Co-efficient Calculations

We have already seen how:-

$$P_n = \prod_{j=1}^{k} b_j . \prod_{j=1}^{k}(Q_j + i) = \prod_{j=1}^{k} b_j . \prod_{j=1}^{k}(Q_{REP} + i) = B_n.(Q_{REP} + i)^n$$

Equation 2.7

where P_n is a Product of Arbitrary Binomial Sums.

It may further be demonstrated that for at least one to eight binomial real multiplicands:-

$$T_n = P_n = B_n . \sum_{k=0}^{n} \binom{n}{k} . R_n^{n-k}$$
Equation 7.9

where:-

$$\binom{n}{k} = {}^nC_k = \frac{n!}{k!\,(n-k)!}$$
Equation 7.10

From Equation 7.9 it is manifest that Co-efficient $C_{n,k}$ is given by:-:-

$$C_{n,k} = \binom{n}{k} . R_n^{n-k}$$
Equation 7.11

Certain Properties of the Coefficient $C_{n,k}$ and the Representative Q-Dividend, R_j

We have seen at an earlier stage that binomial products are capable of being bootstrapped from the subjacent product according to:-

$$P_n = B_n . \left[Q_n \prod_{k=1}^{n-1} (Q_k + i) + i . \prod_{k=1}^{n-1} (Q_k + i) \right]$$
Equation 4.4a

from which it is possible to isolate:-

$$\prod_{k=1}^{n} (Q_k + 1) = (Q_n + 1) \prod_{k=1}^{n-1} (Q_k + 1) = \sum_{j=0}^{n} C_{n,j}$$
Equation 7.12

Because:-

$$B_n \prod_{k=1}^{n} (Q_k + 1) = B_n \sum_{j=0}^{n} C_{n,j}$$
Equation 7.13

Further we may state that:-

$$\prod_{k=1}^{n}(Q_k+1)=(Q_n+1)\prod_{k=1}^{n-1}(Q_k+1)=\sum_{j=0}^{n}C_{n,j}=\left(R_j+1\right)^n$$

Equation 7.14

and that:-

$$C_{n,k}=\binom{n}{k}\cdot R_n{}^{n-k}=\binom{n}{k}\cdot\left(\sqrt[n]{\prod_{j=1}^{n}(Q_j+1)}-1\right)^{n-k}=\binom{n}{k}\cdot\left(\psi^{\frac{1}{n}}-1\right)^{n-k}$$

Equation 7.15

From which the logical (but untested) implication is that:-

$$R_n=\psi^{\frac{1}{n}}-1$$

Equation 7.16

Development for the Quartic Case

The logic of Equation 7.8 may conveniently be demonstrated for the Quartic, which is neither too simple nor too involved to vitiate clarity, but similar elaborations may be performed for any binomial product, and indeed I worked checks for all kinds of expansions from the Linear to the Octic, using random values of Real a and Real b on Mathcad®.

Firstly use long multiplication or whatever to establish:-

$$P_4=\prod_{j=1}^{4}b_j\cdot[Q_1Q_2Q_3Q_4+(Q_1Q_2Q_3+Q_1Q_2Q_4+Q_1Q_3Q_4+Q_2Q_3Q_4)$$

$$+(Q_1Q_2+Q_2Q_3+Q_3Q_4+Q_4Q_1+Q_1Q_3+Q_2Q_4)+\cdot(Q_1+Q_2+Q_3+Q_4)+1]$$

Equation 7.17
after
Equation 4.16

(Note that the product P_4 is tantamount to:-

$$P_4=\prod_{j=1}^{4}b_j\cdot\left[\left(C_{4,0}\right)+\left(C_{4,1}\right)+\left(C_{4,2}\right)+\left(C_{4,3}\right)+\left(C_{4,4}\right)\right]=B_4\cdot\sum_{k=0}^{4}\left(C_{4,k}\right)$$

Equation 7.18

Substitute the Representative Q-Value, R_4, for all the Q in Equation 7.17:-

$$P_4 = \prod_{j=1}^{4} b_j \cdot [R_4R_4R_4R_4 + (R_4R_4R_4 + R_4R_4R_4 + R_4R_4R_4 + R_4R_4R_4)$$

$$+ (R_4R_4 + R_4R_4 + R_4R_4 + R_4R_4 + R_4R_4 + R_4R_4) + (R_4 + R_4 + R_4 + R_4) + 1]$$

Equation 7.19

Simplifying:-

$$P_4 = \prod_{j=1}^{4} b_j \cdot [1.R_4^4 + (4.R_4^3) + (6.R_4^2) + (4.R_4^1) + (1.R_4^0)]$$

Equation 7.20

Substitution and summation then gives:-

$$P_4 = \prod_{j=1}^{4} b_j \cdot [\ (C_{4,0}) + (C_{4,1}) + (C_{4,2}) + (C_{4,3}) + (C_{4,4})] = B_4 \cdot \sum_{k=0}^{4} \binom{4}{k} \cdot R_n^{n-k}$$

Equation 7.21

References

1   Wikipedia contributors. (2020, November 13).
           Argument (complex analysis).
           In Wikipedia, The Free Encyclopedia.
           Retrieved 13:26, November 20, 2020,
           from
https://en.wikipedia.org/w/index.php?title=Argument_(complex_analysis)&oldid=988500252

2   How do you simplify cos(2tan−1x)?
           https://socratic.org/questions/how-do-you-simplify-cos-2-tan-1-x

3   How do you simplify the expression sin(2arctanx)?
           https://socratic.org/questions/how-do-you-simplify-the-expression-sin-
     2-arctan-x

4   Euler's Formula And De Moivre's Theorem
           Byju's The Learning App
           https://byjus.com/maths/eulers-formula-and-de-moivres-theorem/

5   Wikipedia contributors. (2020, November 5).
           Binomial coefficient
           In Wikipedia, The Free Encyclopedia.
           Retrieved 13:05, November 17, 2020,
           from
https://en.wikipedia.org/w/index.php?title=Binomial_coefficient&oldid=987215744

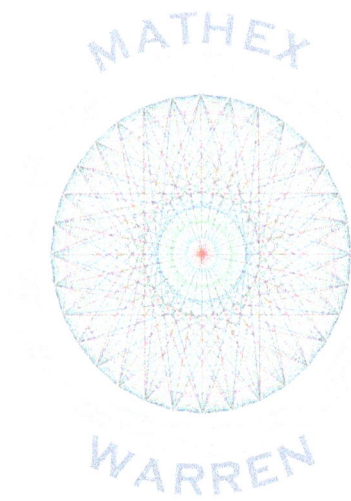

CHAPTER SIX

Polygram Mensuration

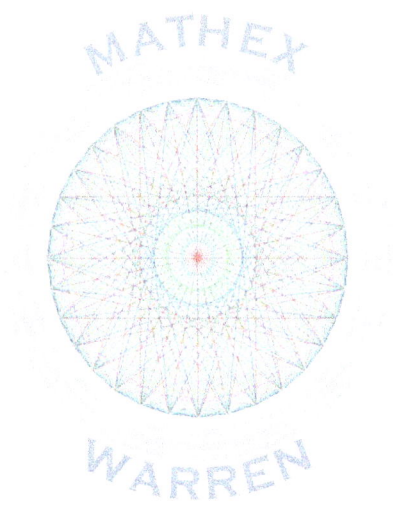

The Regular Polygon and its
First-Order Star Polygram

by
James R Warren BSc MSc PhD PGCE

PART ONE
PRINCIPLES

A regular polygon is a two-dimensional figure with n Sides of equal length. A number of other lineaments may characterise a polygon, and a number of included or excluded Angles. Ultimately, these geometric properties are functions of Sides n and Radius r only, as combined with the Ludolphine Constant π. The Ludolphine Constant is not exactly knowable but has a value near to 3.14159265358979. It is the Ratio of the Circumference of a Circle to the Circle's Diameter.

Technically, a Regular Polygon is an Order (n,1) polygram and its associated First-Order Star Polygon is an Order (n,2) polygram. The Star Polygram is formed from the intersecting chords between alternate (n,1) polygon vertices.

These relations are illustrated for the n = 7 Heptagon and its associated (7,2) polygram in Figure 1.1. I beheld this heptagon and its co-lateral rectangle STUV in a dream at 0400 GMT on the morning of 24 October 2020, and I wondered if there was any relation such that the ratios AP/PC and SB/BT were related. I decided to study this problem in my waking hours, notwithstanding the fogginess of conscious ratiocination.

Though initially developed for the Heptagon, the following analysis is general.

Angular Relationships[1]

The External Angle Δ specifies a particular regular n-gon and is given by:-

$$\Delta = \frac{2\pi}{n}$$
Equation 1.1

whilst the Internal Angle Γ is:-

$$\Gamma = \pi - \Delta = \pi - \frac{2\pi}{n} = \pi\left(1 - \frac{2}{n}\right)$$
Equation 1.2

Of considerable importance to analysis is the Apothemal Angle (Half-Side Subtense) θ shown on the figure as BOQ and defined as:-

$$\theta = \frac{\pi}{n}$$
Equation 1.3

which is associated with the Sector Angle 2θ and the angle $\omega = 3\theta$.
The Internal Co-Angle ψ is defined as:-

$$\psi = \Gamma - \frac{\pi}{2} = \left(\pi - \frac{2\pi}{n}\right) - \frac{\pi}{2} = \pi\left(\frac{1}{2} - \frac{2}{n}\right)$$
Equation 1.4

Basic Lines

The Apical Radius (of the Order (n,1) Polygon) is defined as the scale factor r, which scales the Areas of the forms by r^2. From a mathematical point of view the value of r is arbitrary and to assist computational economy as well as analytical clarity we may as well define it as unity.

At any event the key line segments Apothemal Opposite a = BQ and Apothemal Adjacent b = OQ are respectively:-

$$a = r.\sin\theta$$
Equation 1.5

and:-

$$b = r.\cos\theta$$
Equation 1.6

Therefore the Polygon Side, s, is given by:-

$$s = 2a = 2r.\sin\theta$$
Equation 1.7

and so the Rectangular Extension c is:-

$$c = \frac{s}{\tan\psi}$$
Equation 1.8

It is now clear that the Height of the Rectangular Extension is 2a = SU and its Width is b+c = UV.

<u>The Polygram First Order Transect τ</u>

This chord between alternate vertices defines the (n,2) star polygon or polygram and has length τ = α+β where α is the Major Part and β is the Minor Part. It is inconvenient, but possible, to define either of these parts.

Firstly we require the chord length τ which is easily accessed via the Cosine Rule as:-

$$\tau = \alpha + \beta = s\sqrt{2(1 - \cos\Gamma)}$$
Equation 1.9

As aforementioned, we are especially interested in the ratio Ω = α/β = AP/PC.
Key to knowledge of α and β is the Minor Circle Radius γ = OP.
This is because:-

$$\beta = \sqrt{a^2 + (b - \gamma)^2}$$
Equation 1.10

whilst:-

$$\alpha = \tau - \beta$$
Equation 1.11

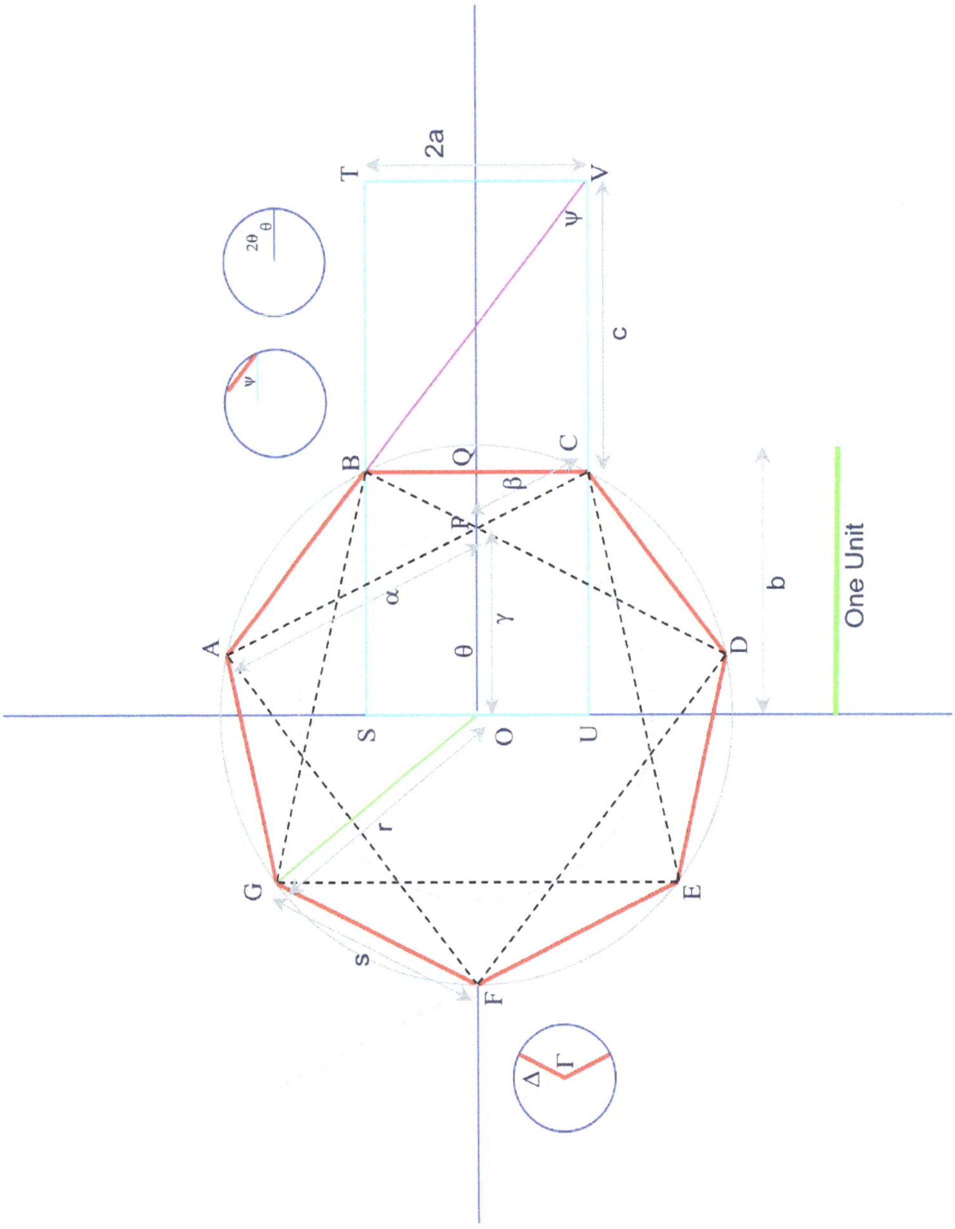

Figure 1.1
The Notation for the Exemplary Heptagon and its
Rectangular Extension

The Minor Circle Radius γ by Linear Interpolation[2]

Allow that the Line t is terminated by the Vertices A = (x1,y1) and C = (x2,y2). Then:-

$$x_1 = r.\cos\omega \qquad y_1 = r.\sin\omega$$
$$x_2 = r.\cos\theta \qquad y_2 = -r.\sin\theta$$

Given that the co-ordinate (x_0,y_0) is that of the desired interpolated point we may note that $\gamma \equiv x_0$ whilst in terms of the figure geometry the attendant y_0 is zero.
Therefore the general interpolative equation:-

$$x_0 = x_1 + (y_0 - y_1).\left(\frac{x_2 - x_1}{y_2 - y_1}\right)$$
Equation 2.1

may be rendered as:-

$$\gamma = x_1 + (0 - y_1).\left(\frac{x_2 - x_1}{y_2 - y_1}\right)$$
Equation 2.2

Appropriate substitutions give:-

$$\gamma = r.\cos\omega + (0 - r.\sin\omega).\left(\frac{r.\cos\theta - r.\cos\omega}{-r.\sin\theta - r.\sin\omega}\right)$$
Equation 2.3

After some lengthy but elementary wrangling this expression resolved to:-

$$\gamma = r.(\tan\theta.\sin 3\theta + \cos 3\theta)$$
Equation 2.4

The Minor Circle Radius γ by Line-Line Intersection[3]

It is also possible, though more protracted, to use a general expression for the intersection of two straight-lines to establish the position of (x_0,y_0). Let us propose the chord AC as our first line and DB as our second line. The two lines visibly intersect at $(\gamma,0) \equiv (x_0,y_0)$.

According to our usual policy we shall use this as an independent check upon our interpolative procedures above.

Trigonometrically, the four required end-points are:-

A	$x_1 = r.\cos\omega$	$y_1 = r.\sin\omega$
C	$x_2 = r.\cos\theta$	$y_2 = -r.\sin\theta$
D	$x_3 = r.\cos\omega$	$y_3 = -r.\sin\omega$
B	$x_4 = r.\cos\theta$	$y_4 = r.\sin\theta$

For efficiency and convenience we may define:-

$$U = x_1 y_2 - y_1 x_2$$
Equation 2.5

$$V = x_3 y_4 - y_3 x_4$$
Equation 2.6

$$W = (x_1 - x_2)(y_3 - y_4) - (y_1 - y_2)(x_3 - x_4)$$
Equation 2.7

and:-

$$Px_0 = \frac{U(x_3 - x_4) - V(x_1 - x_2)}{W}$$
Equation 2.8

$$Py_0 = \frac{U(y_3 - y_4) - V(y_1 - y_2)}{W}$$
Equation 2.9

or:-

$$\gamma = \frac{U(x_3 - x_4) - V(x_1 - x_2)}{W}$$
Equation 2.10

<u>The Development of γ^2</u>

To analyse areas it is often convenient to have an expression for γ^2. By reference to Equation 2.4 we are now in a position to specify:-

$$\zeta = \gamma^2 = r^2.(\tan\theta.\sin 3\theta + \cos 3\theta)^2$$
Equation 2.11

From which the appropriate Triplication of Angle Identities permit the form:-

$$\zeta = \gamma^2 = r^2.\left(\frac{\sin\theta}{\cos\theta}.(3\sin\theta - 4(\sin\theta)^3) + (4(\cos\theta)^3 - 3\cos\theta)\right)^2$$
Equation 2.12

Some protracted derivations allow us to establish:-

$$\zeta = 1 + (\sin\theta)^2.(-3 - 8(\sin\theta)^2 + 16(\sin\theta)^4)$$
$$+ (\tan\theta)^2.(\sin\theta)^2.(9 - 24(\sin\theta)^2 + 16(\sin\theta)^4)$$
Equation 2.13

which clearly includes two quadratic polynomials in $\text{Sin}^2\theta$. Ordinary solution for the quadratic roots allows us to write in the first instance:-

$$-3 - 8(\sin\theta)^2 + 16(\sin\theta)^4 = 16\left((\sin\theta)^2 - \frac{3}{4}\right)\left((\sin\theta)^2 + \frac{1}{4}\right)$$
Equation 2.14

and in the second instance:-

$$9 - 24(\sin\theta)^2 + 16(\sin\theta)^4 = 16\left((\sin\theta)^2 - \frac{3}{4}\right)\left((\sin\theta)^2 - \frac{3}{4}\right)$$
Equation 2.15

or:-

$$9 - 24(\sin\theta)^2 + 16(\sin\theta)^4 = \left[4\left((\sin\theta)^2 - \frac{3}{4}\right)\right]^2$$
Equation 2.16

The appropriate substitutions now allow us to specify $\zeta = \gamma^2$ as:-

$$\zeta = 1 + (\sin\theta)^2.\left(16\left((\sin\theta)^2 - \frac{3}{4}\right)\left((\sin\theta)^2 + \frac{1}{4}\right) + (\tan\theta)^2.\left[4\left((\sin\theta)^2 - \frac{3}{4}\right)\right]^2\right)$$
Equation 2.17

Ratio of Circumscribed and Inscribed Circles

We have seen that:-

$$\zeta = \gamma^2 = 1 + (\sin\theta)^2 . \left\{ 16 \left((\sin\theta)^2 - \frac{3}{4} \right) \left((\sin\theta)^2 + \frac{1}{4} \right) + (\tan\theta)^2 . \left[4 \left((\sin\theta)^2 - \frac{3}{4} \right) \right]^2 \right\}$$

Equation 2.17a

and this allows us immediately to declare the ratio κ of the n-Polygon's Circumscribed Circle, A_r, to that of the Polygram's Minor Circle, A_γ as:-

$$\kappa = \frac{A_r}{A_\gamma} = \frac{\pi r^2}{\pi \gamma^2} = \frac{r^2}{\gamma^2}$$

Equation 3.1

Taking, for convenience, the value of r as unity we may render Equation 3.1 as:-

$$\kappa = \frac{1}{1 + (\sin\theta)^2 . \left\{ 16 \left((\sin\theta)^2 - \frac{3}{4} \right) \left((\sin\theta)^2 + \frac{1}{4} \right) + (\tan\theta)^2 . \left[4 \left((\sin\theta)^2 - \frac{3}{4} \right) \right]^2 \right\}}$$

Equation 3.2

or:-

$$\kappa = \frac{A_r}{A_\gamma} = \frac{\pi r^2}{\pi \gamma^2} = \frac{r^2}{\gamma^2} = \frac{1}{(\tan\theta \sin\theta + \cos\theta)^2}$$

Equation 3.3

The Ratio of α to β in the Chord of the (n,2) Polygram

This ratio is defined by:-

$$\Omega = \frac{\alpha}{\beta}$$

Equation 3.4

By substitution we may write:-

$$\Omega = \frac{\alpha}{\beta} = \frac{(\alpha + \beta) - \beta}{\beta} = \frac{\left(s\sqrt{2(1 - \cos\Gamma)}\right) - \sqrt{a^2 + (b - \gamma)^2}}{\sqrt{a^2 + (b - \gamma)^2}}$$
Equation 3.5

or:-

$$\Omega = \frac{\left(2a\sqrt{2(1 - \cos\Gamma)}\right) - \sqrt{a^2 + (b - \gamma)^2}}{\sqrt{a^2 + (b - \gamma)^2}}$$

$$= \frac{\left(2(r.\sin\theta)\sqrt{2(1 - \cos\Gamma)}\right) - \sqrt{(r.\sin\theta)^2 + \left((r.\cos\theta) - \gamma\right)^2}}{\sqrt{(r.\sin\theta)^2 + \left((r.\cos\theta) - \gamma\right)^2}}$$
Equation 3.6

which reduces to:-

$$\Omega = \frac{\sqrt{8}.\sin\frac{\pi}{n}\sqrt{1 - \cos\left[\pi\left(1 - \frac{2}{n}\right)\right]}}{\sqrt{\left(\frac{\gamma}{r}\right)^2 - 2\left(\frac{\gamma}{r}\right)\cos\frac{\pi}{n} + 1}} - 1$$
Equation 3.7

This simple, real form is best left without the solution of roots in the denominator. Notwithstanding that, simplified and economised renditions are possible and on 9 November 2020 I was able to formulate:-

$$\Omega = \frac{2\cos\left(\frac{2\pi}{n}\right)}{\sqrt{\left(\frac{\gamma}{r}\right)^2 - 2\left(\frac{\gamma}{r}\right)\cos\frac{\pi}{n} + 1}} - 1$$
Equation 3.8

The Ratio of c to b in the Rectangular Extension

This Ratio, φ, is defined by:-

$$\varphi = \frac{c}{b}$$
Equation 3.9

from which appropriate substitutions evolve:-

$$\varphi = \frac{c}{b} = \frac{s}{\tan\psi} \times \frac{1}{b} = \frac{2\tan\left(\frac{\pi}{n}\right)}{\tan\psi} = \frac{2\tan\left(\frac{\pi}{n}\right)}{\tan\left(\Gamma - \frac{\pi}{2}\right)} = \frac{2\tan\left(\frac{\pi}{n}\right)}{\tan\left[\left(\pi - \frac{2\pi}{n}\right) - \frac{\pi}{2}\right]}$$

Equation 3.10

which simplifies to:-

$$\varphi = \frac{2\tan\left(\frac{\pi}{n}\right)}{\tan\left(\pi\left(\frac{1}{2} - \frac{2}{n}\right)\right)}$$

Equation 3.11

A Table of Resulting Values

Table 3.1 is a presentation of the EXCEL® variables and ratios for the regular polygons and their first-order polygrams for the cases n = 5 (the pentagon) to n = 15 (the pentadecagon). The table has been formatted so that values are shown to seven-figure accuracy, except for integers, which are shown native.

In the case of the key ratios $\Omega = \alpha/\beta$ and $\varphi = c/b$ I offer also a plot of the relevant eleven points to which I fitted an EXCEL® quartic polynomial regression that has an R^2 Determination Coefficient of 0.99999745.

The figured quartic is:-

$$\frac{c}{b} = 1.61950125.\left(\frac{\alpha}{\beta}\right)^4 - 16.65252771.\left(\frac{\alpha}{\beta}\right)^3 + 66.68891957.\left(\frac{\alpha}{\beta}\right)^2 - 120.46221262.\left(\frac{\alpha}{\beta}\right)$$
$$+ 85.07832444$$

Equation 3.12

Equation 3.12 has of course no analytic status, but may suggest further avenues of research.

Variable	Symbol	Line Termini	Value 5	Value 6	Value 7	Value 8	Value 9	Value 10	Value 11	Value 12	Value 13	Value 14	Value 15
Number of Sides	n		5	6	7	8	9	10	11	12	13	14	15
Polygon External Angle	Δ		1.2566371	1.0471976	0.8975979	0.7853982	0.6981317	0.6283185	0.5711987	0.5235988	0.4833219	0.4487990	0.4188790
Polygon Internal Angle	Γ		1.8849556	2.0943951	2.2439948	2.3561945	2.4434610	2.5132741	2.5703940	2.6179939	2.6582707	2.6927937	2.7227136
Sector Angle	2θ		1.2566371	1.0471976	0.8975979	0.7853982	0.6981317	0.6283185	0.5711987	0.5235988	0.4833219	0.4487990	0.4188790
Half Side Subtense (Apothemal Angle)	θ		0.6283185	0.5235988	0.4487990	0.3926991	0.3490659	0.3141593	0.2855993	0.2617994	0.2416610	0.2243995	0.2094395
Internal Co-Angle	ψ		0.3141593	0.5235988	0.6731984	0.7853982	0.8726646	0.9424778	0.9995977	1.0471976	1.0874744	1.1219974	1.1519173
Major Circle Radius	r	OB	1	1	1	1	1	1	1	1	1	1	1
Apothemal Opposite	a	BQ	0.5877853	0.5000000	0.4338837	0.3826834	0.3420201	0.3090170	0.2817326	0.2588190	0.2393157	0.2225209	0.2079117
Apothemal Adjacent	b	OQ	0.8090170	0.8660254	0.9009689	0.9238795	0.9396926	0.9510565	0.9594930	0.9659258	0.9709418	0.9749279	0.9781476
Side Length	s	FG	1.1755705	1.0000000	0.8677675	0.7653669	0.6840403	0.6180340	0.5634651	0.5176381	0.4786313	0.4450419	0.4158234
Rectangular Extension	c	CV	3.6180340	1.7320508	1.0881460	0.7653669	0.5739780	0.4490280	0.3621170	0.2988585	0.2512051	0.2143209	0.1851365
Rectangular Length	$c+b$	UV	4.4270510	2.5980762	1.9891149	1.6892464	1.5136706	1.4000845	1.3216100	1.2647843	1.2221469	1.1892488	1.1632841
Extensile Diagonal	d	BV	3.8042261	2	1.3917910	1.0823922	0.8929512	0.7639320	0.6697923	0.5977170	0.5405478	0.4939592	0.4551754
First Order Polygram Transect (Major Chord)	$\alpha+\beta$	AC	1.9021130	1.7320508	1.5636630	1.4142136	1.2855752	1.1755705	1.0812816	1	0.9294463	0.8677675	0.8134733
Major Part of First Order Polygram Transect	α	AP	1.1755705	1.1547005	1.0820883	1	0.9216050	0.8506508	0.7876551	0.7320508	0.6829685	0.6395240	0.6009167
Minor Part of First Order Polygram Transect	β	PC	0.7265425	0.5773503	0.4815746	0.4142136	0.3639702	0.3249197	0.2936265	0.2679492	0.2464779	0.2282435	0.2125566
Minor Circle Radius	γ	OP	0.3819660	0.5773503	0.6920215	0.7653669	0.8152075	0.8506508	0.8767688	0.8965755	0.9119558	0.9241390	0.9339546
Ratio α/β	$\Omega = \alpha/\beta$		1.6180340	2	2.2469796	2.4142136	2.5320889	2.6180340	2.6825071	2.7320508	2.7709121	2.8019377	2.8270909
Ratio c/b	$\varphi = c/b$		4.4721360	2	1.2077509	0.8284271	0.6108146	0.4721360	0.3774046	0.3094011	0.2587231	0.2198325	0.1892726
Area of r-Circle	A_r		3.1415927	3.1415927	3.1415927	3.1415927	3.1415927	3.1415927	3.1415927	3.1415927	3.1415927	3.1415927	3.1415927
Area of Polygon	A_P		2.3776413	2.5980762	2.7364102	2.8284271	2.8925442	2.9389263	2.9735245	3	3.0207006	3.0371862	3.0505248
Area of γ-Circle	A_g		0.4583522	1.0471976	1.5044890	1.8403024	2.0877869	2.2732778	2.4150164	2.5253616	2.6127476	2.6830232	2.7403208
Circles' Ratio	$\kappa = A_r/A_\gamma$		6.8541020	3	2.0881460	1.7071068	1.5047477	1.3819660	1.3008577	1.2440169	1.2024095	1.1709152	1.1464324

Table 3.1
Certain Values of the Angles and Lineaments Characteristic of
Elementary (n,1) Regular Polygons and their Associated (n,2) Polygrams

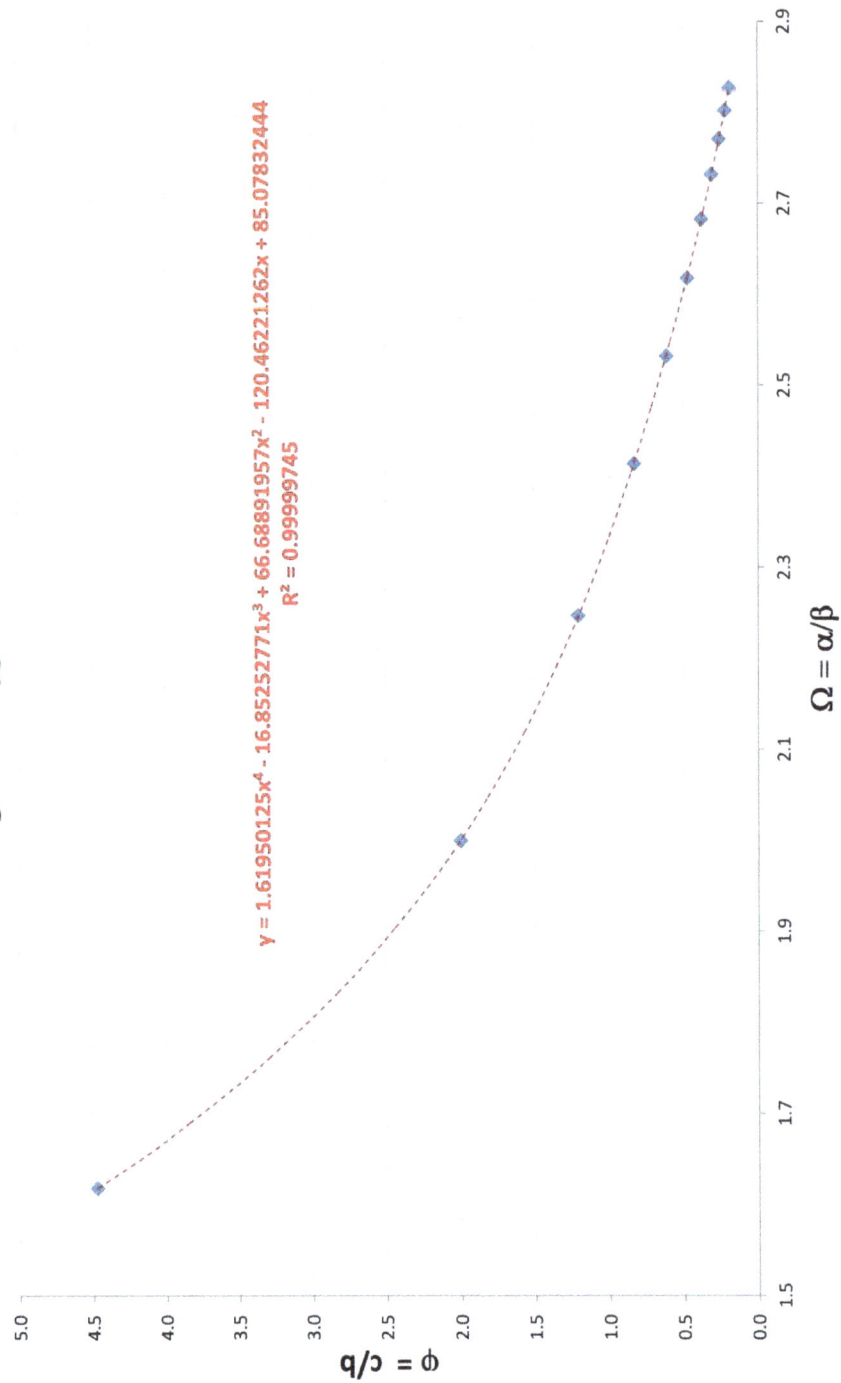

The Rectangular Extension Ratio c/b versus The Chord Ratio α/β for the Regular Polygons n = 5 to 15

$y = 1.61950125x^4 - 16.85252771x^3 + 66.68891957x^2 - 120.46221262x + 85.07832444$
$R^2 = 0.99999745$

$\Omega = \alpha/\beta$

$\phi = c/b$

Figure 3.1
The Chord Ratio φ Plotted versus Rectangular Extension Ratio Ω

References

1   **Elementary Formulae of the Polygon**
     "Handbook of Mathematics"
     IN Bronshtein and KA Semendyayev
     English Translation and Editing by KA Hirsch
     Verlag Harri Deutsch of Thun and Frankfurt-am-Main
     Van Nostrand Reinhold and Company Incorporated of New York
     December 1978
     ISBN 0-442-21171-6
     Page 175 in Section 2.6.1 Plane Geometry
     pp 973

2   **Linear Interpolation**
     "The Penguin Dictionary of Mathematics"
     Edited by John Daintith and RD Nelson
     Penguin Books Limited of Harmondsworth   1989
     ISBN 0-14-051119-9
     Interpolation on Pages 178-179
     pp 350

3   **Line-line Intersection**
     Wikipedia contributors. (2020, September 23).
      Line–line intersection.
     In Wikipedia, The Free Encyclopedia. Retrieved 10:50, October 29, 2020,
     from
     https://en.wikipedia.org/w/index.php?title=Line%E2%80%93line_intersection&oldid=979991801

4   **Trigonometric Constants Expressed in Real Radicals**
     Wikipedia contributors. (2020, October 17).
     Trigonometric constants expressed in real radicals.
     In Wikipedia, The Free Encyclopedia. Retrieved 10:48, October 29, 2020,
     from
     https://en.wikipedia.org/w/index.php?title=Trigonometric_constants_expressed_in_real_radicals&oldid=983985529

5   **University of Surrey Exact Trigonometric Function Values**
     Exact Trigonometric Function Values
     © 1996-2017 Dr Ron Knott
     enquiry@ronknott.com
     updated 30 May 2017
     http://www.maths.surrey.ac.uk/hosted-sites/R.Knott/Fibonacci/simpleTrig

6   **Metallic Mean**
      Wikipedia contributors. (2020, September 30).
      Metallic mean.
      In Wikipedia, The Free Encyclopedia. Retrieved 10:45, October 29, 2020,
      from
      https://en.wikipedia.org/w/index.php?title=Metallic_mean&oldid=981113144

7   **Star Polygons**
      Wikipedia contributors. (2020, October 25).
      Star polygon.
      In Wikipedia, The Free Encyclopedia. Retrieved 10:33, October 29, 2020,
      from
      https://en.wikipedia.org/w/index.php?title=Star_polygon&oldid=985390957

8   **Polygrams**
      Wikipedia contributors. (2019, December 18).
      Polygram (geometry).
      In Wikipedia, The Free Encyclopedia. Retrieved 10:43, October 29, 2020,
      from
https://en.wikipedia.org/w/index.php?title=Polygram_(geometry)&oldid=931388444

CHAPTER SEVEN

Decision Logic

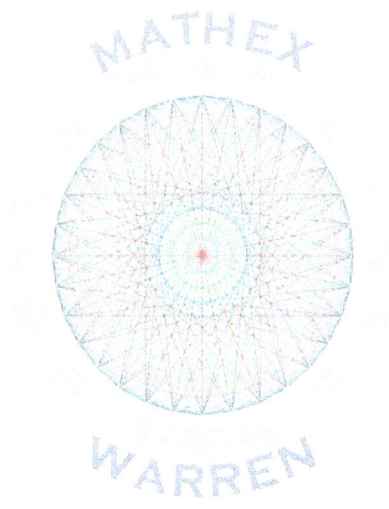

A Boolean Calculus of Moiety
for Two Arguments

By
James R Warren BSc MSc PhD PGCE

PART ONE
ASPECTS OF THE PROCESS ARITHMETIC LOGIC

<u>Predicate</u>

Consider the simplest binary decision structure comprising two parallel multidigit binary numbers A and B that resolve to a decision array D. The Decision Array D issues from operations upon the logical intermediate array C and the array of i BIAS values.

The one-dimensional arrays A, B, C and D are all of equal length i.

A_i, B_i, C_i and D_i are all positional binary digits.

Z is a generalisation of A, B, C or D.

It is accordingly the case that (in denary terms) the largest Z is given by:-

$$Z = 2^n - 1$$
Equation 1.1

For example, a certain A is the binary number 0110 that of course expresses the value $0*2^3+1*2^2+1*2^1+0*2^0 = 6_{denary}$ where A_0 is 0 A_2 is 1 A_3 is 1 and A_4, representing the leftmost $2^3 = 8$ position, is 0. That is to say that for any A, B, C or D:-

$$Z_{denary} = \sum_{i=0}^{n} Z_i \cdot 2^i$$
Equation 1.2

B is a second array of identical length, the binary number 0101 which is obviously 5 in denary terms.

The required digit-wise C is yielded by:-

$$C_i = (A_i - B_i)^2$$
Equation 1.3

It is readily checked that:-
If A_i=0 and B_i=0 then C_i is 0
If A_i=0 and B_i=1 then C_i is 1
If A_i=1 and B_i=0 then C_i is 1
If A_i=1 and B_i=1 then C_i is 0

which constitutes a logical Exclusive Or or XOR structure. Like Binary Addition, this result can be instantiated using simple electronic or hydraulic circuitry. This simple Exclusive Or system may traditionally be summarised with Truth Table XOR.

Furthermore, when $A_i = B_i = $ unity (A and B "agree" positively about Point i) then we wish to record that $D_i = 1$. When $A_i = B_i = $ zero A and B agree in the negative ("reject") Option i; and when A_i and B_i differ the binary outcome D_i is decided by chance conditioned by $BIAS_i$.

The array $BIAS$ is a set of real numbers in the range $0 < BIAS_i < 1$ that bias the selection of either A_i or B_i as the attendant outcome D_i. In our present very basic treatment $BIAS_i$ will always be 0.5 in order to give an equal chance that A_i or B_i will prevail if A_i and B_i differ.

β is the value of the local $BIAS$ (i.e. $BIAS_i$) and for our purposes β will invariably be 0.5. This gives rise to fifth, auxiliary, array BI which we may define using:-

$$BI_j = entier(\beta + rnd(0))$$
Equation 1.4

where rnd(0) is a Uniform Probability Random Number Generator Function operating on the range zero to unity exclusive. It behaves like an unbiased coin-toss with equal probability of being 0 or 1. j is the Criterial Position (the truth tabular combination of arguments A and B).

Accordingly, the required Truth Table for Outcome D is:-

A	B	BI	C	D
0	0	0	0	0
0	0	1	0	0
0	1	0	1	0
0	1	1	1	1
1	0	0	1	0
1	0	1	1	1
1	1	0	0	1
1	1	1	0	1

Truth Table D

Condition:- Positive Agreement of Arguments

The outcome of the arithmetic product $A_i \times B_i$ is zero except for the condition that A_i and B_i are both unity: This is the Boolean AND condition expressed by Truth Table AND:-

A	B	C	A.[1-(A-B)²]	AND(A,B)
0	0	0	0	0
0	1	1	0	0
1	0	1	0	0
1	1	0	1	1

Truth Table AND

This is arithmetically:-

$$AND_j = A_j.\left[1 - \left(A_j - B_j\right)^2\right] = A_j.\left[1 - C_j\right] = A_j.NXOR(A_j, B_j)$$
Equation 1.5

Condition:- Counteridentity of Arguments

When A_i and B_i differ the outcome is unity: When A_i and B_i are the same then the outcome is zero. This is arithmetically equivalent to the C_i expressed by Equation 1.3
This is Boolean Exclusive OR or XOR. The relevant truth table is:-

A	B	C	(A-B)²	XOR(A,B)
0	0	0	0	0
0	1	1	1	1
1	0	1	1	1
1	1	0	0	0

Truth Table XOR

This is arithmetically:-

$$XOR_j = \left(A_j - B_j\right)^2 = C_j$$
Equation 1.6

Condition:- Identity of Arguments

When both A_i and B_i are zero or else both A_i and B_i are unity, then the outcome bit, D_j, is unity: Else the outcome is zero. This is Boolean NOT XOR or NXOR expressed by the following Truth Table NXOR:-

A	B	C	1-(A-B)²	NXOR(A,B)
0	0	0	1	1
0	1	1	0	0
1	0	1	0	0
1	1	0	1	1

Truth Table NXOR

This is arithmetically:-

$$NXOR_j = \left[1 - \left(A_j - B_j\right)^2\right] = \left[1 - C_j\right]$$
Equation 1.7

The Arithmetic Realisation of the Desired Decision D_i

We are now in a position to promulgate the Decision Criterion as:-

$$D_J = C_J \cdot BI_J + A_J\left(1 - C_J\right)^2$$
Equation 1.8

This equation has a number of identities that we may explore later.

Arithmetic Digit Position as an Expression of Criterion Importance

It is of course the case that the value of binary digits doubles at each place to the left, just as denary digits augment ten-fold with each step. (An "order of magnitude").
In principle we could vary the number base at each step and accordingly the utility value of each ordered criterion: But for simplicity's sake we shall forbear this in the current disquisition. We will arbitrarily agree that the value of each sub-decision expressed by a digit-position is double the value of that to its right.

An Example

Arthur and Betty are planning a night out in London.
They have a number of decisions to make: Some are important; some less so. Some are readily agreed; some more contentious. But none are deal-breakers, because Arthur and Betty are in love and some composite compromise shall swiftly be agreed.

Clearly, some criteria are highly contingent: Such as whether to go by tube or to walk will depend upon whether it is raining, or deciding between tube and bus upon whether it is raining and whether they can afford the tube fares.

But we shall leave such complications to statesmen and strategists and examine the simple binary logic.

So in no particular order here are the j decision criteria that offer themselves to Arthur and Betty:-

Criterion	Arthur	Betty
Venue Type	Pictures	Restaurant
Venue Quality	Omnivorous	Vegetarian
Mode of Travel	Bus	Walk
Color of Wine	White	Red
Drink before		
Event	Yes	No
Hamburger	No	No

Table Five
Arthur and Betty's Preferences for their
Night Out in London

This is the First Convention of initial preferences (forgive the pleonasm) before the couple come to any agreement about the relative or absolute importancies of their options; any initial compromises; or any logico-mathematical post-analyses.

As we have seen, some options will be conditioned or even decided by external accidents such as the weather whilst others may be rendered irrelevant or redundant by prior selections: For instance, if they elect to visit a vegetarian restaurant they are unlikely to enjoy a hamburger, though of course one or both of them may partake of a vegetarian meal in a standard omnivores' restaurant. If they go to the pictures then a decision between red and white wine is improbable, except of course they decide to go for a pre-drink in a tavern when that wine criterion re-enters play.

Table Six combines the (descending) order of importance in which Arthur and Betty ordered their six criteria with the Prior Positions the couple agreed (or failed to agree) before Chance played its part. Arthur is coded zero and Betty unity but more importantly we can see that Betty prevailed as regards Mode of Transport since they agreed to walk to their venue: Therefore both Arthur and Betty adopt $A_{j=1}$ as 1 and since A_i and B_i are therefore 1 our process awards D_i the value 1. Thus they determinately agree their most important criterion. At the other end of the scale of importance both are unanimous that they wish No Hamburger. Thus $A_{j=6}$ is 0, $A_i=B_i=0$ and the mechanism sets D_6 as zero.

Where Arthur and Betty cannot agree at the Third Convention, adventitious factors must decide. This might be classical chance where Arthur tosses a coin to decide between pictures and restaurant (I cannot imagine Betty doing anything so unladylike!). Slightly more arcanely, the couple solve the wine dilemma by asking the waiter to decide between red and white for them. *This is tantamount to a stochastic operation.* We know that they are not having fish if they are in a vegetarian restaurant though we have no evidence that they are, but nevertheless the waiter might have recommended red due to some theory or other, unless he could simply charge more for it! Also, the pre-event refoccilation was decided upon the basis of the pubs turning out to be either crowded (no drink) or quiet (drink): They turned-out to be quiet, however that is defined.

In a protracted and more complex negotiation zero-C criteria would be filtered into Treaties of Assent ($A_j=B_j=1$) and Treaties of Repudiation ($A_j=B_j=0$), and further efforts focused upon the $C_j=1$ criteria that depend upon ballot or upon third-party interventions.

Criterion	Arthur Preference	Betty Preference	Agreed Order of Importance	Arthur Prior Position	Betty Prior Position	A_j	B_j	BIAS Decider	C_j	D_j	Outcome
		0	1								
Mode of Travel	Bus	Walk	1 Walk	Walk	1	1		0	1	Walk
Venue Type	Pictures	Restaurant	2 Pictures	Restaurant	0	1	Coin Toss	1	1	Restaurant
Venue Quality	Omnivorous	Vegetarian	3 Vegetarian	Vegetarian	1	1		0	1	Vegetarian
Color of Wine	White	Red	4 White	Red	0	1	Waiter	1	1	Red
Drink before Event	Yes	No	5 No	No	1	0	Crowds	1	1	Drink First
Hamburger	No	No	6 No	No	0	0		0	0	No Hamburger

Table Six
Arthur and Betty's Decisions for their
Night Out in London
In Their Agreed Order of Importance

By now we are in a position to compute denary expressions of the predicate and outcome arrays A, B and D for further mathematical treatments.

These are presented in Table Seven:-

A_j	B_j	C_j	D_j	Outcome	Binary Index	Denary Digit Value	Denary A_j	Denary B_j	Denary C_j	Denary D_j
1	1	0	1	Walk	5	32	32	32	0	32
0	1	1	1	Restaurant	4	16	0	16	16	16
1	1	0	1	Vegetarian	3	8	8	8	0	8
0	1	1	1	Red	2	4	0	4	4	4
1	0	1	1	Drink First	1	2	2	0	2	2
0	0	0	0	No Hamburger	0	1	0	0	0	0
						63	42	60	22	62	Denary Total
						5.977280	5.392317	5.906891	4.459432	5.954196	$\text{Log}_2(\text{Denary Total})$

Table Seven
Denary Expressions of Arthur and Betty's Weighted Decisions
for their Night Out in London
In Their Agreed Order of Importance

Examination of the Intermediate XOR Indicator, C_j

The relevancy of C_j is that it prevents spurious BIAS over-ride.
We may expand C_j in terms of A_j and B_j as:-

$$C_j = \left(A_j - B_j\right)^2 = A_j^2 - 2.A_jB_j + B_j^2$$
Equation 1.9

Therefore the square of the Indicator C_j, which is $C_j{}^2$, may be expressed as:-

$$C_j^2 = \left[\left(A_J - B_J\right)^2\right]^2 = A_J^4 - 4.A_J^3B_J^1 + 6.A_J^2B_J^2 - 4.A_J^1B_J^3 + B_J^4$$
Equation 1.10

All integral powers of the general Binary Digit Z_j are the digit itself (i.e. Z_j).
Accordingly:-

$$C_J^k = \left[\left(A_J - B_J\right)^2\right]^k = A_j - 4.A_jB_j + 6.A_jB_j - 4.A_jB_j + B_j = A_j - 2.A_jB_j + B_j$$
Equation 1.11

Implications for $NXOR_j$

The Negation of Exclusive OR, the logical condition $NXOR(A_j,B_j)$, is yielded by:-

$$NXOR\left(A_j, B_j\right) = 1 - C_j = 1 - \left(A_j^2 - 2.A_jB_j + B_j^2\right)$$
Equation 1.12

or:-

$$NXOR\left(A_j, B_j\right) = 1 - C_j = 1 - A_j^2 + 2.A_jB_j - B_j^2$$

Equation 1.13

therefore the square $NXOR(A_j,B_j)^2$ may be expressed as:-

$$NXOR\left(A_j, B_j\right)^2 = 1 - 2.A_j^2 + 4.A_jB_j - 2.B_j^2 + A_j^4 + B_j^4 - 4.A_j^3B_j + 6.A_j^2B_j^2 - 4.A_jB_j^3$$
Equation 1.14

Since as before a*ll integral powers of the general Binary Digit Z_j are the digit itself (i.e. Z_j)* Equation 1.14 reduces to:-

$$NXOR(A_j, B_j) = 1 - 2.A_j + 4.A_jB_j - 2.B_j + A_j + B_j - 4.A_jB_j + 6.A_jB_j - 4.A_jB_j$$
Equation 1.15

From which simple algebraic addition gives us:-

$$NXOR(A_j, B_j) = 1 - A_j - B_j + 2A_j B_j$$
Equation 1.16

Several Identities of the Decision Function Outcome D$_j$

Certain Decision Functions DEF() may variously be defined in terms of arguments such as A$_j$, B$_j$, C$_j$, and BI$_j$; which for brevities sake will henceforth be called A, B, C and BI. Note that confusion with the arrays of which these are respective elements can be avoided if you remember that the arrays are designated with the upper case Arial font, e.g. A,B,C and BI.

In particular, we would wish to define and examine:-

DEF1(A,B,BI)

and:-

DEF2(A,C,BI)

and the closely-associated:-

DEF3(A,B,β)

Firstly:-

$$DEF1(A, B, BI) = XOR(A, B). entier\left(\frac{1}{2} + BI\right) + NXOR(A, B). A. [1 - (A - B)^2]$$

$$= (A - B)^2. entier\left(\frac{1}{2} + BI\right) + NXOR(A, B). A. [1 - (A - B)^2]$$

$$= (A - B)^2. entier\left(\frac{1}{2} + BI\right) + (1 - XOR(A, B)). A. [1 - (A - B)^2]$$
Equation 1.17

Secondly:-

$$DEF2(A, C, BI) = C. entier\left(\frac{1}{2} + BI\right) + A. (1 - C)^2$$

$$= C. BIAS(\beta) + A. (1 - C)^2$$

$$= (A - 2AB + B). BIAS(\beta) + A. (1 - A - B + 2AB)$$
Equation 1.18

Thirdly:-

$$DEF3(A, B, \beta) = (A - 2AB + B). BIAS(\beta) + AB$$

or by re-arrangement:-

$$DEF3(A, B, \beta) = AB + (A - 2AB + B). BIAS(\beta)$$

$$= AND(A, B) + XOR(A, B). BIAS(\beta)$$

$$= (A - 2AB + B). BIAS(\beta) + A. (1 - A - B + 2AB)$$
Equation 1.19

The Logarithm of Base Two

A logarithm is the inverse of an exponent. For example, the integer number 4 can be specified as 2^2 and 9 can be specified as 3^2. In both cases the exponent is 2: That exponent is the logarithm.

So the Base 2 logarithm of 4 is 2, and the Base 3 logarithm of 9 is 2.

It is possible to interchange the bases of logarithms arbitrarily using this formula:-

$$log_2(x) = \frac{log_e(x)}{log_e(2)}$$

Equation 2.1

or generally:-

$$log_b(x) = \frac{log_e(x)}{log_e(b)}$$

Equation 2.2

In approximate terms, Equation 2.1 may be rendered as:-

$$log_2(x) = 1.44269504088896 \times log_e(x)$$

Equation 2.3

e is The Base of Natural, sometimes called Napierian, Logarithms. Its value is about 2.71828182845905. (No exact knowledge is possible because the constant e is transcendental: In layman's terms the digits of its mantissa can be generated infinitely without falling into any repeating pattern. Many experts think that there are an infinite number of transcendental numbers lurking in the "real number line" continuum).

For example, the denary value of the decision intermediary C in the case of the Arthur and Betty Night-out Plan of Part One turned out to be 22. The Log_2 transformation of 22 can be calculated to as 4.45943161863730.

The original value of 22 can be recovered using the following formula:-

$$C_{denary} = 2^{log_2(C_{denary})} \approx 1.44269504088896 \times log_e(x) \approx 22$$

Equation 2.4

Among the immediate benefits of logarithmic transformations is the fact that the magnitude of the values can be reduced to manageable sizes. An immediate drawback is that no logarithm can be found for a negative number, and the logarithm of unity is zero in any base.

This latter fact will oblige us slightly to change the (maximum) magnitudes of binary numbers from $2^{n+1}-1$ to $2^{n+1}-2$ in order to preserve the finite values of positional numbers.

Numerical Limits

The Maximum Size, M, of an n-digit binary number is given by:-

$$M = 2^{n+1} - 1$$
Equation 2.5

For a predicate of ten binary digits (decisions) the number of array elements n = 9 and accordingly M = 1023.

When we take logarithms (to base 2), however, the technical need to avoid spurious zero values suggests a simple slight augmentation.

Therefore, we shall employ a minimal Anti-Fail Form of M, AFF, as defined by this equation:-

$$AFF = 2^{n+1} + 2$$
Equation 2.6

whose logarithm to base two is n+2.

In practical circumstances in which the predicates A and B have "reasonable length" this revised limit will have little or no numerical impact.

The Heronian Semi-Circuit, s_{AFF}, of a maximal triangle in which a, b and d are all AFF in length is given by:-

$$s_{AFF} = \frac{AFF + AFF + AFF}{2}$$
Equation 2.7

which by substitution becomes:-

$$s_{AFF} = \frac{(2^{n+1} + 2) + (2^{n+1} + 2) + (2^{n+1} + 2)}{2}$$
$$= \frac{3}{2} \cdot 2^{n+1} + 3$$
$$= 3(2^n + 1)$$
Equation 2.8

The square of the semi-perimeter is accordingly:-

$$s_{AFF}^2 = 9(2^{2n} + 2. 2^n + 1) = 9(2^{2n} + 2^{n+1} + 1)$$
Equation 2.9

The relevant Mean Side Length is $\mu_{a,b,d}$ given by:-

$$\mu_{a,b,d} = \frac{(2^{n+1} + 2) + (2^{n+1} + 2) + (2^{n+1} + 2)}{3}$$
Equation 2.10

But it is to be noted that the Mean Uniformly-Distributed Decision Value $\mu_{A,B,D}$ is rather:-

$$\mu_{A,B,D} = 2^n + 1$$
Equation 2.11

The Area A_{AFF} or the relative Euclidean triangle is given by Heron's Formula as:-

$$A_{AFF} = \sqrt{s_{AFF}(s_{AFF} - AFF)\big((s_{AFF} - AFF)s_{AFF} - AFF\big)}$$
Equation 2.12

which by substitution yields:-

$$A_{AFF} = \sqrt{3(2^n + 1)(3(2^n + 1) - AFF)(3(2^n + 1) - AFF)(3(2^n + 1) - AFF)}$$
Equation 2.13

or:-

$$A_{AFF}$$
$$= \sqrt{3(2^n + 1)\big(3(2^n + 1) - (2^{n+1} + 2)\big)\big(3(2^n + 1) - (2^{n+1} + 2)\big)\big(3(2^n + 1) - (2^{n+1} + 2)\big)}$$
Equation 2.14

From which:-

$$A_{AFF} = \sqrt{3(2^n + 1)\big(3(2^n + 1) - (2^{n+1} + 2)\big)^3}$$
Equation 2.15

$$A_{AFF} = \sqrt{3(2^n + 1)(3 \cdot 2^n - 2^{n+1} + 1)^3}$$
Equation 2.16

$$A_{AFF} = \sqrt{3}\sqrt{2^{n+1}}\sqrt{(3 \cdot 2^n - 2^{n+1} + 1)^3}$$
Equation 2.17

The forgoing enables us to define a Heronian Dimensionless Metric, HDM, a characteristic of the decision triangle, as:-

$$HDM = \frac{1}{4} \cdot \frac{A_{AFF}}{S_{AFF}^2}$$
Equation **2.18**

By substitution this becomes:-

$$HDM = \frac{1}{4} \cdot \frac{\sqrt{3}\sqrt{2^{n+1}}\sqrt{(3 \cdot 2^n - 2^{n+1} + 1)^3}}{[3(2^n + 1)]^2}$$
Equation 2.19

Condensation of this surd expression yields:-

$$HDM = \frac{1}{4} \cdot \frac{(3 \cdot 2^n - 2^{n+1} + 1)^{\frac{3}{2}}}{[3(2^n + 1)]^{\frac{3}{2}}}$$

$$= \frac{1}{4} \cdot \frac{(3 \cdot 2^n - 2 \cdot 2^n + 2^0)^{\frac{3}{2}}}{[3(2^n + 1)]^{\frac{3}{2}}}$$

$$= \frac{1}{4} \cdot \frac{(2^n + 1)^{\frac{3}{2}}}{[3(2^n + 1)]^{\frac{3}{2}}}$$

$$= \frac{1}{4} \cdot \frac{1}{3^{\frac{3}{2}}}$$

$$= \frac{1}{4} \cdot \frac{1}{\sqrt{27}}$$

$$= \frac{1}{\sqrt{16}} \cdot \frac{1}{\sqrt{27}}$$

$$\therefore HDM = \frac{1}{\sqrt{432}}$$
Equation 2.20

It is immediately apparent that the HDM does *not* depend upon the scale of the decision arrays, and is indeed a universal constant.

This HDM constant is near to 0.048112522432469.

I developed the EXCEL® spreadsheet TRIGTESTsFailTrapForm512.xlsx to trial 512 iterations of ten-element array predicates with uniformly-random binary digits. Taking the value 0.048112522432469 as the ideal I computed a number of the Percentage Specific Defects.

On several occasions I pressed the compute button for through-calculation of the 512 random triangulations. A typical empirical mean Ω was 0.0459162119 and its population standard deviation 0.0053564714. In this case Percentage Specific Defect was 4.56494571772158.

On the whole, these actual 2^{10}-x uniformly-random trials issue with an empirical Ω within 5% of the ideal reflecting of course the fact that they are not the theoretical maximum values.

If rather than arithmetic values of the geometric forms you treat instead of their logarithms to base two then the following expressions pertain:-

$$\Omega = \frac{1}{4} \cdot \frac{log_2(A_{AFF})}{log_2(s^2_{AFF})}$$
Equation 2.21

and also:-

$$AFF = n + 2$$
Equation 2.22

$$s_{AFF} = \frac{3(n + 2)}{2}$$
Equation 2.23

$$s^2_{AFF} = \frac{9(n + 2)^2}{4}$$
Equation 2.24

$$A_{AFF} = \frac{\sqrt{3}(n + 2)^2}{4}$$
Equation 2.25

Hence:-

$$\Omega = \frac{1}{4} \cdot \frac{\sqrt{3}(n + 2)^2}{4} \cdot \frac{4}{9(n + 2)^2} = \frac{1}{\sqrt{432}}$$
Equation 2.26

Plainly, the Heronian Dimensionless Metric Ω holds good for logarithms to the base two, or I dare say, any base.

The Critical Interangle γ

When the roster of binary digits in A is treated as one premise of a logical argument and the sequence of those in B as the other premise, yielding D as their conclusion: Then we may treat A, B and D as the (denary) sides of a triangle.

The Critical Interangle, γ, lies between A and B and shall in well-conditioned systems be less than one radian, as may readily be established empirically, at least in the case of uniformly random A and **B** elements (digits) A_j and B_j.

To pre-empt possible confusion we shall revert to calling A as (denary) a; and B as denary b, and treat them as ordinary lengths on the Euclidian plane.

The Triangulation of Edges and their Defined Areas

Clearly, since the three sequences of (binary) positional numbers that constitute decision structures all possess magnitude they may potentially constitute triangles.

A word of caution here: These triangles do not necessarily form closed shapes on the Euclidean plane, because one side might not complete the sum of the other two. Indeed the cosine of one or more of the included angles may exceed unity, and that cosine prove complex. No attempt is made in this study to move forward under complex arithmetic.

Note that no tampering will suffice to obviate such a case, and in particular that attempted projections upon non-Euclidean planes are invalid.

Experiments demonstrated that around 1.5% of randomly-generated cases were complex.

But where complex arithmetic occurred, and EXCEL® or other tools required real values only, I took either absolute values or complex moduli.

Not all complex cases reflect absurd decision structures, and not all credible decision structures are real.

This is not to argue that all invalid agreements are arithmetically complex or that tenable outcomes are universally real.

Mathematical tests and comparisons are a guide, not a dogma.

As a rule, complex results occurred where:-

$$d < a + b$$
Inequality 2.27

The favored expression of triangle Area, A_{OBC} or K_{OBC}, was Heron's Formula:-

$$A_{OBC} = \sqrt{s_{OBC}(s_{OBC} - a)(s_{OBC} - b)(s_{OBC} - d)}$$
Equation 2.28

where s_{OBC} is the Heronian Semi-Perimeter; and a, b and d are Triangle Sides representing denary Decision Structures. Equation 2.28 is capable of being irreducibly complex.

s_{OBC} is given by:-

$$s_{OBC} = \frac{a + b + d}{2}$$
Equation 2.29

Sequence of Calculations

The following sequence of calculations was employed to evaluate parameters of interest for plotting:-

Cosine Rule

$$\alpha = cos^{-1}\left(\frac{a^2 + b^2 - d^2}{2bd}\right)$$
Equation 2.30

where α is the First Basal Angle (i.e. Interangle between a and d).
In order to pre-empt a complex cosine I took the absolute value of the argument:-

$$\alpha = cos^{-1}\left(\left|\frac{a^2 + b^2 - d^2}{2bd}\right|\right)$$
Equation 2.31

Experiments confirmed that Equation 2.31 had no effect upon the computational value passed forward.

For the Arthur and Betty Night-Out Data of Part One α was 0.702172428550611 radians, which computed to be 40.2315172193592 degrees. This compared with a measured α of 41.1° giving a Percentage Specific Defect PSD(α°,α°$_{meas}$) of minus 2.15871247386845.

Sine Rule
$$\beta = sin^{-1}\left(\frac{b.\,sin(\alpha)}{a}\right)$$
Equation 2.32

where β is the Second Basal Angle (i.e. the Interangle between b and d).

For the Arthur and Betty Night-Out Data of Part One β was 1.1749809228012 radians, which computed to be 67.3214478848958 degrees. This compared with a measured β of 67.3° giving a Percentage Specific Defect PSD(β°,β°$_{meas}$) of 0.03185891802638.

Angles in a Triangle total π Radians

$$\gamma = \pi - \alpha - \beta$$
Equation 2.33

where γ is The Third Angle, the Critical Angle opposite to Decision Side, d.

For the Arthur and Betty Night-Out Data of Part One γ was 1.26443929223798 radians, which computed to be 72.447034895745 degrees. This compared with a measured γ of 73° giving a Percentage Specific Defect PSD(γ°,γ°$_{meas}$) of minus 0.763268096549032.

The Opposite and Adjacent of Part-Hypotenuse, g

$$x = b \cdot cos(\alpha) \qquad y = a \cdot sin(\beta)$$
Equation 2.34a   **Equation 2.34b**

For the Arthur and Betty Night-Out Data of Part One x was 45.8064516129032 units. This compared with a measured x of 45 giving a Percentage Specific Defect PSD(x,x_{meas}) of 1.7605638028169.

For the Arthur and Betty Night-Out Data of Part One y was 38.7526643036934 units. This compared with a measured y of 39.5 giving a Percentage Specific Defect PSD(y,y_{meas}) of minus 1.92847565382837.

Area of the Critical Triangle, OBC

For this I used the Included Angle Areal Formula:-

$$K1_{OBC} = \frac{1}{2} ab.\sin(\gamma)$$
Equation 2.35

For the Arthur and Betty Night-Out Data of Part One $K1_{OBC}$ was 1201.3325934145 square units.

For comparison, the Heronian A_{OBC} was 1201.3325934145 square units.

Heronian Dimensionless Metric, Ω

$$\Omega_{OBC} = \frac{1}{4} \cdot \frac{A_{OBC}}{s_{OBC}^2}$$
Equation 2.36

For the Arthur and Betty Night-Out Data of Part One Ω_{OBC} was 0.044665845977636 dimensionless units.

Computation of Part-Hypotenuse, g

$$g = \sqrt{x^2 - (b-y)^2}$$
$$= \sqrt{2.b^2.\left(1 - \cos\left(\frac{\pi}{2} - \alpha\right)\right)}$$
$$= \sqrt{2.b^2.\left(1 - \sin(\alpha)\right)}$$
Equation 2.37

For the Arthur and Betty Night-Out Data of Part One, part-hypotenuse g was 50.4943589280702 units. This compared with a measured g of 49.48 units.
The PSD(g,g_{meas}) was 2.00885593876975.

Establishment of Angle δ by Similar Triangles

$$\delta = tan^{-1}\left(\frac{b-y}{x}\right)$$
Equation 2.38

For the Arthur and Betty Night-Out Data of Part One δ was 0.434311944122143 radians, which computed to be 24.8842413903204 degrees. This compared with a measured δ of 24.5° giving a Percentage Specific Defect PSD(δ°,δ°$_{meas}$) of 1.54411534711236.

The Obtuse Angle ABC = η

$$\eta = \pi + \delta - \beta$$
Equation 2.39

For the Arthur and Betty Night-Out Data of Part One η was 1.26443929223798 radians, which computed to be 137.562793505425 degrees. This compared with a measured η of 137° giving a Percentage Specific Defect PSD(η°,η°$_{meas}$) of 0.409117531770973.

Area of Quadrilateral OABC

To establish the Area of the inclusive Quadrilateral OABC we may treat it as the melding of two triangles of a common side AC, that is:-

$$K_{OABC} = K_{AOC} + K_{ABC}$$
Equation 2.40

Equation 2.40 may be expanded as:-

$$K_{OABC} = \frac{1}{2}\cdot b \cdot d \cdot sin\left(\frac{\pi}{2}\right) + \frac{1}{2}\cdot a \cdot g \cdot sin(\eta)$$
Equation 2.41

which in turn expands to:-

$$K_{OABC} = \frac{1}{2}\left(bd + a.sin(\eta)\sqrt{(b.cos(\alpha))^2 + (b - a.sin(\beta))^2}\right)$$
Equation 2.42

or:-

$$K_{OABC} = \frac{1}{2}\cdot b \cdot d \cdot sin\left(\frac{\pi}{2}\right) + \frac{1}{2}\cdot a \cdot \sqrt{2.b^2.(1 - sin(\alpha))}\cdot sin(\eta)$$
Equation 2.41

from which:-

$$K_{OABC} = \frac{1}{2}\left(bd + a.\sin(\eta)\sqrt{2.b^2.(1 - \sin(\alpha))}\right)$$

giving:-

$$K_{OABC} = \frac{1}{2}\left(bd + a.b.\sqrt{2}.\sin(\eta)\sqrt{1 - \sin(\alpha)}\right)$$

or:-

$$K_{OABC} = \frac{b}{2}\left(d + a.\sqrt{2}.\sin(\eta)\sqrt{1 - \sin(\alpha)}\right)$$

Equation 2.42

For the Arthur and Betty Night-Out Data of Part One, Area K_{OABC} was 2575.52614180159 square units.

Length of Great Hypotenuse, h and the Horizon Metric, OD = z

The Great Hypotenuse, h, is given by:-

$$h = \frac{b}{\sin(\delta)}$$

Equation 2.43

whilst the Horizon Metric z is:-

$$z = \sqrt{h^2 - b^2}$$

$$z = \sqrt{\frac{b^2}{\sin^2(\delta)} - b^2}$$

$$z = \sqrt{b^2 \cdot \left(\frac{1}{\sin^2(\delta)} - 1\right)}$$

$$z = b \cdot \sqrt{\frac{1}{\sin^2(\delta)} - 1}$$

Equation 2.44

For the Arthur and Betty Night-Out Data of Part One h was 142.590185376083 units.

For the Arthur and Betty Night-Out Data of Part One z was 129.35208141188 units. This compared with a measured z of 131.5 giving a Percentage Specific Defect PSD(z,z_{meas}) of minus 1.66052108684707.

Area of Triangle OAD

The given numerical values are all for the Arthur and Betty scenario:-

$$K_{OAD} = \frac{bz}{2} = 3880.5624423564 \; square \; units$$
Equation 2.45

Dimensionless Areas' Ratios include:-

$$\frac{K_{OABC}}{K_{OAD}} = 0.663699187955252$$
Equation 2.46

$$\frac{A_{OBC}}{K_{OAD}} = 0.309576926350142$$
Equation 2.47

$$\frac{A_{OBC}}{K_{OABC}} = 0.466441622904343$$
Equation 2.48

Tabulation of the Decision Parameters

Recall the happy circumstances of Arthur and Betty's Night-Out from Part One.
The Decision Parameters computed as above for Arthur and Betty's Night-Out are given in Table 2.1.

The Triangulation of a Productive Decision Structure

It is possible analytically to develop and illustrate this and any valid real-valued decision structure as shown in Figure 2.1.

Table 2.1
Decision Parameters

Variable Name	Variable Symbol	Computed Value	Measured Value	Angle in Degrees	Measured Degrees	Percentage Specific Defect PSD(u_{comp}, u_{meas})
Heronian Semi-Perimeter of the Critical Triangle	S_{OBC}	82				
Heronian Area of the Critical Triangle	A_{OBC}	1201.3325934145				
First Basal Angle	α	0.7021724385850611		40.2315172193592	41.1	-2.1587124738684700
Second Basal Angle	β	1.1749809228012		67.3214478848958	67.3	0.0318589180263796
Critical Angle	γ	1.2644392922379980		72.4470348957450	73	-0.7632680965490320
Adjacent of the Part-Hypotenuse	x	45.8064516129032	45			1.7605633802816900
Complementary Opposite of the Part-Hypotenuse	y	38.7526643036934	39.5			-1.9284756538283700
Area of the Critical Triangle	K_{OBC}	1201.3325934145				
Heronian Dimensionless Metric	Ω	0.0446658459776 36				
Length of Part-Hypotenuse	g	50.4943589280702	49.48			2.0088559387697500
Great Hypotenuse Horizon Angle	δ	0.43431194412214 3		24.8842413903204	24.5	1.5441153471123500
Peripheral Angle ABC	η	2.40092367491073 0		137.5627935054250	137	0.4091175317709730
Area of Inclusive Quadrilateral	K_{OABC}	2575.5261418 0159				
Length of Great Hypotense	h	142.5901853760 83				
Length of Horizon Metric	z	129.3520814118 8	131.5			-1.6605210868470700
Area of Triangle OAD	K_{OAD}	3880.5624423564				
Dimensionless Area Ratio 1	K_{OABC}/K_{OAD}	0.6636991879552 52				
Dimensionless Area Ratio 2	A_{OBC}/K_{OAD}	0.3095769263501 42				
Dimensionless Area Ratio 3	A_{OBC}/K_{OABC}	0.4664416229043 43				

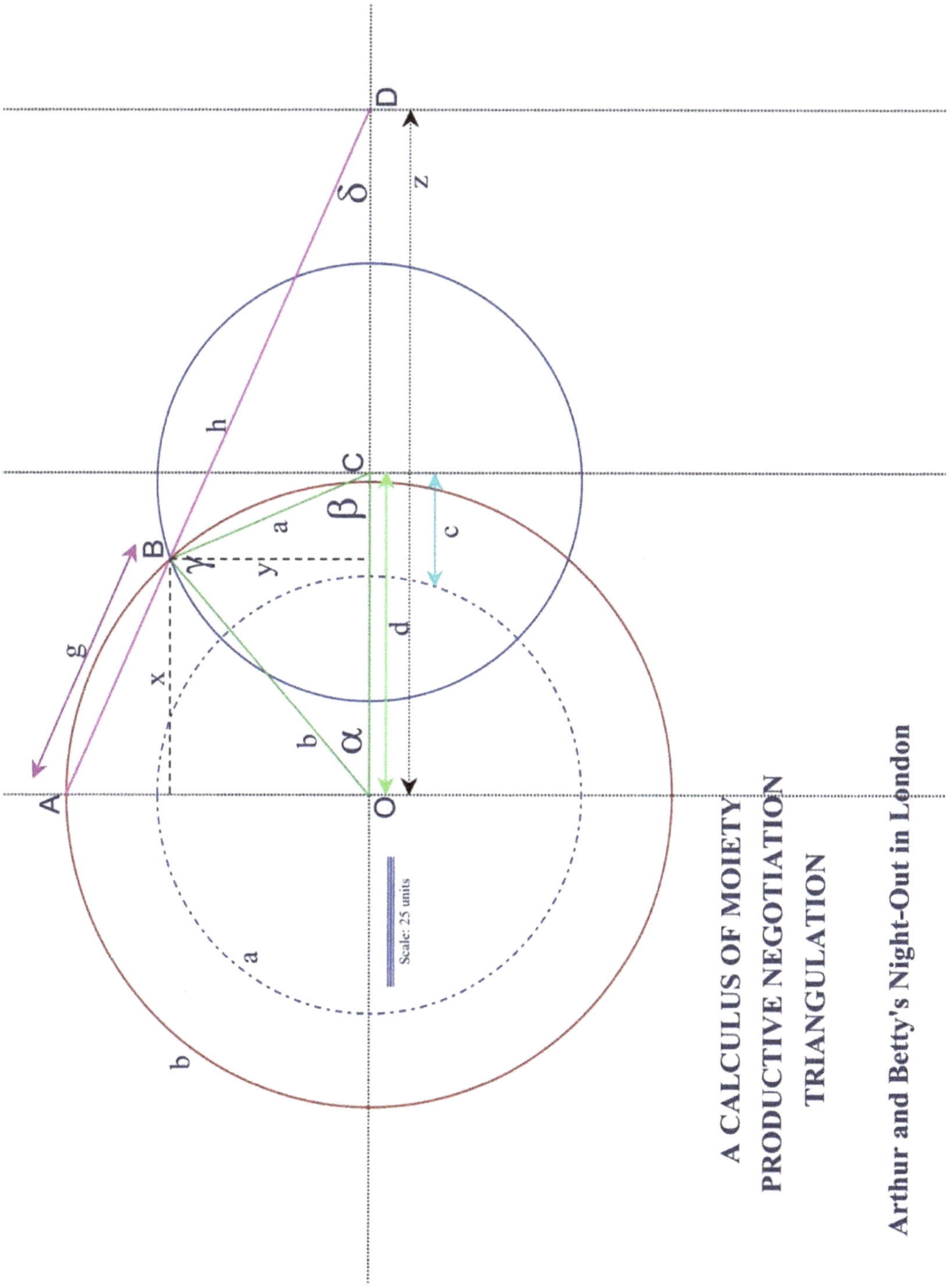

Figure 2.1
Decision Parameter Geometry Plot
For the Data of Arthur and Betty's Night-Out in London

Mathematics and The Higher Nonsense

As a free man or woman we flatter ourselves that we are master or mistress of our fate. But we fail frequently and our final failure is fatal.

Most opinions are meaningless, and most plans illusory or else scenes set for nemesis.

Mathematics helps us to detect nonsense, but does not of course protect us, because it is human art and mirrors the fallibility of its authors. It is a dog that barks its warning but does not bite. Or at best it is one of those tiny piping creatures whose shrill signals can be heard only by the youngest and most astute.

It is important to remember that mathematical methods do not violate the essential irrationality of social interaction. Mathematics can suggest the strength (or weakness) of a tendency: Just as mathematical methods can *describe*, but never *explain*, natural phenomena. Mathematics is language, and as such it can describe a situation, sometimes very graphically. Explanation is beyond art.

If I repeat I shall: This is not to argue that all invalid agreements are arithmetically complex or that tenable outcomes are universally real. *Mathematical tests and comparisons are a guide, not a dogma.*

An Example of a Dysfunctional Decision Structure with Complex Parameters

We do not of course imply that the following scenario is politically or strategically intricate or complicated, though it is both for certain. We use the word "complex" in its technical, algebraic sense to mean "involving use of the square root of minus unity".

Alopecia and Bulimia are two middling powers with a common land frontier and also each has a seaboard. Catatonia and Dysphoria are two smaller countries with no common land border with either Alopecia or Dysphoria. Catatonia is landlocked. Incredibly, Alopecia and Bulimia attempt to negotiate mutual management of their border. Catatonia is "protected" by Euphemia, a major world power.

Lord Allan and his team are treating for Alopecia. Lady Alice is his Assistant Deputy Foreign Relations Secretary. President Frank is a kind of podesta or foreign referee convening the talks. He has been hired from Futility, a neutral country with a long tradition of such things. Bulimia is represented by Mr Bellringer and his team.

Table 2.2 lists, in no particular order, the topics for discussion that the two teams have identified:-

Criterion	Alopecia	Bulimia
Guard Pants	Pink	Blue
Guard Hat	Peaked	Peaked
Anti-Tank Sulcus	Yes	No
Wire Fence	No	Yes
Customs	Yes	Yes
Electric Fence	No	No
Dogs	Yes	Yes
Border Catatonia-Disphoria	No	Yes

Table 2.2
Border Decision Criteria

The color of the uniform trousers worn by the border guards is the most important issue. It is in dispute whether the color should be pink or blue. This is followed by the vexed question of whether there should be an anti-tank fosse. Interestingly, the antagonists have not specified whether the trap ditch should be on the Alopecian or the Bulimian side of the frontier, or one on each side, and though foot sulci are also at issue that sub-question is not provided for in the schedule. You are entitled to think that the uniform issues and even the employment of dogs are trivia that should be delegated to field experts long after the defensive issues have been decided. I digress, and worse opine.

The Allan and Bellringer teams cannot agree about trouser color, so using privilege President Frank produces a deck of cards. He shuffles the cards thoroughly and asks Allan "Red or Black". Allen chooses "Black". Frank passes the whole deck to Lady Alice and invites her to divide and show: The result is the three of spades.

The fosse is also irresolvable so Frank decides the matter with a coin toss and imposes it on both delegations: The decision is to have a fosse, but the attendant implications remain unaddressed.

The wire fence idea is (so to say) ditched when it transpires that neither Alopecia nor Bulimia can provide sufficient material: For many decades Alopecia maintained a wire factory in the city of Jingle so as not to rely upon Bulimian imports, but the uneconomic facility was closed in 2017.

Bulimia agitates for a policed border to be re-established between Catatonia and Dysphoria. This offends the Catatonian ambassador who threatens to send his navy (notwithstanding that it is purely lacustrine). Euphemian diplomacy intervenes to prevent the threat to Catatonian sovereignty.

Alopecia and Bulimia agree that there should be no electric fence, but that they both want Customs activity at the border and the use of guard dogs, as well as peaked hats on the border guards.

You may find my illustration absurd or preposterous, but I have recently read of several international disputes just as stupid, or even worse. Most people simply do not know how to treat, and have never been trained, nor even did they ever suspect that negotiation might be an art. An efficient and optimal agreement should quickly separate the points of decision about which both antagonists agree (that is agree in the positive or the negative) to isolate the arguable clauses.

Today this lack of consensual ability applies almost as forcefully to the privately-educated classes as ill as it does to working people for whom debate was never a livelihood skill.

The ordered criteria and outcomes are summarised in Table 2.3:-

Criterion	Alopecia Preference	Bulimia Preference		Agreed Order of Importance	Alopecia Prior Position	Bulimia Prior Position	A_j	B_j	BIAS Decider	C_j	D_j	Outcome
		0	1									
Guard Pants	Pink	Blue		1 Pink	Blue	0	1	Cards	1	0	Pink Pants
Anti-Tank Fosse*	Yes	No		2 Yes	No	1	0	Coin Toss	1	0	Anti-Tank Fosse
Wire Fence	No	Yes		3 No	Yes	0	1	Lack of Wire	1	0	No Wire Fence
Border Catatonia-Dysphoria	No	Yes		4 No	Yes	0	1	Intrevention by Euphemia	1	0	No Catatonia Border
Guard Hat	Peaked	Peaked		5 Peaked	Peaked	1	1		0	1	Peaked Hats
Customs	Yes	Yes		6 Yes	Yes	1	1		0	1	Customs Emplaced
Electric Fence	No	No		7 No	No	0	0		0	0	No Electric Fence
Dogs	Yes	Yes		8 Yes	Yes	1	1		0	1	Guard Dogs

Table 2.3
Border Criteria and Outcomes

The denary totals of the outcomes in terms of the geometrically-treatable linear factors a=77, b=189, c=240 and d=13 are shown in Table 2.4:-

A_j	B_j	C_j	D_j	Outcome	Binary Index	Denary Digit Value	Denary A_j	Denary B_j	Denary C_j	Denary D_j	
0	1	1	0	Pink Pants	7	128	0	128	128	0	
1	0	1	0	Anti-Tank Fosse	6	64	64	0	64	0	
0	1	1	0	No Wire Fence	5	32	0	32	32	0	
0	1	1	0	No Catatonia Border	4	16	0	16	16	0	
1	1	0	1	Peaked Hats	3	8	8	8	0	8	
1	1	0	1	Customs Emplaced	2	4	4	4	0	4	
0	0	0	0	No Electric Fence	1	2	0	0	0	0	
1	1	0	1	Guard Dogs	0	1	1	1	0	1	
						255	77	189	240	13	Denary Total
						7.994353	6.266787	7.562242	7.906891	3.700440	Log_2(Denary Total)

Table 2.4
Border Denary Totals

The shortness of decision line d relative to the predicates a and b is often "a bad sign" and calculations show that the Critical Angle γ and *every associated angle, lineament and area* are complex.

The value of γ turns out to be 90.9952532427306-16.8554751702702i.

The (planar Euclidean) triangulation of this decision system is impossible as illustrated by Figure 2.2. Be assured that any projection onto a non-Euclidean surface will not do.

The triangle sides cannot connect and the abortive resolution as illustrated by Figure 2.2:-

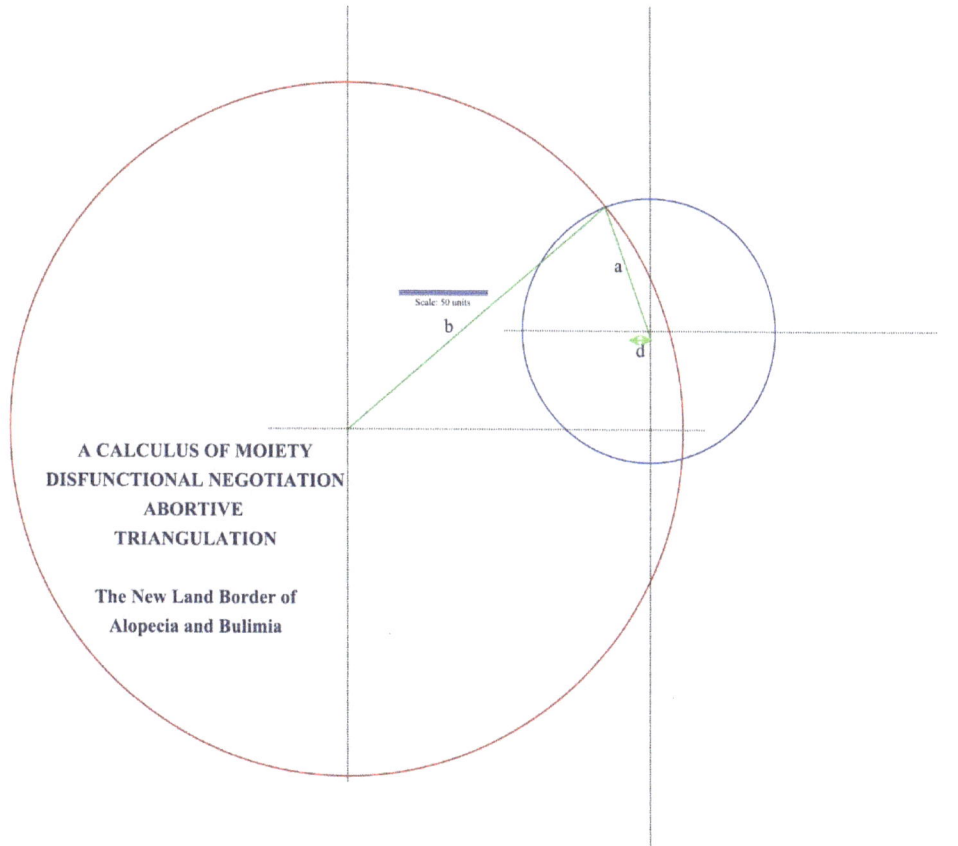

A CALCULUS OF MOIETY
DISFUNCTIONAL NEGOTIATION
ABORTIVE
TRIANGULATION

The New Land Border of
Alopecia and Bulimia

Scale: 50 units

Figure 2.2
The Impossible Real Triangulation of the
Alopecia-Bulimia International Border Negotiation
Illustrated

PART THREE
COMPUTATIONAL EXPERIMENTS

Note that throughout Part Three we shall address the denary length variables a, b and d as their base two logarithms in order to control size managements co-ordinately with the binary arithmetic upon which they are based. As we have seen, this does not affect the relevant relativities, and in particular it does not compromise the values of dimensionless derivatives.

Also note that due to the characteristics of EXCEL® re-calculation procedures *the heights of histogram columns may not necessarily tally* with tabular results in an exact manner. This will not affect any general interpretations we may infer from the experiments.

Table 3.1 presents a summary of the ungrouped and grouped descriptive statistics of some key dimensionless results that issue of the numerical experiments.

In every case, 1024 uniformly randomly generated trilateral side data of values up to $2^{n+1}+2=2^{10}+2=1026$ were computed for a, b and d; and distributed to 32 buckets manually-controlled to accommodate all of the available data.

The Critical Angle, γ

The Critical Angle, γ, subtends outcome side d and thus lies between the input sides a and b such that:-

$$\gamma = cos^{-1}\left(\frac{a^2 + b^2 - d^2}{2ab}\right)$$

Equation 3.1

As we have seen earlier in this disquisition, the expected value of γ is π/3 radians or 60° *when the three angles of the solution trilateral are equal*, and we showed that this was so when a = b = d = 2^x+2. Of course this conclusion is false in general, and in the particular case of random-length a, b and d we demonstrate that the mean γ value is π/6 radians or 30°.

The tabular ungrouped mean γ is 29.345700053383 and the grouped 29.415893554688 whilst Figure 3.1 presents a histogram which helps to confirm that experimentally γ is very near to thirty degrees. A number of experiments suggested that γ is normally distributed about its mean without detectible skew.

The implication of a thirty-degree gamma is that the value of the decision array D is about 51.76% of the (equal) uniformly-random inputs A and B: A little worse than perfect compromise. This is shown be the following application of the Cosine Rule in which $a_{perfect} = b_{perfect}$ = 1 and $\gamma_{perfect}$ = π/6:-

$$d = \sqrt{a_{perfect}^2 + b_{perfect}^2 - 2a_{perfect}b_{perfect}.\cos\gamma_{perfect}} = 0.517638090205041$$

Equation 3.2

Interval Specifiers	Critical Angle γ in Degrees	Trilateral Dimensionless Metric	Heronian Dimensionless Metric	Size-Dependent Box Ratio
Lower Bound	0	0	0	0
Upper Bound	180	0.1	0.15	0.01
Number of Intervals	32	32	32	32
Interval Width	5.625	0.003125	0.0046875	0.0003125
Series Statistics				
Minimum	9.268342788923	0.003109515710	0.005146140389	0.002554273669
Maximum	122.400297014032	0.081306029184	0.099841001832	0.006628906915
Mean	29.345700053383	0.027131808101	0.046063165433	0.003240191902
Pop s	8.167042778184	0.009117013283	0.004739213910	0.000434278278
GFD Mean	29.415893554688	0.027142333984	0.046847534180	0.003246459961
GFD s	8.352975066602	0.009144685854	0.005174786994	0.000432927746
Total Data By Count	1024	1024	1024	1024
Sum of Frequencies	1024	1024	1024	1024
Discrepancies				
Theoretical Value	30	0.027777777778	0.048112522432	0.003084492100
Empirical Mean	29.345700053383	0.027131808101	0.046063165433	0.003240191902
$PSD(Value, \mu_{emp})$	2.180999822056	2.325490837693	4.259508534	-5.047826256

Table 3.1
Statistical Summaries for Grouped Frequency Distributions

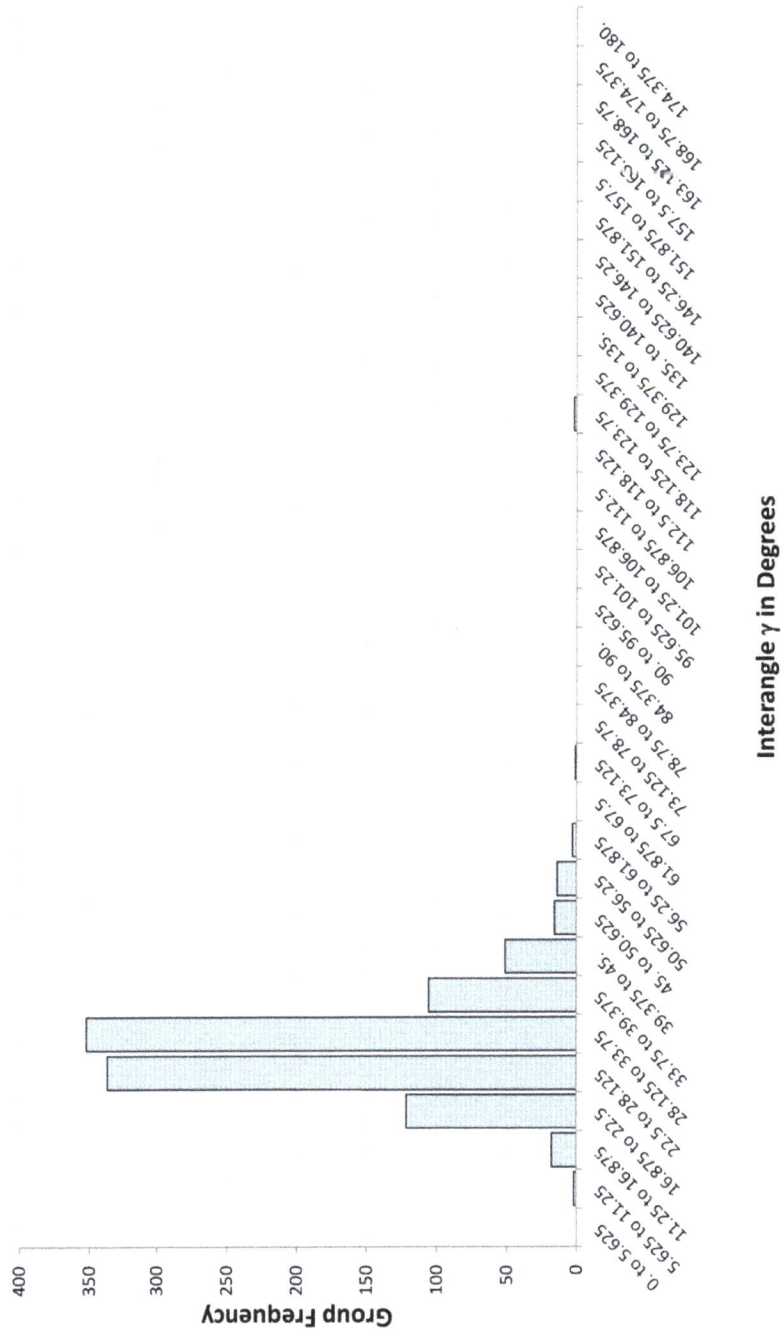

Figure 3.1
A Grouped Frequency Histogram of
The Critical Angle γ is Degrees
for the Moiety Trilateral abd

The Trilateral Dimensionless Metric, TDM

The Trilateral (i.e. triangle OBC) Dimensionless Metric, TDM, is defined by:-

$$TDM = \frac{A_{OBC}}{P_{OBC}^2} = \frac{\frac{1}{2} \cdot ab \cdot \sin\gamma}{(a+b+d)^2}$$

Equation 3.3

where A_{OBC} is the Area of Trilateral OBC and P_{OBC} is its Perimeter. Obviously, area has the dimensions of the square of perimeter.

The tabular ungrouped mean TDM is 0.027131808101 and the grouped 0.027142333984 whilst Figure 3.2 presents an apparent approximate Gaussian Distribution that suggests that modal TDM is near to 0.0265. Note, however, that the distribution has a long tail of sparse values up to 0.0844.

In order to understand this apparently unreasonable result we have to make a leap of faith to assume that the comparative trilateral *does not have* the well-known Euclidean conformation that when a=1, b=1 and subtense d=0.517638090205041 then perforce the subtended angle must be $\pi/6$.

Rather we must believe that even though the sharp angle is $\pi/6$, the sides a, b and d *are all unity*.

We already know that it does not matter whether we use the actual lengths or their logarithms as long as we treat all alike: So bear with me:-

$$
\begin{aligned}
TDM_{perfect} &= \frac{\frac{1}{2} \cdot a_{perfect} b_{perfect} \cdot \sin\gamma_{perfect}}{\left(a_{perfect} + b_{perfect} + d_{perfect}\right)^2} \\
&= \frac{\frac{1}{2} \cdot 1 \cdot 1 \cdot \frac{1}{2}}{(1 + 1 + 1)^2} \\
&= \frac{1}{4} \cdot \frac{1}{9} \\
&= \frac{1}{36} \\
&\approx 0.027777777777778
\end{aligned}
$$

Equation 3.4

Figure 3.2 illustrates the TDM grouped frequency distribution for one particular experiment:-

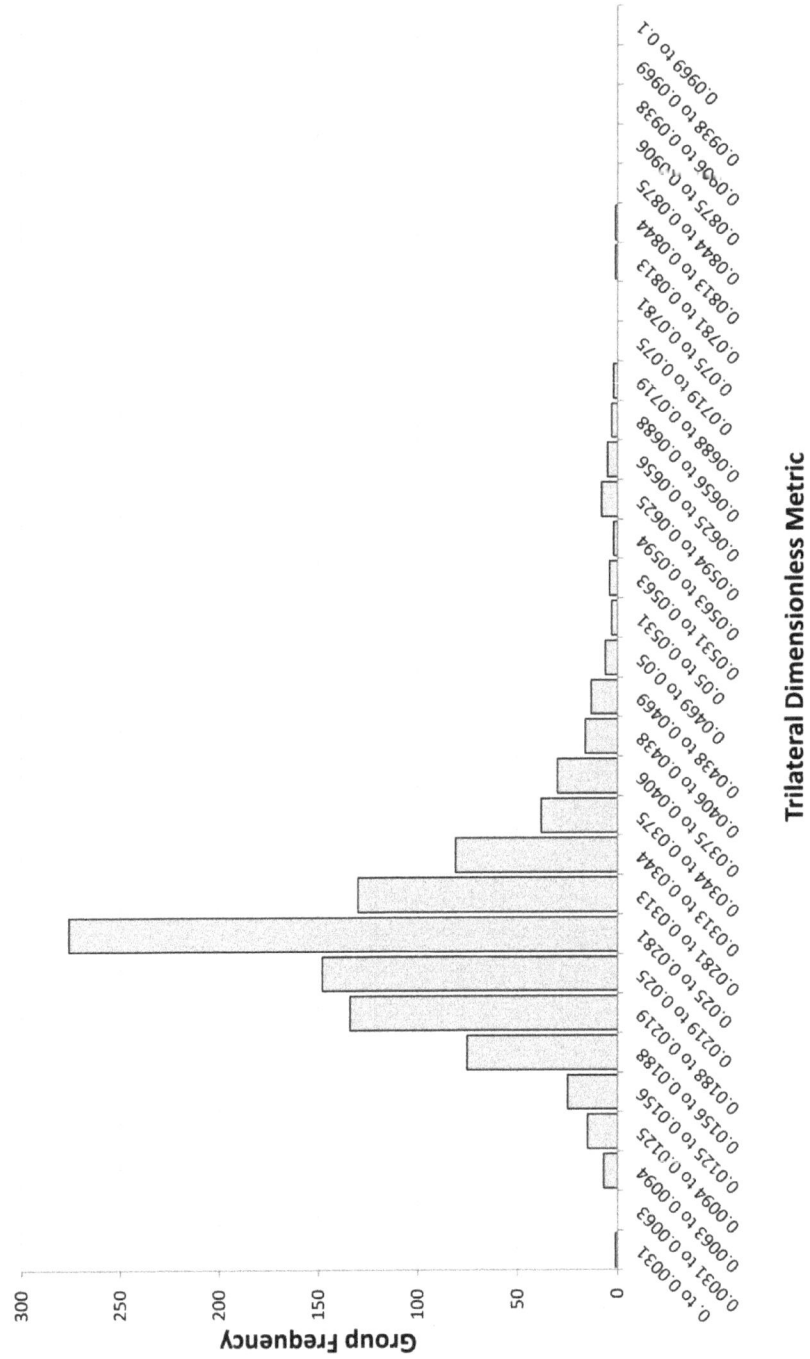

Figure 3.2
A Grouped Frequency Histogram of
The Trilateral Dimensionless Metric
for the Moiety Trilateral abd

<u>The Heronian Dimensionless Metric, HDM</u>

The Heronian Dimensionless Metric, HDM, is given by:-

$$HDM = \frac{A_{OBC}}{s_{OBC}^2} = \frac{\sqrt{s(s-a)(s-b)(s-d)}}{s^2}$$
Equation 3.5

where A_{OBC} is the Area of the Critical Triangle OBC (the same trilateral as for TDM); s_{OBC} is the Semi-Perimeter of Triangle OBC; and s_{OBC} is given by:-

$$s_{OBC} = \frac{a+b+d}{2}$$
Equation 3.6

The HDM is a well-behaved quasi-exponential distribution up to a value near 0.0516 with a handful of the 1024 HDMs outlying in the range 0.0934-0.1051. The valid part of the distribution is accurately representable by a cubic equation where $y = \log_2(\text{frequency})$.

We have seen that the theoretical upper limit of real HDM is:-

$$HDM = \frac{1}{\sqrt{432}} = 0.048112522432$$
Equation 3.7

and Table 3.1 shows that the ungrouped mean is 0.046063165433 whilst the grouped mean is 0.046847534180.

There is a single rogue outlier at 0.10078125: Almost certainly complex.

Figure 3.3 illustrates the HDM grouped frequency distribution for one particular experiment:-

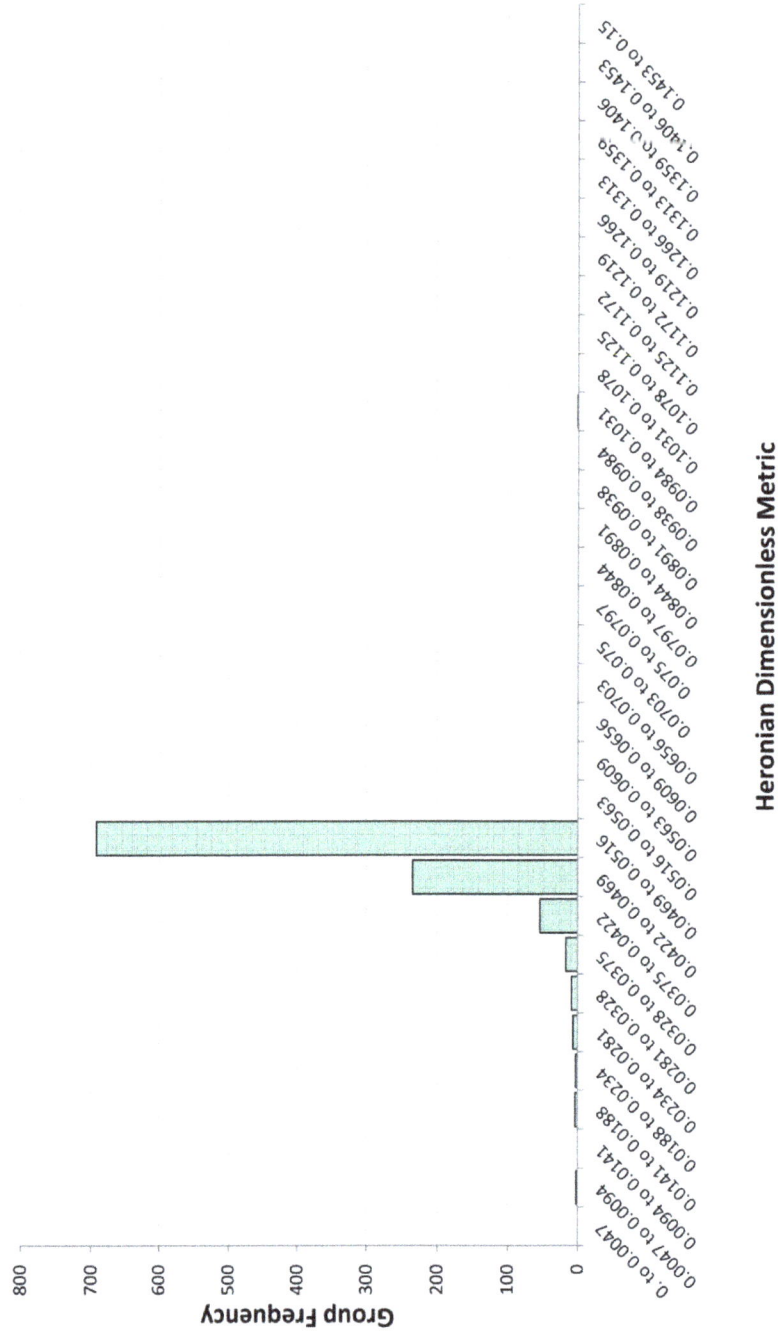

Figure 3.3
A Grouped Frequency Histogram of
The Heronian Dimensionless Metric
for the Moiety Trilateral abd

The Size-Dependent Box Metric

The Size-Dependent Box Metric is defined in these terms:-

$$R_B = \frac{V_{OBC}}{S^2_{OBC}} = \frac{abd}{[2 \cdot (ab + ad + bd)]^2} = \frac{1}{4} \cdot \frac{abd}{(ab + ad + bd)^2}$$

Equation 3.8

where R_B is the Size-Dependant Box Ratio representing the trilateral OBC; V_{OBC} is the Box Volume; and S_{OBC} is the Box Surface Area.

For the maximal condition that $a = b = d = 2^{n+1}+2$:-

$$R_B = \frac{1}{72(2^n + 1)} = \frac{1}{72\mu}$$

Equation 3.9

where μ is the Mean of the Maximal Length AFF=$2^{n+1}+2$. When $n = 9$, 2^n+2 is 513 and R_B = 0.000027073857483. For the illustrated numerical experiment the ungrouped mean is 0.003240191902 and the grouped mean is 0.003246459961. By inspection, the mode is approximately 0.003, from which value the distribution exhibits a long exponentially-decaying tail. If the average R_B is 0.0032, then the indicated n is about 1.75 as the $\log_2$ of a denary number.

Figure 3.4 illustrates the Size-Dependent Box Ratio grouped frequency distribution for one particular experiment.

Complex Values

In the particular experiment illustrated in Part Three six of the 1024 Heronian areas proved complex, representing 0.005859375 of the population. Their statistics were included in terms of the *absolute* value of the Cosine Rule square root.

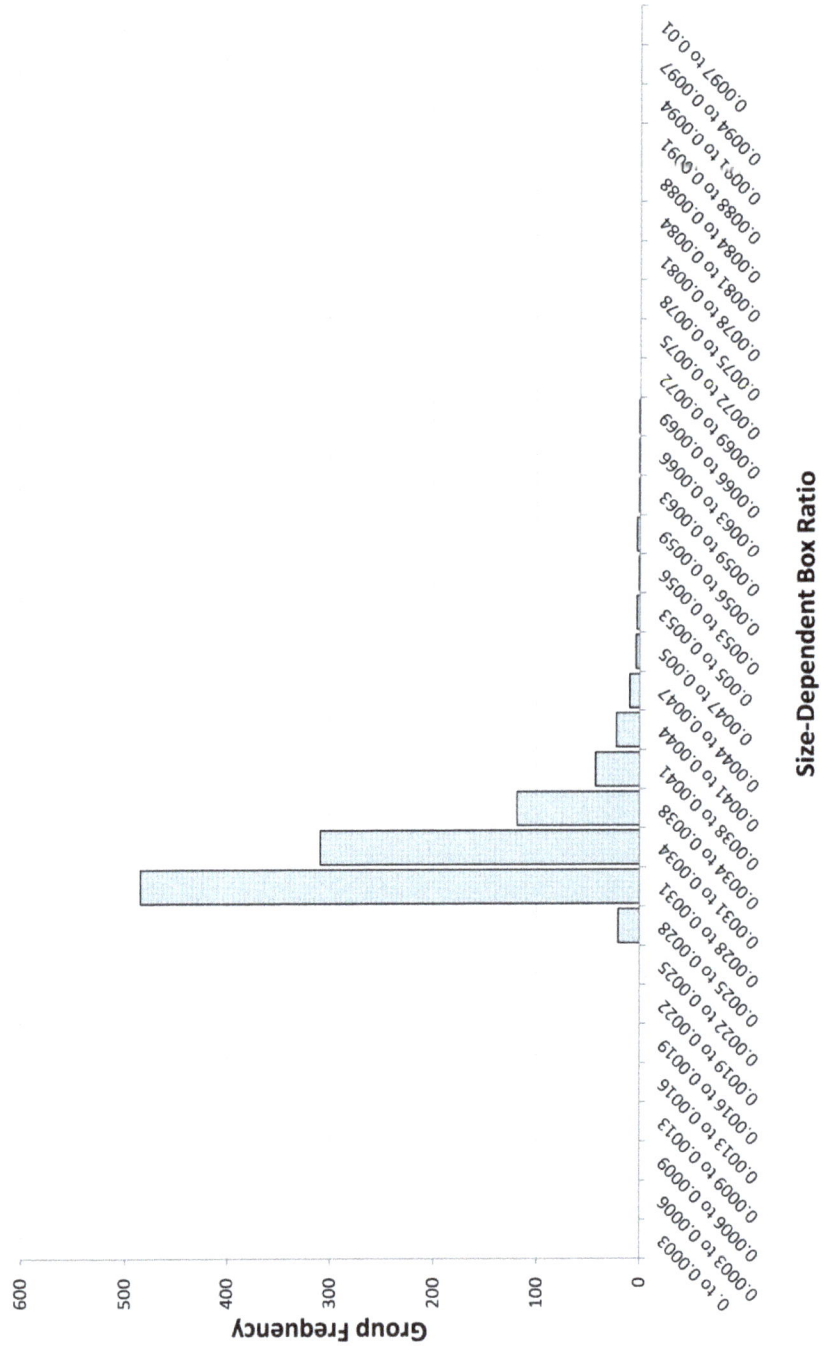

Figure 3.4
A Grouped Frequency Histogram of
The Size-Dependent Box Ratio
for the Moiety Trilateral abd

PART FOUR
ARITHMETICAL FORMULATIONS

To clarify, we shall use the term "arithmetical" to describe formulations that involve the decision arrays A, B and D only, expressed as the Euclidian triangle side lengths a, b and d.

Certainly we shall derive formulae from trigonometrical summaries but will seek formulae involving the cardinal operators, and roots and exponents of length only.

Technology is changing all the time, but as a principle we should eliminate trigonometric functions because they are expensive in time and may introduce error or complexity.

Also, it is more expeditious to design circuitry for simple binary arithmetic rather than anything more elaborate.

Thirty-Degree Relativities

Consider a "perfect" isosceles triangle whose unitary sides a and b include the angle $\gamma = \pi/6 = 30°$.

That is:-

$$a_p = b_p = 1$$
Equation 4.1

and:-

$$\gamma_p = \frac{\pi}{6} = 30°$$
Equation 4.2

Also:-

$$MAX = 2^9 + 2 = 514$$
Equation 4.3

whilst:-

$$LMAX = L2(MAX) = \frac{\ln(MAX)}{\ln(2)} = 9.00562454919388$$
Equation 4.4

Then:-

$$d_p = \sqrt{a_p^2 + b_p^2 - 2.a_p b_p.\cos(\gamma_p)} = 0.517638090205041$$
Equation 4.5

whilst for Trilateral Dimensionless Metric, TDM$_p$, it follows that:-

$$TDM_p = \frac{\frac{1}{2} \cdot a_p b_p \cdot \sin(\gamma_p)}{(a_p + b_p + d_p)^2} = 0.027777777777778$$

Equation 4.6

Furthermore:-

$$TDM_p = \frac{\frac{1}{2} \cdot LMAX \cdot LMAX \cdot \sin(\gamma_P)}{(LMAX + LMAX + LMAX)^2}$$

$$TDM_p = \frac{\frac{1}{2} \cdot LMAX \cdot LMAX \cdot \sin(\gamma_P)}{[3LMAX]^2}$$

$$TDM_p = \frac{1}{4} \cdot \frac{1}{9}$$

$$TDM_p = \frac{1}{36} = 0.027777777777778$$

Equation 4.7

Arrangement of the Critical Triangle OBC

Let the Area of the Critical Triangle OBC be K$_{OBC}$, then:-

$$K_{OBC} = \frac{1}{2} \cdot a \cdot b \cdot \sin(\gamma) = \frac{1}{2} \cdot a \cdot b \cdot \sin(\pi - \alpha - \beta)$$

Equation 4.8

Equation 4.8 may be expanded to the infecund form:-

$$K_{OBC} = \frac{1}{2} \cdot a \cdot b \cdot \sin\left(\pi - \cos^{-1}\left(\frac{b^2 + d^2 - a^2}{2bd}\right) - \sin^{-1}\left(\frac{b}{a} \cdot \sin\alpha\right)\right)$$

Equation 4.9

It is, however, more profitable to develop the area in this way:-

$$K_{OBC} = \frac{1}{2} \cdot a \cdot b \cdot \sin(\gamma) = \frac{1}{2} \cdot a \cdot b \cdot \sin\left(\cos^{-1}\left(\frac{a^2 + b^2 - d^2}{2ab}\right)\right)$$

Equation 4.10

From which, by the identity that the sine of an inverse cosine is equivalent to the root of unity minus the argument, we obtain:-

$$K_{OBC} = \frac{1}{2} \cdot a \cdot b \cdot \sqrt{1 - \left(\frac{a^2 + b^2 - d^2}{2ab}\right)}$$

Equation 4.11

Equation 4.11 may be expanded as:-

$$K_{OBC} = \frac{1}{2} \cdot a \cdot b \cdot \sqrt{1 - \left(\frac{a^4 + b^4 \mp d^4 + 2a^2b^2 - 2a^2d^2 - 2d^2b^2}{4a^2b^2}\right)}$$

Equation 4.12

which in turn may be expressed as:-

$$K_{OBC} = \frac{1}{2} \cdot a \cdot b \cdot \sqrt{1 - \left(\frac{a^4}{4a^2b^2} + \frac{b^4}{4a^2b^2} \mp \frac{d^4}{4a^2b^2} + \frac{2a^2b^2}{4a^2b^2} - \frac{2a^2d^2}{4a^2b^2} - \frac{2d^2b^2}{4a^2b^2}\right)}$$

Equation 4.13

which by cancellation becomes:-

$$K_{OBC} = \frac{1}{2} \cdot a \cdot b \cdot \sqrt{1 - \left(\frac{a^4}{4a^2b^2} + \frac{b^4}{4a^2b^2} \mp \frac{d^4}{4a^2b^2} + \frac{2}{4} - \frac{2d^2}{4b^2} - \frac{2d^2}{4a^2}\right)}$$

Equation 4.14

Equation 4.14 may be grouped into:-

$$K_{OBC} = \frac{1}{2} \cdot a \cdot b \cdot \sqrt{1 - \frac{1}{4} \cdot \left\{\frac{1}{4a^2b^2}(a^4 + b^4 + d^4) + 2\left(1 - d^2\left[\frac{1}{b^2} + \frac{1}{a^2}\right]\right)\right\}}$$

Equation 4.15

Arrangement of the Heron's Formula Semi-Perimeter and its Square

During this stage we shall use the Arthur and Betty Night-Out Data to assist our illustrations.

Recall that = 41, b = 60, and d = 62.

Now the Heronian Semi-Perimeter, s, is given by:-

$$s = \frac{a + b + d}{2} = \frac{a}{2} + \frac{b}{2} + \frac{d}{2} = f + g + h$$

Equation 4.16

Accordingly the intermediates are f = 21, g = 30 and h = 31. Therefore, the Semi-Perimeter is 62 and its square s^2 is 6724.

That square is given by:-

$$s^2 = (f + g + h)^2 = \frac{(a + b + d)^2}{4}$$

Equation 4.17

which develops as:-

$$s^2 = f^2 + g^2 + h^2 + 2(fg + fh + gh)$$
$$s^2 = f^2 + g^2 + h^2 + 2[(f + h) \cdot (g + h) - h^2]$$
$$s^2 = \left(\frac{a}{2}\right)^2 + \left(\frac{b}{2}\right)^2 + \left(\frac{d}{2}\right)^2 + 2\left[\frac{(a + d)}{2} \cdot \frac{(b + d)}{2} - \left(\frac{d}{2}\right)^2\right]$$
$$s^2 = \left[\frac{a^2 + b^2 + d^2}{4}\right] + 2\left[\frac{(a + d)}{2} \cdot \frac{(b + d)}{2} - \left(\frac{d}{2}\right)^2\right]$$
$$s^2 = \left[\frac{a^2 + b^2 + d^2}{4}\right] + 2\left[\frac{(a + d) \cdot (b + d)}{4} - \frac{d^2}{4}\right]$$
$$s^2 = \left[\frac{a^2 + b^2 + d^2}{4}\right] + \left[\frac{(a + d) \cdot (b + d)}{2} - \frac{d^2}{2}\right]$$
$$s^2 = \left[\frac{a^2 + b^2 + d^2}{4}\right] + \left[\frac{d^2 + d(a + b) + ab}{2} - \frac{d^2}{2}\right]$$
$$s^2 = \left[\frac{a^2 + b^2 + d^2}{4}\right] + \left[\frac{d(a + b) + ab}{2}\right]$$
$$s^2 = \frac{1}{2}\left[\frac{a^2 + b^2 + d^2}{2} + d(a + b) + ab\right]$$

Equation 4.18

The Arrangement of the Heronian Dimensionless Metric, HDM

Using the Heronian Form of the Area of a Triangle, A_{OBC}, we may write:-

$$A_{OBC} = \sqrt{s(s - a)(s - b)(s - d)}$$
Equation 4.19

where s is the Heronian Semi-Perimeter as defined by Equation 4.16. Equation 4.19 may be quoted as:-

$$A_{OBC} = \sqrt{(f + g + h) \cdot (-f + g + h) \cdot (f - g + h) \cdot (f + g - h)}$$
Equation 4.20

or:-

$$A_{OBC} = \sqrt{\frac{(a+b+d)}{2} \cdot \frac{(-a+b+d) \cdot (a-b+d) \cdot (a+b-d)}{8}}$$

Equation 4.21

from which:-

$$A_{OBC} = \frac{1}{4}\sqrt{(a+b+d) \cdot (-a+b+d) \cdot (a-b+d) \cdot (a+b-d)}$$

Equation 4.22

It is manifest that any two sides must exceed the third: Or else the area is complex. Because the Heronian Dimensionless Metric, HDM_{OBC}, is given by:-

$$HDM_{OBC} = \frac{1}{4} \cdot \frac{A_{obc}}{s_{obc}^2}$$

Equation 4.23

we may insert Equations 4.18 and 4.22 to give:-

$$HDM_{OBC} = \frac{1}{4} \cdot \frac{\frac{1}{4}\sqrt{(a+b+d) \cdot (-a+b+d) \cdot (a-b+d) \cdot (a+b-d)}}{\frac{1}{2}\left[\frac{a^2+b^2+d^2}{2} + d(a+b) + ab\right]}$$

Equation 4.24

From which:-

$$HDM_{OBC} = \frac{1}{8} \cdot \frac{\sqrt{(a+b+d) \cdot (-a+b+d) \cdot (a-b+d) \cdot (a+b-d)}}{\frac{a^2+b^2+d^2}{2} + d(a+b) + ab}$$

Equation 4.25

$$HDM_{OBC} = \frac{1}{4} \cdot \frac{\sqrt{(a+b+d) \cdot (-a+b+d) \cdot (a-b+d) \cdot (a+b-d)}}{a^2+b^2+d^2 + 2[d(a+b) + ab]}$$

Equation 4.26

Given the Arthur and Betty data the value of HDM_{OBC} is 0.044665845977636, not far shy of its Ω limiting value of 0.048112522432.

PART FIVE
ARITHMETICAL FORMULATIONS:
ARRANGEMENTS FOR THE QUADRILATERAL OABC

Insofar as numerical validation of our derivations in Part Five are required we shall use the Arthur and Betty data a = 42, b = 60, c = 22 and d = 62.

Some Convenient Auxiliaries

To save some repetitions and to facilitate derivations we shall define the following auxiliary variables at the outset:-

$$t = \left(\frac{b^2 + d^2 - a^2}{2bd}\right)^2$$

Equation 5.1

$$u = \sqrt{1 - t}$$

Equation 5.2

$$v = \frac{1 - u}{\sqrt{t}}$$

Equation 5.3

$$w = \frac{v}{\sqrt{1 + v^2}}$$

Equation 5.4

From these definitions this system of identities may immediately be formulated:-

$$1 - t = 4\left(w^2 - \frac{1}{2}\right)\left(w^2 - \frac{1}{2}\right) = (2w^2 - 1)^2$$

Equation 5.5

$$t = 1 - (2w^2 - 1)^2 = 4w^2(1 - w^2)$$

Equation 5.6

$$\sqrt{1 - t} = \sqrt{4\left(w^2 - \frac{1}{2}\right)\left(w^2 - \frac{1}{2}\right)} = 1 - 2w^2$$

Equation 5.7

Furthermore:-

$$w^2 = \frac{v^2}{1 + v^2} = \frac{\left(1 - \sqrt{1 - t}\right)^2}{t + \left(1 - \sqrt{1 - t}\right)^2}$$
Equation 5.8

and:-

$$\sqrt{1 - w^2} = \frac{1}{\sqrt{2}}\sqrt{1 + \sqrt{1 - t}}$$
Equation 5.9

The Arrangement of the Area of the Quadrilateral OABC

We have seen earlier how a quadrilateral's area is simply the sum of the areas of two component triangles: Which two component triangles is a question of convenience.

Given that the obtuse angle η is $\pi + \delta - \beta$, reference to the Figure 2.1 master diagram confirms that the following definition of the Area of OABC is a possibility:-

$$K_{OABC} = A_{AOC} + A_{ABC}$$
Equation 5.10

which may conveniently be elaborated as:-

$$K_{OABC} = \frac{1}{2} \cdot b \cdot d \cdot \sin\left(\frac{\pi}{2}\right) + \frac{1}{2} \cdot a \cdot g \cdot \sin(\eta)$$
Equation 5.11

which expands to:-

$$K_{OABC} = \frac{bd}{2} + \frac{1}{2} \cdot a \cdot \sqrt{2b^2(1 - \sin(\alpha))} \cdot \sin(\pi + \delta - \beta)$$
Equation 5.12

or by substitution for angle α:-

$$K_{OABC} = \frac{bd}{2} + \frac{1}{2} \cdot a \cdot \sqrt{2b^2\left(1 - \sin\left(acos\left(\frac{b^2 + d^2 - a^2}{2bd}\right)\right)\right)} \cdot \sin(\pi + \delta - \beta)$$
Equation 5.13

Noting the convenient identity:-

$$\sin\left(acos\left(\frac{b^2 + d^2 - a^2}{2bd}\right)\right) = \sqrt{1 - \left(\frac{b^2 + d^2 - a^2}{2bd}\right)^2}$$

Equation 5.14

we may render Equation 5.13 as:-

$$K_{OABC} = \frac{bd}{2} + \frac{1}{2} \cdot a \cdot \sqrt{2b^2\left(1 - \sqrt{1 - \left(\frac{b^2 + d^2 - a^2}{2bd}\right)^2}\right)} \cdot \sin(\pi + \delta - \beta)$$

Equation 5.15

Furthermore, in order to abridge the very long and intricate equations that subsequently arise from substitutions for the three angles π, δ and β we may usefully invoke the above Equation 5.1 through 5.4 for t, u, v and w to give:-

$$K_{OABC} = \frac{bd}{2} + \frac{1}{2} \cdot a \cdot \sqrt{2b^2(1 - u)} \cdot \sin\left(\pi + \tan^{-1} v - \sin^{-1}\frac{b}{a}u\right)$$

Equation 5.16

Because $\sin(\pi) = 0$, the Quadrilateral Area may be re-expressed as:-

$$K_{OABC} = \frac{bd}{2} + \frac{1}{2} \cdot a \cdot \sqrt{2b^2(1 - u)} \cdot \sin\left(-\tan^{-1} v + \sin^{-1}\frac{b}{a}u\right)$$

Equation 5.17

whose trigonometrical terms may be re-formulated to yield:-

$$K_{OABC} = \frac{bd}{2} + \frac{1}{2} \cdot a \cdot \sqrt{2b^2(1 - u)} \cdot \sin\left(\sin^{-1}\frac{b}{a}u - \sin^{-1} w\right)$$

Equation 5.18

Noting the standard identity:-

$$\sin^{-1}\frac{b}{a}u - \sin^{-1} w = \sin^{-1}\left(\frac{b}{a}u\sqrt{1 - w^2} - w\sqrt{1 - \left(\frac{b}{a}u\right)^2}\right)$$

Equation 5.19

We may re-configure Equation 5.18 as:-

$$K_{OABC} = \frac{bd}{2} + \frac{1}{2} \cdot a \cdot \sqrt{2b^2(1-u)} \cdot \left[\frac{b}{a} u \sqrt{1-w^2} - w \sqrt{1 - \left(\frac{b}{a}u\right)^2} \right]$$

Equation 5.20

An involved sequence of substitutions and simplifications then enables us to arrive at:-

$$K_{OABC} = \frac{bd}{2} + \left\{ \frac{b\sqrt{1-\sqrt{1-t}}}{\sqrt{2}} \right.$$
$$\left. \cdot \left[b \cdot \sqrt{1-t} \cdot \frac{1}{\sqrt{2}} \cdot \sqrt{1+\sqrt{1-t}} - a \cdot \sqrt{\frac{1-\sqrt{1-t}}{2}} \cdot \sqrt{1 - \left(\frac{b}{a} \cdot \sqrt{1-t}\right)^2} \right] \right\}$$

Equation 5.21

At this juncture it is helpful to note that if Equation 5.21 is:-

$$K_{OABC} = \frac{bd}{2} + [z_1 - z_2]$$

Equation 5.22

then:-

$$z_1 = \frac{b\sqrt{1-\sqrt{1-t}}}{\sqrt{2}} \times b \cdot \sqrt{1-t} \cdot \frac{1}{\sqrt{2}} \cdot \sqrt{1+\sqrt{1-t}} = \frac{b^2}{2} \cdot \sqrt{1-t} \cdot \sqrt{t}$$

Equation 5.23

and:-

$$z_2 = a \cdot b \cdot \frac{\sqrt{1-\sqrt{1-t}}}{\sqrt{2}} \cdot \frac{\sqrt{1-\sqrt{1-t}}}{\sqrt{2}} \cdot \sqrt{1 - \left(\frac{b}{a} \cdot \sqrt{1-t}\right)^2}$$
$$= a \cdot b \cdot \frac{1-\sqrt{1-t}}{2} \cdot \sqrt{1 - \left(\frac{b}{a} \cdot \sqrt{1-t}\right)^2}$$

Equation 5.24

Substitution of the simplified z_1 and z_2 into Equation 5.22 then permits:-

$$K_{OABC} = \frac{bd}{2} + \left[\frac{b^2}{2} \cdot \sqrt{1-t} \cdot \sqrt{t} - a \cdot b \cdot \frac{1 - \sqrt{1-t}}{2} \cdot \sqrt{1 - \left(\frac{b}{a} \cdot \sqrt{1-t}\right)^2} \right]$$

Equation 5.25

from which further manipulations allow:-

$$K_{OABC} = \frac{b}{2} \cdot \left\{ d + \left[b \cdot u \cdot \sqrt{t} - a \cdot (1-u) \cdot \sqrt{1 - \left(\frac{b}{a} u\right)^2} \right] \right\}$$

Equation 5.26

Observe that the major square root of Equation 5.26 constitutes a quadratic equation in the quotient b/a. Accordingly:-

$$\sqrt{1 - \left(\frac{b}{a} u\right)^2} = \sqrt{(t-1) \cdot \left(\frac{b}{a} - \frac{\sqrt{1-t}}{t-1}\right) \cdot \left(\frac{b}{a} + \frac{\sqrt{1-t}}{t-1}\right)}$$

Equation 5.27

from which we may establish the real result:-

$$\sqrt{1 - \left(\frac{b}{a} u\right)^2} = \sqrt{(t-1) \cdot \left(\frac{b}{a} - \frac{1}{\sqrt{1-t}}\right) \cdot \left(\frac{b}{a} + \frac{1}{\sqrt{1-t}}\right)} = \sqrt{(t-1) \cdot \left(\frac{b^2}{a^2} - \frac{1}{1-t}\right)}$$

Equation 5.28

or as a numerical negative:-

$$\sqrt{1 - \left(\frac{b}{a} u\right)^2} = \sqrt{(t-1) \cdot \left(\frac{b}{a} - \frac{1}{\sqrt{1-t}}\right) \cdot \left(\frac{b}{a} + \frac{1}{\sqrt{1-t}}\right)} = \sqrt{t-1} \sqrt{\frac{b^2}{a^2} - \frac{1}{1-t}}$$

Equation 5.29

Whilst the expressions Equation 5.27, 5.28 and 5.29 are all real the two product components:-

$$\sqrt{t-1}$$
and

$$\sqrt{\frac{b^2}{a^2} - \frac{1}{1-t}}$$

are both complex.
Indeed the area expression:-

$$K_{OABC} = \frac{b}{2} \cdot \left\{ d + \left[b \cdot u \cdot \sqrt{t} - a \cdot (1-u) \cdot u \cdot \sqrt{\frac{b^2}{a^2} - \frac{1}{1-t}} \right] \right\}$$

Equation 5.30

is complex, but a real-arithmetic utility formula can easily be offered as:-

$$K_{OABC} = \frac{b}{2} \cdot \left\{ d + \left[b \cdot u \cdot \sqrt{t} - a \cdot (1-u) \cdot u \cdot \sqrt{abs\left(\frac{b^2}{a^2} - \frac{1}{1-t}\right)} \right] \right\}$$

Equation 5.31

which of course offers up the required numerical area value, which for the stated Arthur and Betty Night-Out data is 2575.52614180159 square units.

Equation 5.31 can further be condensed to the complex expression:-

$$K_{OABC} = \frac{b}{2} \cdot \left\{ d + \left[b \cdot u \cdot \sqrt{t} - a \cdot (u - u^2) \cdot \sqrt{\left(\frac{b^2}{a^2} - \frac{1}{1-t}\right)} \right] \right\}$$

Equation 5.32

whose numerical value is 2747.56102114911-172.034879347514i.
On that basis:-

$$K_{OABC} = \frac{b}{2} \cdot \left\{ d + \left[b \cdot u \cdot \sqrt{t} - a \cdot (u - u^2) \cdot \sqrt{abs\left(\frac{b^2}{a^2} - \frac{1}{1-t}\right)} \right] \right\}$$

Equation 5.33

is of course a real-arithmetic utility formulation.

www.ingramcontent.com/pod-product-compliance
Lightning Source LLC
Chambersburg PA
CBHW040142200326
41458CB00025B/6348